Probability and Risk Analysis

Igor Rychlik   Jesper Rydén

Probability and Risk Analysis

An Introduction for Engineers

With 46 Figures and 7 Tables

 Springer

Dr. Jesper Rydén
School of Technology and Society, Malmö University,
Ö Varvsg 11A,
SE-20506 Malmö, Sweden
e-mail: jesper.ryden@ts.mah.se

Prof. Igor Rychlik
Dept. of Mathematical Statistics,
Lund University,
Box 118,
22100 Lund, Sweden
e-mail: igor@maths.lth.se

ISBN-13 978-3-642-06346-6   e-ISBN-13 978-3-540-39521-8

Springer is a part of Springer Science+Business Media.

springer.com

© Springer-Verlag Berlin Heidelberg 2010
Printed in Germany

Cover design: Estudio Calamar, Viladasens

Printed on acid-free paper

Preface

The purpose of this book is to present concepts in a statistical treatment of risks. Such knowledge facilitates the understanding of the influence of random phenomena and gives a deeper knowledge of the possibilities offered by and algorithms found in certain software packages. Since Bayesian methods are frequently used in this field, a reasonable proportion of the presentation is devoted to such techniques.

The text is written with student in mind – a student who has studied elementary undergraduate courses in engineering mathematics, may be including a minor course in statistics. Even though we use a style of presentation traditionally found in the math literature (including descriptions like definitions, examples, etc.), emphasis is put on the understanding of the theory and methods presented; hence reasoning of an informal character is frequent. With respect to the contents (and its presentation), the idea has not been to write another textbook on elementary probability and statistics — there are plenty of such books — but to focus on applications within the field of risk and safety analysis.

Each chapter ends with a section on exercises; short solutions are given in appendix. Especially in the first chapters, some exercises merely check basic concepts introduced, with no clearly attached application indicated. However, among the collection of exercises as a whole, the ambition has been to present problems of an applied character and to a great extent real data sets have been used when constructing the problems.

Our ideas have been the following for the structuring of the chapters: In Chapter 1, we introduce probabilities of events, including notions like independence and conditional probabilities. Chapter 2 aims at presenting the two fundamental ways of interpreting probabilities: the frequentist and the Bayesian. The concept of intensity, important in risk calculations and referred to in later chapters, as well as the notion of a stream of events is also introduced here. A condensed summary of properties for random variables and characterisation of distributions is given in Chapter 3. In particular, typical distributions met in risk analysis are presented and exemplified here. In Chapter 4 the most important notions of classical inference (point estimation, confidence intervals) are discussed and we also provide a short introduction to bootstrap methodology. Further topics on probability are presented in Chapter **5**, where notions like covariance, correlation, and conditional distributions are discussed.

The second part of the book, Chapters **6-10**, are oriented at different types of problems and applications found in risk and safety analysis. Bayesian methods are further discussed in Chapter **6**. There we treat two problems: estimation of a probability for some (undesirable) event and estimation of the mean in a Poisson distribution (that is, the constant risk for accidents). The concept of conjugated priors to facilitate the computation of posterior distributions is introduced.

Chapter **7** relates to notions introduced in Chapter **2** – intensities of events (accidents) and streams of events. By now the reader has hopefully reached a higher level of understanding and applying techniques from probability and statistics. Further topics can therefore be introduced, like lifetime analysis and Poisson regression. Discussion of absolute risks and tolerable risks is given. Furthermore, an orientation on more general Poisson processes (e.g. in the plane) is found.

In structural engineering, safety indices are frequently used in design regulations. In Chapter **8**, a discussion on such indices is given, as well as remarks on their computation. In this context, we discuss Gauss' approximation formulae, which can be used to compute the values of indices approximately. More generally speaking, Gauss' approximation formulae render approximations of the expected value and variance for functions of random variables. Moreover, approximate confidence intervals can be obtained in those situations by the so-called delta method, introduced at the end of the chapter.

In Chapter **9**, focus is on how to estimate characteristic values used in design codes and norms. First, a parametric approach is presented, thereafter an orientation on the POT (Peaks Over Threshold) method is given. Finally, in Chapter **10**, an introduction to statistical extreme-value distributions is given. Much of the discussion is related to calculation of design loads and return periods.

We are grateful to many students whose comments have improved the presentation. Georg Lindgren has read the whole manuscript and given many fruitful comments. Thanks also to Anders Bengtsson, Oskar Hagberg, Krzysztof Nowicki, Niels C. Overgaard, and Krzysztof Podgórski for reading parts of the manuscript; Tord Isaksson and Colin McIntyre for valuable remarks; and Tord Rikte and Klas Bogsjö for assistance with exercises. The first author would like to express his gratitude to Jeanne Wéry for her long-term encouragement and interest in his work. Finally, a special thanks to our families for constant support and patience.

Lund and Malmö,   *Igor Rychlik*
March, 2006   *Jesper Rydén*

Contents

1

Basic Probability

Different definitions of what *risk* means can be found in the literature. For example, one dictionary[1] starts with:

> "A quantity derived both from the probability that a particular hazard will occur and the magnitude of the consequence of the undesirable effects of that hazard. The term risk is often used informally to mean the probability of a hazard occurring."

Related to risk are notions like risk analysis, risk management, etc. The same source defines *risk analysis* as:

> "A systematic and disciplined approach to analyzing risk – and thus obtaining a measure of both the probability of a hazard occurring and the undesirable effects of that hazard."

Here, we study the parts of risk analysis concerned with computations of probabilities closer. More precisely, what is the role of probability in the fields of risk analysis and safety engineering? First of all, identification of failure or damage scenarios needs to be done (what can go wrong?); secondly, chances for these and their consequences have to be stated. Risk can then be *quantified* by some measures, often involving probabilities, of the potential outputs. The reason for quantifying risks is to allow coherent (logically consistent) actions and decisions, also called risk management.

In this book, we concentrate on mathematical models for randomness and focus on problems that can be encountered in risk and safety analysis. In that field, the concept (and tool) of probability often enters in two different ways. Firstly, when we need to describe the uncertainties originating from incomplete knowledge, imperfect models, or measurement errors. Secondly, when a representation of the genuine variability in samples has to be made, *e.g.* reported temperature, wind speed, the force and location of an earthquake, the number of people in a building when a fire started, etc. Mixing of these

---

[1] *A Dictionary of Computing*, Oxford Reference.

two types of applications in one model makes it very difficult to interpret what the computed probability really measures. Hence we often discuss these issues.

We first present two data sets that are discussed later in the book from different perspectives. Here, we formulate some typical questions.

Example 1.1 (Periods between earthquakes). The time intervals in days between successive serious earthquakes world-wide have been recorded. "Serious" means a magnitude of at least 7.5 on the Richter scale or more than 1000 people killed. In all, 63 earthquakes have been recorded, *i.e.* 62 waiting times. This particular data set covers the period from 16 December 1902 to 4 March 1977.

In Figure 1.1, data are shown in the form of a histogram. Simple statistical measures are the sample mean (437 days) and the sample standard deviation (400 days). However, as is evident from the figure, we need more sophisticated probabilistic models to answer questions like: "How often can we expect a time period longer than 5 years or shorter than one week?" Another important issue for allocation of resources is: "How many earthquakes can happen during a certain period of time, *e.g.* 1 year?". Typical probabilistic models for waiting times and number of "accidents" are discussed in Chapter 7.

(This data set is presented in a book of compiled data by Hand *et al.* [34].)

□

Example 1.2 (Significant wave height). Applications of probability and statistics are found frequently in the fields of oceanography and offshore technology. At buoys in the oceans, the so-called *significant wave height* H_s is recorded, an important factor in engineering design. One calculates H_s as the

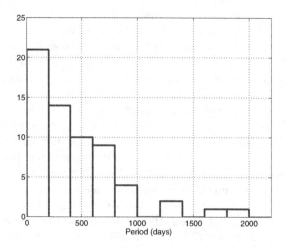

Fig. 1.1. Histogram: periods in days between serious earthquakes 1902-1977.

average of the highest one-third of all of the wave heights during a 20-minute sampling period. It can be shown that H_s^2 is proportional to average energy of sea waves.

In Figure 1.2, measurements of H_s from January to December 1995 are shown in the form of a time series. The sampling-time interval is one hour, that is, H_s is reported every hour. The buoy was situated in the North East Pacific. We note the seasonality, *i.e.* waves tend to be higher during winter months.

One typical problem in this scientific field is to determine the so-called 100-year significant wave (for short, the 100-year wave): a level that H_s will exceed on average only once over 100 years. The 100-year wave height is an important parameter when designing offshore oil platforms. Usually, 100 years of data are not recorded, and statistical models are needed to estimate the height of the 100-year wave from available data.

Another typical problem is to estimate durations of storms (time periods with high H_s values) and calm periods. For example, transport of large cargos is only allowed when longer periods of calmer weather can be expected. In Chapters 2 and 10 we study such questions closer.

(The data in this example are provided by the National Data Buoy Center and are accessible on the Internet.)   □

In this chapter a summary of some basic properties of probabilities is given. The aim is to give a review of a few important concepts: sample space, events, probability, random variables, independence, conditional probabilities, and the law of total probability.

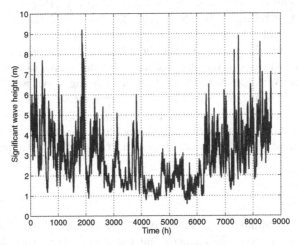

Fig. 1.2. Time series: significant wave height at a buoy in the East Pacific (Jan 1995 – Dec 1995).

1.1 Sample Space, Events, and Probabilities

We use the term *experiment* to refer to any process whose outcome is not known in advance. Generally speaking, probability is a concept to measure the uncertainty of an outcome of an experiment. (Classical simple experiments are to flip a coin or roll a die.) With the experiment we associate a collection (set) of all possible outcomes, call it *sample space*, and denote it by $\mathcal{S}$. An element s in this set will be denoted by $s \in \mathcal{S}$ and called a *sample point*. Intuitively, an *event* is a statement about outcomes of an experiment. More formally, an event A is a collection of sample points (a subset of $\mathcal{S}$, written as $A \subset \mathcal{S}$) for which the statement is true. Events will be denoted by capital letters A, B, C; sometimes we will use indices, *e.g.* A_i, $i = 1, \ldots, k$, to denote a collection of k events.

Random variables

We now introduce the fundamental notion of a *random variable* (r.v.), which is a number determined by the outcome of an experiment.

> **Definition 1.1.** *A **random variable** is a real-valued function defined on a sample space.*

In many experiments, only finitely many results need to be considered and hence the sample space is also finite. For illustration of some basic concepts we often use the already-mentioned experiments: "flip a coin" and "roll a die." The sample space of flipping a coin is $\mathcal{S} = \{\text{"heads"}, \text{"tails"}\}$. We write 0 if heads is shown, and 1 for tails; in this situation the sample space is $\mathcal{S} = \{0, 1\}$. Example of an event could be "The coin shows heads" with a truth set $A = \{0\}$. For an experiment of rolling a die, $\mathcal{S} = \{1, 2, 3, 4, 5, 6\}$, and the event "The die shows an odd number" is equivalent to the set $A = \{1, 3, 5\}$.

Let N be a number shown by the die. Clearly, N is a numerical function of an outcome of the experiment of rolling a die and serves as a simple example of a random variable. Now the statement "The die shows an odd number" is equivalent to "N is odd." We also use an experiment of rolling a die twice; then $\mathcal{S} = \{(1, 1), (1, 2), \ldots, (6, 6)\} = \{(i, j) : i, j = 1, 2, \ldots, 6\}$. Here it is natural to define two random variables to characterize the properties of outcomes of the experiment: N_1, the result of the first roll, and N_2, the result of the second roll.

Probabilities

Probabilities are numbers, assigned to statements about an outcome of an experiment, that express the chances that the statement is true. For example, for the experiment of rolling a fair die,

$$P(\text{"The die shows odd number"}) = P(A) = \frac{1}{2}.$$

Verbal statements and logical operations defining events are often closer to the practical use of probabilities and easier to understand. However, they lead to long expressions and hence are not convenient when writing formulae. Consequently it is more common to use sets, *e.g.* the statement "The die shows odd number" gives a set $A = \{1, 3, 5\}$, where the statement is true. Here we use both methods: the more intuitive $\mathsf{P}("N$ is odd") and the more formal $\mathsf{P}(\{1, 3, 5\})$, or simply $\mathsf{P}(A)$.

We assume that basic facts (definitions) of set theory are known; for example, that for two events A, B, the symbol $A \cup B$, which is a sum of two sets, means that A or B or both are true, while $A \cap B$ means A and B are true simultaneously. Two events (statements) are excluding if they cannot be true simultaneously, which transfers into the condition on the sets that $A \cap B = \emptyset$ (the empty set). For any event A, denote by A^c its complement, *i.e.* $A \cup A^c = S$ and $A \cap A^c = \emptyset$.

Probability is a way to assign numbers to events. It is a measure of the chances of an event to occur in an experiment or a statement about a result to be true. As a measure, similarly for volume or length, it has to satisfy some general rules in order to be called a probability. The most important is that

$$\mathsf{P}(A \cup B) = \mathsf{P}(A) + \mathsf{P}(B), \quad \text{if} \quad A \cap B = \emptyset. \tag{1.1}$$

Furthermore, for any event A, $0 \leq \mathsf{P}(A) \leq 1$. The statements that are always false have probability zero, similarly, always-true statements have probability one.

One can show that

$$\mathsf{P}(A \cup B) = \mathsf{P}(A) + \mathsf{P}(B) - \mathsf{P}(A \cap B). \tag{1.2}$$

The definition of probability just discussed is too wide, we need to further limit the class of possible functions P that can be called probability to such that satisfy the following more restrictive version of Eq. (1.1).

Definition 1.2. *Let $A_1, A_2, \ldots$ be an infinite sequence of statements such that at most one of them can be true (A_i are mutually excluding); then*

$$\mathsf{P}("At \text{ } least \text{ } one \text{ } of \text{ } A_i \text{ } is \text{ } true") = \mathsf{P}(\cup_{i=1}^{\infty} A_i) = \sum_{i=1}^{\infty} \mathsf{P}(A_i). \tag{1.3}$$

Any function P satisfying (1.3), taking values between zero and one and assigning value zero to never-true statements (impossible events) and value one to always-true statements (certain events) is a correctly defined probability.

Obviously, for a given experiment with sample space S, there are plenty of such functions P, which satisfy the condition of Eq. (1.3). Hence, an important problem is how to choose an adequate one, *i.e.* well measuring the uncertainties one has to consider. In the following we present the classical example how to define probabilities.

Example 1.3 (Classical definition). An important example of a probability P defined for events in $\mathcal{S}$ with a finite number of sample points is the following one, sometimes referred to as the "classical" definition of probability:

$$P(A) = \frac{N_A}{N_\mathcal{S}} \tag{1.4}$$

where N_A is the number of sample points that belong to the event A, $N_\mathcal{S}$ is the total number of sample points in the sample space $\mathcal{S}$.

The probability defined by Eq. (1.4) is a proper model for situations when each individual output of the random experiment has equal chance to occur. Then (1.1) states that (1.4) is the only possible probability on $\mathcal{S}$. For example, it is clear that for the experiment "roll a fair die," all outcomes have the same chance to occur. Then

$$P(\text{"The die shows odd number"}) = \frac{3}{6} = \frac{1}{2}.$$

$\square$

Generally for a countable sample space, *i.e.* when we can enumerate all possible outcomes, denoted by $\mathcal{S} = \{1, 2, 3, \ldots\}$, it is sufficient to know the probabilities

$$p_i = P(\text{"Experiment results with outcome } i\text{"})$$

in order to be able to define a probability of any statement. These probabilities constitute the *probability-mass function*. Simply, for any statement A, Eq. (1.3) gives

$$P(A) = \sum_{i \in A} p_i, \tag{1.5}$$

i.e. one sums all p_i for which the statement A is true; see Eq. (1.6).

Example 1.4 (Rolling a die). Consider a random experiment consisting of rolling a die. The sample space is $\mathcal{S} = \{1, 2, 3, 4, 5, 6\}$. We are interested in the likelihood of the following statement: "The result of rolling a die is even". The event corresponding to this statement is $A = \{2, 4, 6\}$. If we assume that the die is "fair", *i.e.* all sample points have the same probability to come up, then, by Eq. (1.4)

$$P(A) = \frac{3}{6} = 0.5.$$

However, if the die was not fair and showed 2 with probability $p_2 = 1/4$ while all other results were equally probable ($p_i = 3/20$, $i \neq 2$), then by Eq. (1.5)

$$P(A) = p_2 + p_4 + p_6 = \frac{11}{20} = 0.55. \tag{1.6}$$

The probability-mass functions for the two cases are shown in Figure 1.3.

The question of whether the die is "fair" or how to find the numerical values for the probabilities p_i is important and we return to it in following chapters. Here we only indicate that there are several methods to *estimate* the values of p_i. For example:

- One can assume that any values for p_i are possible. In order to find them, one can roll the die many times and record the frequency with which the six possible outcomes occur. This method would require many rolls in order to get reliable estimates of p_i. This is the *classical* statistical approach.
- Another method is to use our experience from rolling different dice. The experience can be quantified by probabilities (or odds), now describing "degree of belief," which values p_i can have. Then one can roll the die and modify our opinion about the p_i. Here the so-called *Bayesian* approach is used to update the experience to the actual die (based on the observed outcomes of the rolls).
- Finally, one can assume that the die is fair and wait until the observed outcomes contradict this assumption. This approach is referred to as *hypothesis testing*.

□

In many situations, one can assume (or live with) the assumption that all possible outcomes of an experiment are equally likely. However, there are situations when assigning equal probabilities to all outcomes is not obvious. The following example, sometimes called the Monty Hall problem, serves as an illustration.

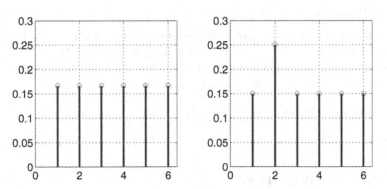

Fig. 1.3. Probability-mass functions. *Left:* Fair die; *Right:* Biased die.

Example 1.5 ("Car or Goat?"). In an American TV show, a guest (called "player" below) has to select one of three closed doors. He knows that behind one of the doors is a prize in the form of a car, while behind the other two are goats. For simplicity, suppose that the player chooses No. 1, which he is not allowed to open. The host of the show opens one of the remaining doors. Since he knows where the car is, he always manages to open a door with a goat behind. Suppose the host opened door No. 3.

We have two closed doors, where No. 1 has been chosen by the player. Now the player gets the possibility to open his door and check whether it is a car behind it or to abandon his first choice and open the second door. The question is which strategy is better: to switch and hence open No. 2, or stick to the original choice and check what is hidden behind No. 1.

Often people believe their first choice is a good one and do not want to switch, others think that their odds are 1:1 to win, regardless of switching. However, the original odds for the car to be behind door No. 1 was 1:2. Thus the problem is whether the odds should be changed to 1:1 (or other values) when one knows that the host opened door No. 3. A solution employing Bayes' formula is given in Example 2.2.

Note that if odds are unchanged, this would mean that the probability that the car is behind door No. 1 is independent of the fact that the host opens door No. 3 (see Remark 1.1, page 13).

(This problem has been discussed in an article by Morgan *et al.* [55].)

$\square$

1.2 Independence

Another important concept that is used to compute (construct) more complicated probability functions P is the notion of *independence*. We illustrate it using an experiment: roll a die twice. It is intuitively clear that the results of the two rolls of the die (if performed in a correct way) should give independent results.

As before, let the sample space of this experiment be

$$\mathcal{S} = \{(1,1),(1,2),\ldots,(6,6)\}.$$

We shall now compute the probability of the statements $A_1 =$ "The first roll gave odd number" and $A_2 =$ "The second roll gave one (1)". If the die is fair and the rolls have been performed correctly, then any of the sample points in $\mathcal{S}$ are equally probable. Now using Eq. (1.4), we have that

$$\mathsf{P}(A_1 \cap A_2) = \mathsf{P}(\{(1,1),(3,1),(5,1)\}) = \frac{3}{36} = \frac{1}{12}.$$

Similarly, we obtain that $\mathsf{P}(A_1) = 1/2$ while $\mathsf{P}(A_2) = 1/6$ and the following equality follows

$$\mathsf{P}(A_1 \cap A_2) = \mathsf{P}(A_1) \cdot \mathsf{P}(A_2). \tag{1.7}$$

This is not by accident but an evidence that our intuition was correct, because the definition of independence requires that (1.7) holds. The definition of independence is given now.

Definition 1.3. *For a sample space* S *and a probability measure* P, *the events* $A, B \subset S$ *are called* **independent** *if*

$$P(A \cap B) = P(A) \cdot P(B). \tag{1.8}$$

Two events A *and* B *are* **dependent** *if they are not independent, that is,*

$$P(A \cap B) \neq P(A) \cdot P(B).$$

Observe that independence of events is not really a property of the events but rather of the probability function P. We turn now to an example of events where we have little help from intuition to decide whether the events are independent or dependent.

Example 1.6 (Rolling a die). Consider a random experiment consisting of rolling a die. The sample space is $S = \{1, 2, 3, 4, 5, 6\}$. We are interested in two statements: "The result of rolling a die is even" and "The result is 2 or 3". The events corresponding to this statements are $A = \{2, 4, 6\}$ and $B = \{2, 3\}$. Can one directly by intuition say whether A and B are independent or dependent?

Let us check it by using the definition. If we assume that the die is "fair", *i.e.* all sample points have the same probability to come up, then

$$P(A \cap B) = \frac{1}{6} = P(A) \cdot P(B) = \frac{3}{6} \cdot \frac{2}{6}.$$

So the events A and B are independent. Observe that if the die was not fair and showed 2 with probability $1/4$ while all other results were equally probable, then the events A and B become dependent (check it). (Solution: $1/4 \neq (1/4 + 3/20 + 3/20) \cdot (1/4 + 3/20)$). $\qquad\square$

The conclusion of the last example was that the question whether two specific events are dependent or not may not be easy to answer using only intuition. However, the important application of the concept of independence is to define probabilities. Often we construct probability functions P so that independence of some events is obvious or assumed, as we see in the following simple example. The specific of that example is that we will compute probabilities of some events without first specifying the sample space S.

Example 1.7 (Rescue station). At a small rescue station, one has observed that the probability of having at least one emergency call a given day is 0.15. Assume that emergency calls from one day to another are independent in the statistical sense. Consider one week; we want to calculate the probability of emergency calls (i) on Monday; (ii) on Monday and Tuesday; (iii) on Monday, Tuesday, and Wednesday.

The probability wanted in (i) is simply 0.15. By independence, we get the probabilities asked for in (ii): $0.15 \cdot 0.15 = 0.0225$ and (iii): $0.15^3 = 0.0034$.

Consider now the statement A: "There will be exactly one day with emergency calls in a week". Then $P(A) = 7 \cdot 0.15 \cdot 0.85^6$, which can be motivated as follows: Let A_i, $i = 1, 2, \ldots, 7$ be the statement "Emergency on the ith day of the week and no calls the remaining six days." Obviously, the statements A_i are mutually excluding, $i.e.$ only one of them can be true. Since $A = A_1 \cup A_2 \cup \ldots \cup A_7$, we obtain by Eq. (1.3)

$$P(A) = P(A_1) + P(A_2) + \cdots + P(A_7).$$

Now, any of the probabilities $P(A_i) = 0.15 \cdot 0.85^6$, because of the assumed independence.   □

The reasoning in the last example is often met in applications, as shown in the following subsection.

1.2.1 Counting variables

Special types of random variables are the so-called *counting variables*, which are related to statements or questions of the type "how many"; an example is found in Example 1.7. Three commonly used types of counting variables in applications are now discussed: binomial, Poisson, and geometric.

Binomial probability-mass function

Suppose we are in a situation where we can perform an experiment n times in an independent manner. Let A be a statement about the outcome of an experiment. If A is true we say that the experiment leads to a success and denote by $p = P(A)$ the probability for "success" in each trial; it is then interesting to find the probability for the number of successes $K = k$ out of n trials. One can derive the following probability (see [25], Chapter VI, or any textbook on elementary probability, $e.g.$ [70], Chapter 3.4):

$$P(K = k) = p_k = \binom{n}{k} p^k (1 - p)^{n-k}, \quad k = 0, 1, 2, \ldots, n \qquad (1.9)$$

where with $n! = 1 \cdot 2 \cdot \ldots \cdot n$,

$$\binom{n}{k} = \frac{n!}{k!\,(n-k)!}.$$

For the random variable K taking values $k = 0, \ldots, n$, the sequence of probabilities $p_k = P(K = k)$ given by Eq. (1.9) is called the *binomial probability-mass function* for K. A shorthand notation is $K \in \text{Bin}(n, p)$.

Example 1.8. The total number of days with at least one call during one week at the rescue station in Example 1.7 can be described by an r.v. $K \in$ Bin$(7, 0.15)$. Hence,

$$P(K = k) = \binom{7}{k} p^k (1 - p)^{7-k}, \quad k = 0, 1, \ldots, 7$$

where $p = 0.15$. For example, the probability of exactly three days with calls is

$$P(K = 3) = \binom{7}{3} 0.15^3 \cdot 0.85^4 = 0.062. \tag{1.10}$$

□

Poisson probability-mass function

The *Poisson distribution* is often used in risk analysis to model the number of rare events. A thorough discussion follows in Chapters 2 and 7. For convenience, we present the probability mass function at this moment:

$$P(K = k) = e^{-m} \frac{m^k}{k!}, \qquad k = 0, 1, 2, \ldots \tag{1.11}$$

The shorthand notation is $K \in$ Po(m). Observe that now the sample space $S = \{0, 1, 2, \ldots\}$ is the set of all non-negative integers, which actually has an infinite number of elements. (All sets that have as many elements as the set of all integers are called *countable sets*, e.g. the set of all rational numbers is countable. Obviously not all sets are countable (for instance, the elements in the set $\mathbb{R}$ of all real numbers cannot be numbered); such sets are called *uncountable*.) Under some conditions, given below, the Poisson probability-mass function can be used as an approximation to the binomial probability mass.

Poisson approximation of Binomial probability-mass function.
If an experiment is carried out by n independent trials and the probability for "success" in each trial is p, then the number of successes K is given by the binomial probability-mass function:

$K \in$ Bin(n, p).

If $p \to 0$ and $n \to \infty$ so that $m = n \cdot p$ is constant, we have approximately that

$K \in$ Po(np).

The approximation is satisfied if $p < 0.1$, $n > 10$. It is occasionally called the *law of small numbers*, following von Bortkiewicz (1898).

Example 1.9 (Poisson approximation). Consider a power plant. For a given month, the probability of no interruptions (stops in production) is 0.95. Denote by K the number of months with at least one interruption during one year. Clearly, $K \in \text{Bin}(n,p)$, with $n = 12$, $p = 0.05$. We investigate the validity of the Poisson approximation, $i.e.$ $K \in \text{Po}(0.6)$, since $np = 12 \cdot 0.05 = 0.6$. The following table results:

k	0	1	2	3	4
Binomial, $P(K = k)$	0.5404	0.3413	0.0988	0.0173	0.0021
Poisson, $P(K = k)$	0.5488	0.3293	0.0988	0.0198	0.0030

Repeating the calculation with a smaller probability, $p = 0.01$, we have $K \in \text{Po}(0.012)$ and obtain

k	0	1	2	3	4
Binomial, $P(K = k)$	0.8864	0.1074	0.0060	$2.01 \cdot 10^{-4}$	$4.57 \cdot 10^{-6}$
Poisson, $P(K = k)$	0.8869	0.1064	0.0064	$2.55 \cdot 10^{-4}$	$7.66 \cdot 10^{-6}$

Clearly, the lower the value of p is, the better the approximation works.   □

Geometric probability-mass function

Consider again the power plant in Example 1.9. Suppose we start a study in January (say) and are interested in the following random variable

$K = $ "The number of months before the first interrupt".

Using assumed independence, we find

$$P(K = k) = 0.05(1 - 0.05)^k, \quad k = 0, 1, 2, \dots .$$

Generally a variable K such that

$$P(K = k) = p(1 - p)^k, \quad k = 0, 1, 2, \dots \tag{1.12}$$

is said to have a geometric probability-mass function. If p is the probability of success then K is the time of the first success.

1.3 Conditional Probabilities and the Law of Total Probability

We begin with the concept of conditional probability. We wish to know the likelihood that some statement B is true when we know that another statement A, say, is true. (Intuitively, the chance that B is true should not be changed if we know that A is true and that the statements A and B are independent.)

For example, consider again an experiment of rolling a die. Let N be the number showed by the die. We can ask, what is the probability that $N = 1$ if we *know* that the result is an odd number, which we denote

$$p_1 = P(N = 1 \mid N \text{ is odd}). \tag{1.13}$$

Since all outcomes are equally probable, it is easy to agree that $p_1 = 1/3$. Obviously, we also have

$$p_2 = P(N = 2 \mid N \text{ is odd}) = 0.$$

We may ask what is the probability that $N < 3$ if N is odd. By Eq. (1.1), we get

$$P(N < 3 \mid N \text{ is odd}) = P(\, N = 1 \text{ or } N = 2 \mid N \text{ is odd})$$
$$= P(N = 1 \mid N \text{ is odd}) + P(N = 2 \mid N \text{ is odd }\,)$$
$$= p_1 + p_2 = \frac{1}{3}.$$

We turn now to the formal definition of conditional probability.

Definition 1.4 (Conditional probability). *The conditional probability of B given A such that $P(A) > 0$ is defined as*

$$P(B|A) = \frac{P(A \cap B)}{P(A)}. \tag{1.14}$$

Note that the conditional probability as a function of events B, A fixed, satisfies the assumptions of Definition 1.2, i.e. is a probability itself.

The conditional probability can now be recomputed by direct use of Eq. (1.14),

$$\frac{P(N < 3 \text{ and } N \text{ is odd})}{P(N \text{ is odd})} = \frac{P(N = 1)}{P(N \text{ is odd})} = \frac{1/6}{1/2} = \frac{1}{3},$$

i.e. the same result as obtained previously.

Remark 1.1. Obviously, if A and B are independent then

$$P(B|A) = \frac{P(A \cap B)}{P(A)} = \frac{P(A) \cdot P(B)}{P(A)} = P(B),$$

so in that case, knowledge that A occurred has not influenced the probability of occurrence of B. □

We turn now to a simple consequence of the fundamental Eq. (1.1). For a sample space $\mathcal{S}$ and two excluding events $A_1, A_2 \subset \mathcal{S}$ (that means $A_1 \cap A_2 = \emptyset$), if A_2 is a complement to A_1, *i.e.* if $A_1 \cup A_2 = \mathcal{S}$, then

$$P(A_1 \cup A_2) = P(A_1) + P(A_2) = 1,$$

A_1, A_2 is said to be a *partition* of $\mathcal{S}$, see the following definition. (Obviously $A_2 = A_1^c$.)

Definition 1.5. *A collection of events* $A_1, A_2, \ldots, A_n$ *is called a* partition *of* $\mathcal{S}$ *if*

(i) The events are mutually excluding, i.e.

$$A_i \cap A_j = \emptyset \quad \text{for } i \neq j$$

(ii) The collection is exhaustive, i.e.

$$A_1 \cup A_2 \cup \ldots \cup A_n = \mathcal{S},$$

that is, at least one of the events A_i *occurs.*

For a partition of $\mathcal{S}$,

$$P(A_1 \cup A_2 \cup \ldots \cup A_n) = P(A_1) + P(A_2) + \cdots + P(A_n) = 1.$$

Using the formalism of statements one can say that we have n different hypotheses about a sample point such that any two of them cannot be true simultaneously but at least one of them is true. Partitions of events are often used to compute (define) the probability of a particular event B, say. The following fundamental result can be derived:

Theorem 1.1 (Law of total probability). *Let* $A_1, A_2, \ldots, A_n$ *be a partition of* $\mathcal{S}$. *Then for any event* B

$$P(B) = P(B|A_1)P(A_1) + P(B|A_2)P(A_2) + \cdots + P(B|A_n)P(A_n).$$

Proof. Obviously, we have

$$B = (B \cap A_1) \cup (B \cap A_2) \cup \ldots \cup (B \cap A_n)$$

and since the events $B \cap A_i$ are mutually excluding we obtain

$$P(B) = P(B \cap A_1) + P(B \cap A_2) + \cdots + P(B \cap A_n). \tag{1.15}$$

Now from Eq. (1.14) it follows that

$$P(B|A)P(A) = P(B \cap A). \tag{1.16}$$

Combining Equations (1.15) and (1.16) gives the law of total probability. $\square$

The law of total probability is a useful tool if the chances of B to be true depend on which of the statements A_i are true. Obviously, if B and $A_1, \ldots, A_n$ are independent then nothing is gained by splitting B into n subsets, since

$$P(B) = P(B)P(A_1) + \cdots + P(B)P(A_n).$$

Example 1.10 (Electrical power supply). Assume that we are interested in the risk of failure of an electric power supply in a house. More precisely, let the event B be "Errors in electricity supply during a day". From experience we know that in the region errors in supply occurs on average once per 10 thunder storms, 1 per 5 blizzards, and 1 per 100 days without any particular weather-related reasons. Consequently, one can consider the following partition of a sample space:

$A_1 = $ "A day with thunder storm", $A_2 = $ "A day with blizzard,"

$A_3 = $ "Other weather".

Obviously the three statements A_1, A_2, and A_3 are mutually exclusive but at least one of them is true. (We ignore the possibility of two thunderstorms in one day.)

From the information in the example it seems reasonable to estimate $P(B|A_1) = 1/10$, $P(B|A_2) = 1/5$, and $P(B|A_3) = 1/100$. Now in order to compute the probability that day one has no electricity supply, we need to compute the probabilities (frequencies) of days with thunder storm, blizzard. Assume that we have on average 20 days with thunderstorms and 2 days with blizzards during a year, then

$$P(B) = 0.1 \cdot \frac{20}{365} + 0.2 \cdot \frac{2}{365} + 0.01 \cdot \left(1 - \frac{20}{365} - \frac{2}{365}\right) = 0.016.$$

$\square$

1.4 Event-tree Analysis

Failure of a complicated engineering system can lead to different damage scenarios. The consequence of a particular failure event may depend on a sequence of events following the failure. The means for systematic identification of the possible event sequences is the so-called *event tree*. This is a visual representation, indicating all events that can lead to different scenarios. In the following example, we first identify events. Later on, we show how conditional probabilities can be applied to calculate probabilities of possible scenarios.

Example 1.11 (Information on fires). Consider an initiation event A, fire ignition reported to a fire squad. After the squad has been alarmed and has done its duty at the place of accident, a form is completed where a lot of information about the fire can be found: type of alarm, type of building, number of staff involved, and much more. We here focus on the following:

- The condition of the fire at the arrival of the fire brigade. This is described by the following statement

    E_1: "Smoke production without flames"

and the complement

E_1^c: "A fire with flames (not merely smoke production)".

- The place where the fire was extinguished, described by the event

    $E_2 =$ "Fire was extinguished in the item where it started"

and the complement

    $E_2^c =$ "Fire was extinguished outside the item".

For an illustration, see Figure 1.4.   ☐

Let us consider one branch of an event tree, starting with the failure event A_1 and the following ordered events of consequences $A_2, \ldots, A_n$. It is natural to compute or estimate from observations, the conditional probabilities $P(A_2|A_1)$, $P(A_3|A_2$ and $A_1)$, etc. We turn to a formula for the probability of a branch "A_1 and A_2 and $\ldots A_n$."

Using the definition of conditional probabilities Eq. (1.14), we have that for $n = 2$

$$P(A_1 \cap A_2) = P(A_2|A_1)P(A_1).$$

Similarly for $n = 3$ we have that

$$P(A_1 \cap A_2 \cap A_3) = P(A_3|A_2 \cap A_1)P(A_2 \cap A_1) = P(A_3|A_2 \cap A_1)P(A_2|A_1)P(A_1).$$

Repeating the same derivation n times we obtain the general formula

$$P(A_1 \cap A_2 \cap \ldots \cap A_n) = P(A_n|A_{n-1} \cap \ldots \cap A_1)$$
$$\cdots \cdot P(A_3|A_2 \cap A_1) \cdot P(A_2|A_1)P(A_1). \quad (1.17)$$

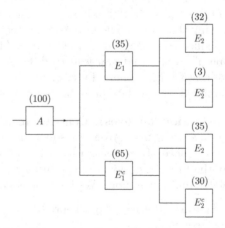

Fig. 1.4. The event tree discussed in Examples 1.11 and 1.12. The numbers within parentheses indicate the number of cases observed after 100 fire ignitions.

The derived Eq. (1.17) is a useful tool to calculate the probability for a "chain" of consequences. Often in applications, events can be assumed to be independent and the probability for a specific scenario can then be calculated. If $A_1, \ldots, A_n$ are independent, then

$$P(A_i \,|\, A_{i-1}, \ldots, A_1) = P(A_i)$$

and $P(A_1 \cap \ldots \cap A_n) = P(A_1) \cdot \ldots \cdot P(A_n)$. In applications with many branches the computations may be cumbersome, and approximate methods exist; see [3], Chapter 7.5. We now return to our example from fire engineering.

Example 1.12 (Information on fires). From statistics for fires in industries in Sweden (see Figure 1.4), we can assign realistic values of the probabilities, belonging to the events in the event tree:

$$P(E_1) = \frac{35}{100} = 0.35, \quad P(E_2|E_1) = \frac{32}{35} = 0.91, \quad P(E_2|E_1^c) = \frac{35}{65} = 0.54.$$

Within an event tree, obviously some outcomes are more interesting than others with respect to the potential damage, the more the number of serious damages, the higher the costs. Consider in our simple example the scenario that there was a fire with flames at the arrival *and* that the fire was extinguished outside the item where it started. We calculate probabilities according to Eq. (1.17) and have $A_1 = E_1^c$ and $A_2 = E_2^c$; hence the probability is given as

$$P(E_1^c \cap E_2^c) = P(E_2^c|E_1^c) \cdot P(E_1^c) = (1 - 0.54) \cdot (1 - 0.35) = 0.30.$$

(Note that this probability could be directly obtained from Figure 1.4: $P(E_1^c \cap E_2^c) = 30/100$). □

Problems

1.1. A student writes exams in three subjects in one week. Let A, B, and C denote events of passing each of the subjects. The probabilities for passing are $P(A) = 0.5$, $P(B) = 0.8$, and $P(C) = 0.2$, respectively. Assume that A, B, and C are independent and the total number of exams that the student will pass is denoted by X.

(a) What are the possible values of X, or in other words, give the sample space.
(b) Calculate $P(X = 0)$, $P(X = 1)$.
(c) Calculate $P(X < 2)$.
(d) Is it reasonable to assume independence?

1.2. Demonstrate that for any events A and B

$$P(A \cup B) = P(A) + P(B) - P(A \cap B).$$

1.3. Consider two independent events A and B with $P(A) > 0$, $P(B) > 0$. Are the events excluding?

1.4. For any event A, denote by A^c its complement. Can the events A and A^c be independent?

1.5. For a given month, the probability of at least one interruption in a powerplant is 0.05. Assume that events of interrupts in the different months are independent. Calculate

(a) The probability of exactly three months with interruptions during a year,
(b) The probability for a whole year without interruptions.

1.6. In an office, there are 110 employees. Using a questionnaire, the number of vegetarians has been found. The following statistics are available:

	Vegetarians	Nonvegetarians
Men	25	35
Women	32	18

One of the employees is chosen at random (any person has the same probability to be selected).

(a) Calculate the probability that the chosen person is a vegetarian.
(b) Suppose one knows that a woman was chosen. What is the probability that she is a vegetarian?
(c) Are the events "woman is chosen" and "vegetarian is chosen" independent? Explain your reasoning.

1.7. The lifetime of a certain type of light bulb is supposed to be longer than 1000 hours with probability 0.55. In a room, four light bulbs are used. Find the probability that at least one light bulb is functioning for more than 1000 hours. Assume that the lifetimes of the different light bulbs are independent.

1.8. Consider the circuit in Figure 1.5.

The components A_1 and A_2 each function with probability 0.8. Assuming independence, calculate the probability that the circuit functions. *Hint:* The system is working as long as one of the components is working.

1.9. Consider the lifetime of a certain filter. The probability of a lifetime longer than one year is equal to 0.9, while the probability of a lifetime longer than five years is 0.1. Now, one has observed that the filter has been functioning longer than one year. Taking this information into account, what is the probability that it will have a lifetime longer than five years?

1.10. Consider a chemical waste deposit where some containers with chemical waste are kept. We investigate the probability of leakage during a time period of five years, that is, with

B = "Leakage during five years"

the goal is to compute $P(B)$.

Due to subterranean water, *corrosion* of containers can lead to leakage. The probability of subterranean water flow at the site during a time period of five years

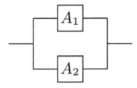

Fig. 1.5. Circuit studied in Problem 1.8.

is $P(A_1) = 0.04$ and the probability of leakage under these conditions is $P(B \mid A_1) = 0.6$. The other important reason for leakage is *thermal expansion* due to chemical reactions in the container. The probability of conditions for thermal expansion is $P(A_2) = 0.01$ and $P(B \mid A_2) = 0.9$. Leakage can also occur for other reasons than the two mentioned, $P(B \mid A_1^c \cap A_2^c) = 0.01$.

Based on this information, compute $P(B)$, the probability for leakage of a container at the site during a five-year period. (Discussion on environmental problems and risk analysis is found in a book by Lerche and Paleologos [50]).

1.11. Color blindness is supposed to appear in 4 percent of the people in a certain country. How many people need to be tested if the probability to find at least one colour blind person is to be 0.95 or more? Note that for simplicity we allow to test a person several times, i.e., people are chosen with replacement. *Hint:* Use suitable approximation.

1.12. A manufacturer of a certain type of filters for use in powerplants claims that on average one filter out of thousand has a serious fault. At a powerplant with 200 installed filters, 2 erroneous filters have been found, which rather indicates that one filter out of hundred is of low quality.

The management of the powerplant wants to claim money for the filters and want to calculate, based on the information from the manufacturer, the probability of more than two erroneous filters out of 200. Calculate this probability (use suitable approximation).

2

Probabilities in Risk Analysis

In the previous chapter, we introduced conditions that a function P has to satisfy in order to be called probability, see Definition 1.2. The probability function is then used as a measure of the chances that a statement about an outcome of an experiment is true. This measure is intended to help in decision making in situations with uncertain outcomes.

In order to be able to model uncertainties in a variety of situations met in risk analysis, we need to further elaborate on the notion of probability. The following four common usages of the concept of probability are discussed in this chapter:

(**1**) To measure the present state of knowledge, *e.g.* the probability that a patient tested positively for a disease is really infected, or that the detected tumour is malignant. "The patient is infected or not", "the tumour is malignant or benign" — we just do not know which of the statements is true. Usually further studies or tests will give exact answer to the question, see also Examples 2.2 and 2.3.

(**2**) To quantify the uncertainty of an outcome of a non-repeatable event, for instance the probability that your car will break down tomorrow and you miss an important appointment, or that the flight you took will land safely. Here again the probability will depend on the available information, see Example 2.4.

(**3**) To describe variability of outcomes of repeatable experiments, *e.g.* chances of getting "Heads" in a flip of a coin; to measure quality in manufacturing; everyday variability of environment, see Section 2.4.

(**4**) In the situation when the number of repetitions of the experiment is uncertain too, *e.g.* the probability of fire ignition after lightning has hit a building. Here we are mostly interested in conditional probabilities of the type: given that a cyclone has been formed in the Caribbean Sea, what are the chances that its centre passes Florida. Obviously, here nature controls the number of repetitions of the experiment.

If everybody agrees with the choice of P, it is called an *objective proba-bility*. This is only possible in a situation when we use mathematical models. For example, under the assumption that a coin is "fair" the probability of getting tails is 0.5. However, there are probably no fair coins in reality and the probabilities have to be estimated. It is a well-known fact that measurements of physical quantities or estimation of probabilities done by different laboratories will lead to different answers (here we exclude the possibility of arithmetical errors). This happens because different approaches, assumptions, knowledge, and experience from similar problems will lead to a variety of estimates. Especially for the problems that have been described in (**1**) and (**2**), the probability incorporates often different kinds of information a person has when estimating the chances that a statement A, say, is true. One then speaks of *subjective probabilities*. As new information about the experiment (or the outcome of the experiment) is gathered there can be some evidence that changes our opinions about the chances that A is true. Such modifications of the probabilities should be done in a coherent way. Bayes' formula, which is introduced in Section 2.1, gives a means to do it.

Sections 2.4–2.6 are devoted to a discussion of applications of probabilities for repeatable events, as described in (**3**) and (**4**). In this context, it is natural to think of how often a statement is true. This leads to the interpretation of probabilities as frequencies, which is discussed in Section 2.4. However, often even the repetition of experiments happens in time in an unpredictable way, at random time instants. This aspect has to be taken into account when modelling safety of systems and is discussed in Sections 2.5 and 2.6, respectively. Concepts presented in those sections, in particular the concept of a *stream of events*, will be elaborated in later chapters.

2.1 Bayes' Formula

We next present *Bayes' formula*, attributed to Thomas Bayes (1702–1761). Bayes' formula is valid for any properly defined probability P; however, it is often used when dealing with subjective probabilities in cases (**1-2**). These types of applications are presented in the following two subsections.

> **Theorem 2.1 (Bayes' formula).** *Let $A_1, A_2, \ldots, A_k$ be a partition of S, see Definition 1.5, and B an event with $P(B) > 0$. Then*
>
> $$P(A_i \mid B) = \frac{P(A_i \cap B)}{P(B)} = \frac{P(B \mid A_i)P(A_i)}{P(B)}.$$

In the framework of Bayes' formula, we deal with a collection of alternatives $A_1, A_2, \ldots, A_n$, for which one and only one is true: we want to deduce which one. The function $L(A_i) = P(B|A_i)$ is called the *likelihood* and measures how likely the observed event is under the alternative A_i. Note that for an event B,

$$P(B) = P(B|A_1)P(A_1) + \cdots + P(B|A_n)P(A_k),$$

by the law of total probability.

Often a version of Bayes' formula is given, which particularly puts emphasis on the role of $P(B)$ as a normalization constant:

$$P(A_i|B) = c\,P(B|A_i)P(A_i), \tag{2.1}$$

where $c = 1/P(B)$ is a normalization constant. In practical computations, all terms $P(B|A_i)P(A_i)$ are first evaluated, then added up to derive c^{-1}. Actually, this approach is particularly convenient when *odds* are used to measure chances that alternative A_i is true (see the following subsection). Then the constant c does not have to be evaluated; any value could be used.

2.2 Odds and Subjective Probabilities

Consider a situation with two events; for example, the odds for $A_1 =$ "A coin shows heads" and $A_2 =$ "A coin shows tails" when flipping a fair coin is usually written 1:1. In this text we define the odds for events A_1 and A_2, to be any positive numbers q_1, q_2 such that $q_1/q_2 = P(A_1)/P(A_2)$. Knowing probabilities, odds can always be found. However, the opposite is not true: odds do not always give the probabilities of events. For instance, the odds for $A_1 =$ "A die shows six" against $A_2 =$ "A die shows one" for a fair die are also 1:1. However, if one knows that A_1, A_2 form a partition, *e.g.* $A_2 = A_1^c$, the probabilities $P(A_1)$ and $P(A_2)$ are given by

$$P(A_1) = \frac{q_1}{q_1 + q_2}, \quad P(A_2) = \frac{q_2}{q_1 + q_2},$$

respectively. In the following theorem we generalize this result to more than two events.

Theorem 2.2. *Let $A_1, A_2, \ldots, A_k$ be a partition of S, having odds q_i, i.e. $P(A_j)/P(A_i) = q_j/q_i$. Then*

$$P(A_i) = \frac{q_i}{q_1 + \cdots + q_k}. \tag{2.2}$$

Example 2.1. Consider an urn with balls of three colours. 50 % of the balls are red, 30 % black, and the remaining balls green. The experiment is to draw a ball from the urn. Clearly A_1, A_2, and A_3, defined as the ball being red, black, or green, respectively, forms a partition. It is easy to see that the odds for A_i are 5:3:2. Now by Theorem 2.2 we find, for instance $P(A_2)$, the probability that a ball picked at random is black:

$$P(A_2) = \frac{3}{5+3+2} = 0.3.$$

$\square$

We now present Bayes' formula for odds. Consider again any two statements A_i and A_j having odds $q_i : q_j$, which we call *a priori* odds and also denote as $q_i^{\text{prior}} : q_j^{\text{prior}}$. Next, suppose that one knows that a statement B about the result of the experiment is true. Knowledge that B is true may influence the odds for A_i and A_j, and lead to *a posteriori* odds, any positive numbers $q_i^{\text{post}}, q_j^{\text{post}}$ such that $q_i^{\text{post}}/q_j^{\text{post}} = P(A_i|B)/P(A_j|B)$. Now Bayes' formula can be employed to compute the *a posteriori* odds:

$$q_i^{\text{post}} = P(B \mid A_i)q_i^{\text{prior}}, \tag{2.3}$$

for any value of i. (Obviously, $q_i^{\text{post}} = cP(B \mid A_i)q_i^{\text{prior}}$, for any positive c, are also the *a posteriori* odds, since the ratio $q_i^{\text{post}}/q_j^{\text{post}}$ remains unchanged.)

The notions *a priori* and *a posteriori* are often used when applying Bayes' formula. These are known from philosophy, and serve, in a general sense, to make a distinction among judgements, concepts, ideas, arguments, or *kinds of knowledge*. The *a priori* is taken to be independent of sensory experience, which *a posteriori* presupposes, being dependent upon or justified by reference to sensory experience. The importance of Bayesian views in science has been discussed for instance by Gauch [27].

Example 2.2 ("Car or goat?"). Let us return to the Monty Hall problem from Example 1.5 and compute the posterior odds for a car being behind door No. 1.

As before, let us label the doors No. 1, No. 2, and No. 3, and suppose that the player chooses door No. 1, and that the following statement

$B =$"The host opens door No. 3"

is true. The player can now decide to open door No. 1 or No. 2. The prior odds for a car being behind No. 1 against it being not was 1:2. Now, he wishes to base his decision on the posterior odds, *i.e.* rationally he will open door No. 1 if this has the highest odds to win the car.

In order to find the odds let us first introduce the following three alternatives:

$A_1 =$ "The car is behind No. 1", $A_2 =$ "The car is behind No. 2",

$A_3 =$ "The car is behind No. 3".

Let $q_1^{\text{prior}}, q_2^{\text{prior}}, q_3^{\text{prior}}$ be the odds for A_1, A_2, A_3, respectively. Here the odds are denoted as *a priori* odds since their values will be chosen from knowledge of the rules of the game and experience from similar situations. It seems reasonable to assume that the prior odds are 1:1:1. However, since B is true the player wishes to use this information to compute the *a posteriori* odds. In order to be able to use Eq. (2.3) to compute the posterior odds he needs to know the likelihood function $L(A_i)$, *i.e.* the probabilities of B conditionally that the alternative A_1 (or A_2) is true: $P(B|A_1)$ and $P(B|A_2)$. The assigned values for the probabilities reflect his knowledge of the game.

Since the player chooses door No. 1 a simple consequence of the rules is that $P(B|A_2) = 1$. He turns now to the second probability $P(B|A_1)$; if A_1 is true (the car is behind the door No. 1) then the host had two possibilities: to open door No. 2 or No. 3. If one can assume that he has no preferences between the doors then $P(B|A_1) = 1/2$, which the player assumes, leading to the following posterior odds by Eq. (2.3)

$$ q_1^{\text{post}} = P(B \mid A_1)q_1^{\text{prior}} = \frac{1}{2} \cdot 1 = \frac{1}{2}, \qquad q_2^{\text{post}} = P(B \mid A_2)q_2^{\text{prior}} = 1 \cdot 1 = 1. $$

Since $q_3^{\text{post}} = 0$, the posterior odds for a car being behind No. 1 is still 1:2. Hence a rational decision is to open door No. 2. (Note that the odds would be 1:1 if the host opens door No. 3 whenever he can, since then $P(B|A_1) = 1$.)

□

Bayes' formula in the formulation in Eq. (2.3) is often used in the case when A_i are interpreted as alternatives. For example, in a courtroom, one can have

$$ A_1 = \text{``The suspect is innocent''}, \quad A_2 = A_1^c = \text{``The suspect is guilty''} $$

while B is the *evidence*, for example

$$ B = \text{``DNA profile of suspect matches the crime sample''}. $$

Using modern DNA analysis, it can often be established that the conditional probability $P(B|A_2)$ is very high while $P(B|A_1)$ very low. However, what is really of interest are the posterior odds for A_1 and A_2 *conditionally* the evidence B, which are given by Eq. (2.3), *i.e.* $P(B|A_1)q_1^{\text{prior}} : P(B|A_2)q_2^{\text{prior}}$. Here the prior odds summarizes the strength of all the other evidences, which can be very hard to estimate (choose) and quite often, erroneously, taken as 1:1.

We end this section with an example of a typical application of Bayes' formula, where the prior odds dominates the conditional probabilities $P(B|A_i)$. The values for various probabilities used in the example are hypothetical and probably not too realistic. This is an important example, illuminating the role of priors, which is often erroneously ignored, cf. [39], pp. 52-54.

Example 2.3 (Mad cow disease). Suppose that one morning a newspaper reports that the first case of a suspected mad cow (BSE infected cow) is found. "Suspected" means that a test for the illness gave positive result. Since this information can influence shopping habits, a preliminary risk analysis is desired. The most important information is the probability that a cow, positively tested for BSE, is really infected.

Let us introduce the statements

$$A = \text{"Cow is BSE infected"} \quad \text{and} \quad B = \text{"Cow is positively tested for BSE"}.$$

The posterior odds for $A_1 = A$ and $A_2 = A^c$ given that one knows that B is true are of interest. These can be computed using Bayes' formula (2.3), if the *a priori* odds $q_1^{\text{prior}}, q_2^{\text{prior}}$ and the likelihood function, *i.e.* the conditional probabilities $P(B|A_1)$ and $P(B|A_2)$, are known.

Selection of prior odds. Suppose that one could find, *e.g.* on the Internet, a description of how the test for BSE works. The important information is that the frequency of infected cows that pass the test, *i.e.* are not detected, is 1 per 100 (here human errors, like mixing the samples etc, are included), while a healthy cow can be suspected for BSE in 1 per 1000 cases. This implies that $P(B|A_1) = 0.99$ while $P(B|A_2) = 0.001$. Assume first that the odds that a cow has BSE are 1:1 (half of the population of cows is "mad"). Then the posterior odds are

$$q_1^{\text{post}} = 0.99 \cdot 1 = 0.99, \qquad q_2^{\text{post}} = 0.001 \cdot 1 = 0.001,$$

in other words 990:1 in favour that the cow has BSE. Many people erroneously neglect estimating the prior odds, which leads to the "pessimistic" posterior odds 990:1 for a cow to be BSE infected.

In order to assign a more realistic value to the prior odds, the problem needs to be further investigated. Suppose that the reported case was observed on a cow chosen at random. Then the reasonable odds for A and A^c would be

"Number of BSE infected cows" : "Number of healthy cows".

Note that the numbers are unknown! In such situations one needs to rely on the experience and has to ask an expert for his opinion.

Prior odds: Expert's opinion. Suppose an expert claims that there can be as many as 10 BSE infected cows in a total population of ca 1 million cows. This results in the priors $q_1^{\text{prior}} = 1$, $q_2^{\text{prior}} = 10^5$ leading to the posterior odds

$$q_1^{\text{post}} = 0.99, \qquad q_2^{\text{post}} = 0.001 \cdot 10^5,$$

which can be also written as $1 : 100$ in favour of that the cow is healthy.

Finally, suppose one decides to test *all* cows and as a consumer one should be interested in the odds that a cow that passed the test is actually infected, *i.e.* $P(A_1|B^c)$. Again we start with the conditional probabilities

$$P(B^c \mid A_1) = 1 - 0.99 = 0.01, \qquad P(B^c \mid A_2) = 1 - 0.001 = 0.999,$$

and then using the expert's odds for A_1 and A_2, $1 : 10^5$, Bayes' formula gives the following posterior odds

$$q_1^{\text{post}} = 0.01 \cdot 1, \qquad q_2^{\text{post}} = 0.999 \cdot 10^5,$$

which (approximately $1 : 10^7$) is clearly a negligible risk, if one strongly believes in the expert's odds.   □

2.3 Recursive Updating of Odds

In many practical situations the new information relevant for risk estimation is
collected (or available) in different time instances. Hence the odds are changing
in time with new received information. Again, Bayes' formula is the main tool
to compute the new, updated, priors for truth of statements A_i.

Sequences of statements

Before giving an example let us formalize the described process of updating of
the odds. Suppose one is interested in the odds for a collection of statements
$A_1, \ldots, A_k$, which form a partition, *i.e.* these are mutually excluding and
always one of them is true, (see Definition 1.5). Let q_i^0 denote the *a priori*
odds for A_i. Let $B_1, \ldots, B_n, \ldots$ be the sequence of statements (evidences)
that become available with time and let q_i^n be the *a posteriori* odds for A_i
with the knowledge that $B_1, \ldots, B_n$ are true is included. Obviously, Bayes'
formula (2.3) can be used to compute q_i^n, if the likelihood function $L(A_i)$,
i.e. the conditional probability $\mathsf{P}(\text{all } B_1, \ldots, B_n \text{ are true} \,|\, A_i)$, is known. The
formula simplifies if it can be assumed that given that A_i is true $B_1, \ldots, B_n$
are independent. For $n = 2$ this means that

$$\mathsf{P}(B_1 \cap B_2 \,|\, A_i) = \mathsf{P}(B_1 \,|\, A_i)\mathsf{P}(B_2 \,|\, A_i).$$

This property will be called *conditional independence*.

Theorem 2.3. *Let* $A_1, A_2, \ldots, A_k$ *be a partition of* S, *and* $B_1, \ldots, B_n, \ldots$
a sequence of true statements (evidences). If the statements B *are conditionally independent of* A_i *then the* a posteriori *odds after receiving the*
n th *evidence*

$$q_i^n = \mathsf{P}(B_n \,|\, A_i)q_i^{n-1}, \qquad n = 1, 2, \ldots, \tag{2.4}$$

where q_i^0 *are the* a priori *odds.*

The last theorem means that each time a new evidence B_n, say, is available
the posterior odds for A_i, A_j are computed using Bayes' formula (2.3) and
then the prior odds are *updated*, *i.e.* replaced by the posterior odds. This recursive estimation of the odds for A_i is correct only if the evidences $B_1, B_2, \ldots$
are conditionally (given A_i is true) independent.

In the following example, presenting an application actually studied with
Bayesian techniques by von Mises in the 1940s [54], we apply the recursive

Bayes' formula to update the odds. The example represents a typical applications of Bayes' formula and the subjective probabilities[1] (odds) in safety
analysis.

Example 2.4 (Waste-water treatment). A new unit at a biological waste-
water treatment station has been constructed. The active biological substances
can work with different degree of efficiency, which can vary from day to day,
due to variability of waste-water chemical properties, temperature, etc. This
uncertainty can be measured by means of the probability that a chemical
analysis of the processed water, done once a day or so, satisfies a required
standard and can be released. We write this as $p = P(B)$ where

$B = $ "The standard is satisfied".

Since p is the frequency of water releases, the higher the value of p, the more
efficient the waste-water treatment is.

The constant p is needed in order to make a decision whether a new
bacterial culture has to be used to treat the waste water or a change of the
oxygen concentrations should be made. Under stationary conditions one can
assume that the probability is constant over time and, as shown in the next
section, using rules of probabilities, one can find the value p if an infinite
number of tests were performed: simply, this is the fraction of times B were
true. However, this is not possible in practice since it would take infinitely
long time and require not-negligible costs. Consequently, the efficiency of the
unit needs to be evaluated based on a finite number of tests during a trial
period.

Subjective probabilities. By experience from similar stations we claim that for
a randomly chosen bacterial culture, the probability p can take values 0.1,
0.3, 0.5, 0.7, and 0.9, which means that we here have $k = 5$ alternatives to
choose between

$A_1 = $ "$p = 0.1$", ..., $A_5 = $ "$p = 0.9$"

about the quality of bacterial culture, *i.e.* the ability to clean the waste water.
(Note that if A_5 is true, the volume of cleaned water is $0.9/0.1 = 9$ times
higher than if A_1 were true.) Mathematically, if the alternative A_i is true
then $P(B|A_i) = p$, that is

$P(B|A_1) = 0.1$, $P(B|A_2) = 0.3$, $P(B|A_3) = 0.5$, $P(B|A_4) = 0.7$,

$P(B|A_5) = 0.9$,

furthermore $P(B^c|A_i) = 1 - P(B|A_i)$. However, we do not know which of the
alternatives A_i is correct. The ignorance about possible quality (the p value)
of the bacterial culture can be modelled by means of odds q_i for which of A_i
is true.

---

[1] A formalization of the notion of subjective probabilities was made in a classical
paper by Anscombe and Aumann [4], often referred to in economics when expected
utility is discussed.

Selection of prior odds. Suppose nothing is known about the quality of the bacterial culture, *i.e.* any of the values of p are equally likely. Hence the prior odds, denoted by q_i^0, are all equal, that is, $q_i^0 = 1$.

Computing posterior odds for A_i. Denote by B_n the result of the nth test, *i.e.* B or B^c is true, and let the odds for the alternative A_i be q_i^n (including all evidences $B_1, \ldots, B_n$). The posterior odds will be computed using the recursive Bayes' formula (2.4). This is a particularly efficient way to update the odds when the evidences B_n become available at different time points[2].

Suppose the nth measurement results in that B is true; then, by Theorem 2.3, the posterior odds

$$q_i^n = P(B|A_i)q_i^{n-1}, \qquad n > 0,$$

and $q_i^0 = 1$, while if instead the nth measurement resulted in B^c being true

$$q_i^n = P(B^c|A_i)q_i^{n-1} = \left(1 - P(B|A_i)\right)q_i^{n-1}.$$

Note that the odds are defined up to a factor c. In the following example we choose to use $c = 10$.

Suppose the first 5 measurements resulted in a sequence $B \cap B^c \cap B \cap B \cap B$, which means the tests were positive, negative, positive, positive, and positive. Let us again apply the recursion to update the uniform prior odds. Let us choose $c = 10$; then, each time the standard is satisfied the odds $q_1, \ldots, q_5$ are multiplied by $1, 3, 5, 7, 9$, respectively, while in the case of negative test result one should multiply the odds by the factors $9, 7, 5, 3, 1$. Consequently, starting with uniform odds as the results of tests arrive, the odds are updated as follows

	A_1 $p = 0.1$	A_2 $p = 0.3$	A_3 $p = 0.5$	A_4 $p = 0.7$	A_5 $p = 0.9$
prior	1	1	1	1	1
B	1	3	5	7	9
B^c	9·1	7·3	5·5	3·7	1·9
B	1·9	3·21	5·25	7·21	9·9
B	1·9	3·63	5·125	7·147	9·81
B	1·9	3·189	5·625	7·1029	9·729

We note that after this particular sequence, the highest likelihood is given for $p = 0.7$ as

$$P(p = 0.7) = \frac{7 \cdot 1029}{1 \cdot 9 + 3 \cdot 189 + 5 \cdot 625 + 7 \cdot 1029 + 9 \cdot 729} = 0.41$$

using Eq. (2.2). (Note that the observed frequency of positive test is $4/5$, *i.e.* between alternatives A_4 and A_5.)   □

---

[2] However, in order to be able to use the formula one needs to assume that B_n are conditionally independent if A_i is true. This can be a reasonable assumption if one uses tests separated by long enough periods of time. Let us assume this.

The previous example is further investigated below, where the efficiency of the cleaning is introduced through properties of p.

Example 2.5 (Efficiency of cleaning). As already mentioned the probability p is only needed to make a decision to keep or replace the bacterial culture in the particular waste-water cleaning station. For example, suppose on basis of economical analysis it is decided that the bacterial culture is called *efficient* if $p \geq 0.5$, *i.e.* on average cleaned water is released at least once in two days. Hence our rational decision, whether to keep or replace the bacterial culture, will be based on odds for

$A =$ "Bacterial culture is efficient"

against A^c.

We have that A is true if A_3, A_4, or A_5 are true while A^c is true if A_1 or A_2 are true. Hence, since A_i are excluding, we have

$$\mathsf{P}(A) = \mathsf{P}(A_3) + \mathsf{P}(A_4) + \mathsf{P}(A_5) \quad \text{while} \quad \mathsf{P}(A^c) = \mathsf{P}(A_1) + \mathsf{P}(A_2).$$

For the odds, we have $q_A/q_{A^c} = \mathsf{P}(A)/\mathsf{P}(A^c)$ and thus the odds for A against A^c are computed as

$$q_A = q_3^n + q_4^n + q_5^n, \quad q_{A^c} = q_1^n + q_2^n.$$

The same sequence of measurements as in the previous example, B, B^c, B, B, B, results in the posterior odds in favour for A (the bacterial culture is efficient) being $16889 : 567 = 29.8 : 1$. The posterior probability that A is true after receiving results of the first 5 tests is $\mathsf{P}(A) = 29.8/(1+29.8) = 0.97$. □

In the last example, the true probability $p = \mathsf{P}(B)$ can be only one of the five possibilities; this is clearly an approximation. In Chapter 6 we will return to this example and present a more general analysis where p can be any number between zero and one.

Remark 2.1 (Selection of information). It is important to use *all* available information to update priors. A biased selection of evidences for A (against A^c) that supports the claim that A is true will obviously lead to wrong posterior odds. Consider for example the situation of the courtroom, discussed in page 25: imagine situations and information that when omitted could change the posterior odds. □

2.4 Probabilities as Long-term Frequencies

In previous sections of this chapter, we studied probabilities as used in situations (1-2), *e.g.* we have non-repeatable scenarios and wish to measure uncertainties and lack of knowledge to make decisions whether statements are true or not. In this section, we turn to a diametrically different setup of repeatable events.

Frequency interpretation of probabilities

In Chapter 1, some basic properties of probabilities were exemplified by using two simple experiments: flip a coin and roll a die. Let us concentrate on the first one and denote its sample space $\mathcal{S} = \{0, 1\}$, which represents the physically observed outcomes $\mathcal{S} = \{$"Heads", "Tails"$\}$. Next, let us flip the coin many times, in the independent manner, and denote the results of the ith flip by X_i. (The random variables X_i are independent.)

If the coin is fair then $\mathsf{P}(X_i = 1) = \mathsf{P}(X_i = 0) = 1/2$. In general, a coin can be biased. Then there is a number p, $0 \leq p \leq 1$, such that $\mathsf{P}(X_i = 1) = p$ and, obviously, $\mathsf{P}(X_i = 0) = 1 - p$. (For example, $p = 1$ means that the probability for getting "Tails" is one. This is only possible for a coin that has "Tails" on both sides.) Finding the exact value of p is not possible in practice. However, using suitable statistical methods, estimates of p can be computed. One type of estimation procedure is called the *Frequentist Approach*. This is motivated by the fundamental result in theory of probability, "Law of Large Numbers" (LLN), given in detail in Section 3.5. The law says that the fraction of "tails" observed in the first n independent flips converges to p as n tends to infinity:

$$\bar{X} = \frac{1}{n}(X_1 + X_2 + \cdots + X_n) \to p, \quad \text{as } n \to \infty \tag{2.5}$$

since $\sum_{i=1}^{n} X_i$ is equal to the number of times "tails" is shown in n flips. Thus we can interpret p as "long-term frequency" of "tails" in an infinite sequence of flips. (Later on in Chapter 6 we will also present the so-called *Bayesian Approach* to estimate p.)

Practically, one cannot flip a coin infinitely many times. Consequently, we may expect that in practice $\bar{X} \neq p$ and it is important to study[3] the error $\mathcal{E} = p - \bar{X}$ or relative error $|p - \bar{X}|/p$. Obviously errors will depend on the particular results of a flipping series and hence are random variables themselves. A large part of Chapter 4 will be devoted to studies of the size of errors. Here we only mention that (as expected) larger n should give on average smaller errors. An interesting question is how large n should be so that the error is sufficiently small (for the problem at hand).

In Chapter 9 we will show that for a fair coin ($p = 0.5$) about 70 flips are needed in order to have $0.4 < \bar{X} < 0.6$, *i.e.* relative error less than 20%, with high probability (see Problem 9.5). A result from a computer simulation is shown in Figure 2.1. Of hundred such simulations, on average 5 would fail to satisfy the bound. In the more interesting case when the probability p is small, $100/p$ flips are required approximately in order to have a "reliable" estimate of the unknown value of p.

As shown next, $\bar{X}$ can also be used to estimate general probabilities $p = \mathsf{P}(A)$ of a statement A about an outcome of an experiment that can be performed infinitely many times in an independent manner.

---

[3]Note that the value of p is also unknown.

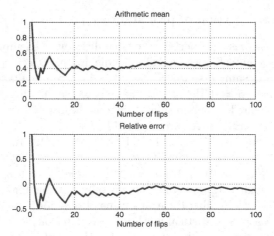

Fig. 2.1. Simulation, tosses of a fair coin. *Top*: Arithmetic mean $\bar{x} = \frac{1}{n}(x_1 + x_2 + \cdots + x_n)$. *Bottom*: Relative error $(\bar{x} - p)/p$.

Long-term frequencies. Define a sequence of independent random variables X_i as follows

$$X_i = \begin{cases} 1, & \text{if the statement } A \text{ about the outcome of the } i\text{th experiment} \\ & \text{is true,} \\ 0, & \text{otherwise.} \end{cases}$$

Again, by *LLN*, $\bar{X} = \frac{1}{n}(X_1 + X_2 + \cdots + X_n) \to p$, where $p = \mathsf{P}(A)$.

Here we interpret the probability $\mathsf{P}(A)$ as observed *long-term frequencies* when the statement A about a result of an experiment is true. In most computations of risks, one wishes to give probabilities interpretations as frequencies of times when A is true. However, this is not always possible as discussed in the previous section.

An approach to construct the notion of probability based on long-term frequencies (instead of the axiomatic approach given in Definition 1.2) was suggested by von Mises in the first decades of the 20th century (cf. [16] for discussion). However, the approach leads to complicated mathematics, hence the axiomatic approach (presented by Kolmogorov in 1933 [44]), see Definition 1.2, is generally used at present. Nevertheless the interpretation of probabilities as frequencies is intuitively very appealing and is important in engineering applications.

We end this subsection with some remarks on practical conditions when $\bar{X}$ can be used as an estimator of the unknown probability $\mathsf{P}(A)$.

Remark 2.2. Often, in practice, the assumption that results of experiments are independent cannot be checked (or is not appropriate). If the assumption of independence cannot be motivated, then one checks whether experiments are *stationary*, which means that properties of the experiment do not change with time. Under the assumption of stationarity, $\bar{X}$ converges but not necessarily to $P(A)$. What is really needed is that the sequence is *ergodic*, see [17]. Then the long-term frequencies will converge to the probabilities.   □

2.5 Streams of Events

Earthquakes, storms, floods, drafts, fire ignitions in dwellings, forest fires, train collisions, etc. can be regarded as results (outcomes) of experiments, which can result from environment and/or human activities. Some of these outcomes can be called accidents or catastrophes if their impacts on society are particularly harmful, but generally we will treat them as "initiation" events that can lead to hazards. The risk for storms, floods, etc. can be measured by means of frequencies; fractions of days with storms or years with floods. In this section we formalize these measures of risk so that it satisfies the assumptions of Definition 1.2 and can be called probabilities. Let us first define a stream of events.

Definition 2.1. *If an event A is true at times $0 < S_1 < S_2 < \ldots$ and fails otherwise, then the sequence of times S_i, $i = 1, 2, \ldots$ will be called a* **stream of events** A.

An important common property of the streams mentioned above is that the exact times S_i, when A is true, are unknown and may vary in an unpredictable way. We turn now to the definition of probability of the event A.

Definition 2.2. *For a stream A, i.e. a sequence of times S_i, $i = 1, 2, \ldots$ when A is true, let (for $t > 0$)*

$$N_A(t) = \text{Number of times } A \text{ occurred in the interval } [0, t]$$

and denote the probability of at least one event in $[0, t]$ by

$$P_t(A) = P(N_A(t) > 0).$$

Further, for fixed s and t,

$$N_A(s, t) = \text{Number of times } A \text{ occurred in the interval } [s, s + t]$$

and $P_{st}(A) = P(N_A(s, t) > 0)$.

Note that $P_t(A)$ means the probability that the event A occurs *at least once* in the time interval $[0, t]$. Again, $\bar{X}$ can be used to estimate $P_t(A)$ as shown next. Let t be one time unit, year say, and define a sequence of random variables X_i as follows

$$X_i = \begin{cases} 1, & \text{if } A \text{ occurred in } i\text{th year} \\ 0, & \text{otherwise.} \end{cases}$$

If the events that A occurred in different years are independent then again by *LLN*,

$$\bar{X} = \frac{1}{n}(X_1 + X_2 + \cdots + X_n) \to P_t(A). \tag{2.6}$$

Clearly the definition of $P_t(A)$ easily modifies to other time periods t. The subscript t is needed since the value assigned to the probability of A depends on t. A shorter t means a lower probability while for longer periods t, $P_t(A)$ can take values close to one.

Remark 2.3. In some risk and safety management documents, for example BSI [78], probabilities $P_t(A)$ are called *frequencies* of events (accidents) A, *e.g.* frequency of fires. Some authors call these frequencies simply probabilities, with adjective "per year" if t is one year, see *e.g.* Ramachandran [65].   □

We turn now to two examples of streams to which we will return on several occasions.

Example 2.6. Consider alarm systems for floods. A warning is issued if the water level at some measuring station exceeds a critical threshold u^{crt}. Now, with $A =$"Warning for flood is issued", and t one year, the yearly probability of flood warnings $P_t(A)$ is the frequency of years in which at least one warning was issued. Actually the probability is also equal to

P("Maximal water level during one year exceeds u^{crt} ");

the last chapter will be devoted to computations of this probability.   □

Example 2.7 (Fire ignition). Probabilities of ignitions have been studied intensively in fire-safety literature and formulae have been proposed for different types of buildings as well as different geographical locations. Here we use the following formula, see [65], t equal to one year:

$P_t(A) = \exp(\beta_0 + \beta_1 \cdot \ln a)$,

where a is the total floor area of a building while β_0 and β_1 are constants that vary between types of activities, geographical location, country, etc. For textile industry in Great Britain the proposed values are $\beta_0 = -5.84$, $\beta_1 = 0.56$, while for hospitals $\beta_0 = -7.1$, $\beta_1 = 0.75$. (Note that for extremely large a, the last formula can give probabilities exceeding one which obviously is not allowed.)

Suppose now a textile industry has a total area of $10\,000$ m^2. Then $P_t(A_2) = \exp(-5.84 + 0.56 \cdot \ln(10000)) = 0.506$.   □

Finally, note that in some situations t is not time but a region in space. For example, if we are interested in the frequency of corrosion damages on a pipeline, t is measured in metres (or km) while the frequency of infected trees in a forest depends on t that has unit m^2 (or km^2).

Initiation events and scenarios

Let us consider a stream of events A, *e.g.* "fire is detected", "warning for flood is issued", or "failure of a pump". Obviously not all times when A occur need to cause hazard for harm or economical losses for people. In order for A to develop an accident or catastrophe, some other unfortunate circumstances, described by events B, have to take place. We call A an "initiating event", B a "scenario". The description of the event B can be very complex and contain both event trees and fault trees. In risk evaluations, it is the stream $A \cap B$ and probability $\mathsf{P}_t(A \text{ and } B)$, which are of interest. In this subsection we examine this closer.

Remark 2.4. We do not in this book discuss consequences of $A \cap B$ for society in terms of financial or human losses, etc. but refer to the literature. For instance, a discussion of problems related to transport and storage of hazardous materials from an economic perspective is given in [50].   □

It is important to note that B has the role to describe a scenario when an initiating event A has occurred. For example, if A is "fire ignition" B could be "failure of sprinkler system", so that if A and B are true one may expect larger economical losses. One can ask why we do not directly consider a stream of A where $A =$ "fire ignition and sprinklers out of order". Obviously this could be done, but in some situations it can be more convenient to separate scenarios from initiating events. Description of different risk-reduction measures, taken in order to avoid the hazard, is often included in B. For instance, B tells us how the systems preventing the hazard can fail. Consequently B can be modified until the acceptable measure of risk for hazard is reached, while the definition of the initiation event A remains unchanged.

Independence

The probability $\mathsf{P}_t(A \text{ and } B)$ is the final goal. Since this is hopefully very small it may be difficult to estimate from historical data. The problem of computing $\mathsf{P}_t(A \text{ and } B)$ is in general very complex and here we treat only the case when B can be assumed to be independent of the stream of A.

A formal definition of independence between the stream of A and the scenario B is somewhat technical and is not given here. Intuitively, the conditional probability that B is true at time $S_n = s$ does not depend on our knowledge of the stream and whether B occurred or not up to time s and that $S_n = s$. The value of the probability $\mathsf{P}(B)$ is a limiting fraction of times s_i when B is true. In Figure 2.2, an estimate of $\mathsf{P}(B)$ is given as $3/6$.

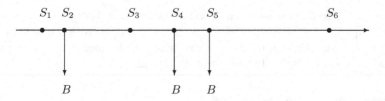

Fig. 2.2. Stream of events A at times S_i with related scenarios B.

In order to give some intuition for the concepts of the stream of events A and an independent scenario B, a mathematical random mechanism generating B independent of a stream is given in the following example.

Example 2.8. This is an artificial example of B which is independent of any stream A. Suppose that a biased coin, giving "heads" with probability p, is available. At each time S_i, when A occurs, the coin is flipped in an independent manner. If "heads" comes up, one decides that B is true (activated), otherwise B is false (is disconnected). Such defined B is independent of the stream and $P(B) = p$.   □

Inspired by Eq. (1.8), we would like to be able to write $P_t(B \text{ and } A) = P(B) \cdot P_t(A)$ when B is independent of the stream A. However, such a formula is usually not correct, except for the situation that only one event A can occur during the period t. However, still the following *approximation* is often used: if B is independent of the stream of events A, then

$$P_t(B \text{ and } A) \approx P(B) \cdot P_t(A). \tag{2.7}$$

Example 2.9. (Continuation of Example 1.11.) Suppose we want to estimate the probability p of at least one "serious" fire during one year, that is, the scenario $B = E_1^c \cap E_2^c$ happening at least once and A the stream of fire ignition. For the stream of initiation events, a realistic value is $P_t(A) = 0.5$. In Example 1.12, we found $P(B) = 0.30$. If the scenario is assumed to be independent of the stream, we have by Eq. (2.7)

$$p \approx P(B)P_t(A) = 0.30 \cdot 0.5 = 0.15.$$

□

An example of estimation of $P(B)$

Estimation of a probability $P(B)$ can be difficult since it is a fraction of times S_i when B is true, which hopefully occurs very rarely. Hence $P(B)$ is often computed by means of laws of probabilities under different assumptions, mixtures of experts' opinions, experiences from similar situations, some data of recorded failures of components, etc. For example $P(B)$ can be taken as a fraction of times when B is true when checked at fixed time points according to some schedule chosen in advance, see the following example. The final result, the probability $P_t(A) \cdot P(B)$, although a useful measure of risk, is usually not equal to $P_t(A \text{ and } B)$.

Example 2.10. Let us discuss a simple scenario for hazard due to fire in a textile industry, where $A = $"Fire starts" defines a stream of fire ignitions. As in Example 2.7, consider a building with total area $8\,000$ m^2; hence the probability of ignition per year

$$P_t(A) = \exp(-5.84 + 0.56 \cdot \ln 8000) = 0.446.$$

We define a "scenario" B, which increases risk for hazard of harm for employees as

$B = $ "At least one of the evacuation doors cannot be opened"

and assume that B is independent of the stream of fire ignitions. A proper way for estimation of $P(B)$ is to use historical data and check how often it happens that exit doors were malfunctioning when a fire started in a textile industry. Even if such data existed, the estimated frequency of failures would be very uncertain. An alternative method is now presented.

Suppose the safety regulations require periodic tests of functionality of exit doors. From reports, one estimates that on average in 1 per 100 inspections (experts' opinion) not all the doors could be opened for different reasons, which gives $P(B) = 0.01$; consequently, by Eq. (2.7), we find

$$P_t(A \text{ and } B) \approx P_t(A)P(B) = 0.01 \cdot 0.45 = 0.0045.$$

□

2.6 Intensities of Streams

In the previous section, the notion of a stream of events was presented and used to define probabilities. For example, for a stream of events A and a fixed period of time t

$$P_t(A) = P(\text{"}A \text{ occurs at least once in the time interval of length } t\text{"}).$$

The probability had frequentistic interpretation, see Eq. (2.6). Some technical difficulties in using a so-defined probability to measure risks resulted in the sign $\approx$ in Eq. (2.7) instead of equality. These difficulties have their origins in the possibility of multiple occurrences of A in the period[4]. A natural solution

---

[4]For example, if there were three fire ignitions in a building during a specific year, each time there is a risk for unfortunate development leading to hazard of harm for people. However, $P_t(A \cap B)$ measures risk for the hazard assuming that only one fire will occur during the period. Clearly the risk is underestimated.

to the problem would be to use smaller periods t so that the possibility of more than one accident in the period can be neglected. (Equivalently one can always use $t = 1$ but change units in which t is measured to the smaller ones, e.g. from years to days, hours, seconds, etc.) In this section we formalize this idea by introducing a concept of *intensity* λ_A of a stream A. Intuitively, for $t = 1$ measured in such units that one can neglect the possibility of occurrence of more than one A in the interval, the intensity λ_A is approximately equal to the probability $P_t(A)$. The formal definition follows:

Definition 2.3. *For a stationary stream of events A (mechanism creating events is not changing in time) the **intensity of events** λ_A and its inverse T_A, called the **return period of** A, are defined as*

$$\lambda_A = \lim_{t \to 0} \frac{P_t(A)}{t}, \qquad T_A = \frac{1}{\lambda_A}.$$

Note that λ_A has a *unit*; for instance, if $t = 1$ day, $\lambda_A \approx P_t(A) = 10^{-3}$, then λ_A with unit $[\text{year}^{-1}]$ is approximately 0.365.

Remark 2.5. In Definition 2.3, stationarity of a stream was required. That concept was not precisely defined since the definition is very technical. A necessary condition for stationarity is that $P_{st}(A) = P_t(A)$ for any value of s, where $P_{st}(A)$ was defined in Definition 2.2. □

We shall demonstrate later on that intensity is a very useful tool in risk analysis and hence estimation of λ_A will be one of the main problems discussed in this book. One of the methods is introduced next.

Suppose we have access to historical data. Then a short period of time t could be chosen and Eq. (2.6) be used to estimate $P_t(A)$, hence $\lambda_A \approx P_t(A)/t$. However, some seconds of reflection and some calculations show that such an estimate is equal to $N_A(T)/T$, where T is the time span of the historical record. This is an intuitive motivation for the following important result; if the mechanism generating accidents is ergodic then

$$\lambda_A = \lim_{T \to \infty} \frac{N_A(T)}{T}, \tag{2.8}$$

where $N_A(T)$ is the number of times A happened in the time interval $[0, T]$. We do not discuss what ergodicity means and merely use it as a term for an assumption sufficient for Eq. (2.8) to be true. In this book we only consider stationary streams that are ergodic. (Note that not all stationary streams are ergodic.) Two very useful properties of intensities are given next.

Theorem 2.4. *Suppose there are* n *stationary-independent streams where* A_i *happen with intensity* λ_{A_i}. *Let* A *be an event that any of* A_i *occurs* $(A = A_1 \cup A_2 \cup \ldots \cup A_n)$. *Then the stream of* A *is stationary and its intensity* λ_A *is given by*

$$\lambda_A = \sum \lambda_{A_i}. \tag{2.9}$$

Consider a scenario B *that can happen when* A *occurs. If* B *is independent of the stream* A, *then the stream of events when* A *and* B *are true simultaneously has the intensity*

$$\lambda_{A \cap B} = \lambda_A \cdot \mathsf{P}(B). \tag{2.10}$$

A consequence of Eq. (2.9) is that even if intensities of accidents A_i are small, chances are likely that any of A_i will occur. For instance, consider the intensity of fire in a flat i in a building, λ_{A_i}, which is small. However, since there are many buildings in a country, the intensity $\lambda_A = \sum \lambda_{A_i}$ of fires in any of the buildings in a country is much higher.

In the following example, we illustrate the important problem of estimating the intensity given data.

Example 2.11 (Accidents in mines). Consider a data set with information on serious accidents in coal mines in United Kingdom, starting from 1851, see [40]. (The data set is also presented in the book by Hand *et al.* [34].) Let $A =$ "Accident in a coal mine happens"; then

$N_A(s, t)$: The number of accidents occurring in the time interval $[s, s+t]$.

The function $N_A(t) = N_A(s, s+t)$, $s = 1851$, is shown in Figure 2.3, left panel, from which by inspection we find for instance $N_A(10) = 30$, $N_A(30) = 100$.

The intensity λ_A is estimated by means of (2.8)

$$\lambda_A \approx \frac{N_A(T)}{T} = \frac{120}{40} = 3, \quad [\text{year}^{-1}].$$

The probability $\mathsf{P}_t(A)$ for $t = 1$ month can be approximated, using the defining equation of the intensity in Definition 2.3, $\mathsf{P}_t(A) \approx t\lambda_A = 3/12$.

Now let us assign to each accident a measure of how severe the accident was; for example, by means of the number of deaths in an accident K, say. Let a "scenario" defining a catastrophe be $B = K > u^{\text{crt}}$, where u^{crt} is some critical number, say 75. In the present case of accidents in mines, one has also access to the number of deaths in each accident, see Figure 2.3, right panel, and hence $\mathsf{P}(B)$ can be estimated. Since there were 17 accidents with more

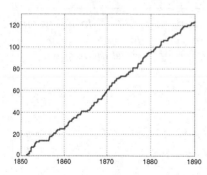
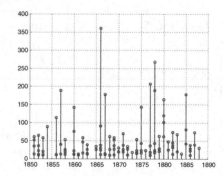

Fig. 2.3. *Left:* Number of accidents $N_A(t)$ in coal mines in United Kingdom ($N_A(1851) = 0$). *Right:* Number of those who died in accidents in coal mines in United Kingdom.

than 75 deaths the probability of B is estimated by $17/120$. Assume that the number of perished K is independent of the stream. Then the intensity $\lambda_{A \cap B} = \lambda_A P(B) \approx 0.43$ year^{-1}.

Suppose one wishes to increase the threshold u^{crt} to a much higher level, for example to 400 deaths. Now there are no data to estimate $P(B) = P(K > 400)$ and hence mathematical modelling is needed to estimate the probability. Methods to estimate this type of probabilities will be discussed in Chapter 9.

$\square$

2.6.1 Poisson streams of events

In the previous section we assumed that $P_t(A) \approx \lambda_A \cdot t$. For large t, $\lambda_A \cdot t > 1$ and hence cannot be used as an approximation of $P_t(A)$. In the following theorem conditions are given when the intensity λ_A defines uniquely $P_t(A)$ for all values of t. First conditions are given that are sufficient for a stream to be called Poisson. Further properties of Poisson streams will be studied in Chapter 7.

(I) More than one event cannot happen simultaneously, i.e. at exactly the same time. (Let the event A define the stream. If $A =$"An aeroplane crashes", the possibility that two aeroplanes crash at the same instance is negligible and **(I)** holds. However, if $A =$"A person dies in an aeroplane accident" then **(I)** is not satisfied since usually several persons die in the same accident.)

(II) The expected number of events observed in any period of time is finite. (The concept of expected value will be described in Chapter 3. For ergodic streams **(II)** means that intensity λ_A is finite.)

(III) The number of events that occur in disjoint intervals are independent. (This is a crucial assumption that has to be motivated in each case studied.)

Theorem 2.5 (Poisson stream of events). *For a stationary stream of event A, if conditions (I) and (II) hold then one has the following bound*

$$P_t(A) \leq \lambda_A t = \frac{t}{T_A}, \tag{2.11}$$

where λ_A is the intensity, T_A the return period of A, see Definition 2.3. If in addition condition (III) is satisfied then the number of events $N_A(s,t)$ observed in a time interval of length t, $[s, s+t]$, $N_A(s,t) \in Po(m)$, where $m = \lambda_A \cdot t$, viz.

$$P(N_A(s,t) = n) = e^{-\lambda_A t} \frac{(\lambda_A t)^n}{n!}, \qquad n = 0, 1, 2, \dots . \tag{2.12}$$

Consequently, the probability of at least one accident in $[0,t]$ is given by

$$P_t(A) = 1 - P(N_A(t) = 0) = 1 - e^{-\lambda_A t}. \tag{2.13}$$

(The proof of the theorem can be found in [17], Chapter 3, where a weaker assumption that $P(N_A(s,t) = n)$ depends only on t and n, instead of required stationarity of the stream, is used.)

It is easy to see that for a stream A and scenario B, if B is independent of the stream and the stream A is Poisson then also the stream of $A \cap B$ is Poisson. The intensity of the stream is given by Eq. (2.10), $\lambda_{A \cap B} = \lambda_A \cdot P(B)$, and hence the number of times that both A and B occur simultaneously in the period of time with length t

$$N_{A \cap B}(s,t) \in Po(m), \quad m = \lambda_A \cdot P(B) \cdot t. \tag{2.14}$$

The last equation is a very useful result that will be used frequently in Chapter 7.

Example 2.12 (Accidents in mines, continuation). In Example 2.11 we measured risk for accidents by $P_t(A)$, $t = 1$ month. This probability was estimated by $1/4$. Now, suppose that we wish to know the probability of more than one accident during the month, *i.e.* $P(N_A(t) > 1)$.

In order to use Theorem 2.5 to compute the probability, one needs to check that assumptions (I-III) hold for the stream. There is no problem in accepting (I-II), only (III) needs to be checked. We have no tools to do this yet and hence we just find it reasonable to assume that the number of accidents between different years is independent and hence assume that (III) is also true. Consequently, the probability that there will be more than one accident in one month is by Eq. (2.12) equal to

$$P(N_A(1/12) > 1) = 1 - P(N_A(1/12) = 0) - P(N_A(1/12) = 1)$$

$$= 1 - e^{-\lambda_A/12} - \frac{\lambda_A}{12} e^{-\lambda_A/12} \approx 0.027,$$

since $\lambda_A \approx 3 \ [\text{year}]^{-1}$.

Finally, consider the scenario B introduced in Example 2.11, *i.e.* $B =$ "$K > 75$", where K is the number of deaths in an accident, assumed to be independent of the stream. Then the stream of catastrophes, *i.e.* accidents when B is true, is Poisson too. Now, since[5] $P(B) \approx 17/120$, the probability of a serious accident during one month is

$$P_t(A \cap B) = 1 - e^{-\lambda_A P(B)/12} \approx \frac{17}{40 \cdot 12},$$

i.e. not negligible. (We have used that $1 - \exp(-x) \approx x$ for small x.) The probability of more than one catastrophe during one month is

$$P(N_{A \cap B}(1) > 1) = 1 - e^{-\lambda_A P(B)/12} - \frac{\lambda_A P(B)}{12} e^{-\lambda_A P(B)/12} \approx 6.1 \cdot 10^{-4}.$$

□

In Chapters 6, 7, and 9 we will return to the problems discussed here, give further applications and methods to estimate λ_A and $P(B)$ from data.

Return period of an event — 100-year waves

Consider a stream of events A. We now give a typical application of the concept of return period T_A of A, met in reliability and safety analysis where one often talks about 100-year waves or 50-year wind speeds. Several non-equivalent definitions of the 100-year value exist. Here we present two of them by means of an example where $A =$ "Water level exceeds u^{crt}" defines the stream. Both definitions extend easily to any stationary stream.

(1) If for $t = 1$ year $P_t(A) = 1/100$, then u^{crt} is called a 100-year water level (or wave). One could also say that A is a 100-year event.
(2) For stationary streams another approach is often used, namely: A is a 100-year event (u^{crt} a 100-year level) if its return period $T_A = 100$ years.

Do these two approaches give different heights for 100-year levels? We answer this question next.

Consider a stationary stream, let $t = 1$ year and u^{crt} be chosen so that $P_t(A) = 1/100$. Since $P_t(A) \leq 1/T_A$, u^{crt} is somewhat smaller than derived by means of Method (2), *i.e.* by solving the equation $T_A = 100$ for A, *i.e.* u^{crt}. However, if the stream is Poisson then

$$P_t(A) = 1 - e^{-1/T_A} \approx \frac{1}{T_A}$$

and the difference is very small and not important in practice. The true advantage of the first definition is that it can be used even for non-stationary

---

[5]The sign $\approx$ means that the value of the probability is estimated and hence uncertain.

streams exhibiting seasonal variability. In addition, in a non-stationary situation (*e.g.* caused by climate change) the 100-year value can be computed by solving $P_{st}(A) = 1/100$ and the critical level is updated as conditions change.

Finding the magnitudes of 100-year levels is an important problem that will be discussed in Chapter 10, where we shall use (1) as the definition of 100-year values.

Example 2.13 (Design of sea walls). When designing protection against high sea level, one speaks about 100-year or 10 000-year storms, which means, if Method (1) is used, that the probability of observing a storm stronger than 100- or 10 000-year storm in one year is $1/100$ and $1/10000$, respectively. Here the stream of storms will be identified with the inception times when water level exceeds a critical level u^{crt} at some specified observation point.

We discuss two examples of design of sea walls: at Ribe in Denmark (at the North Sea) and in the Netherlands. In Denmark, one chose in the 1970s a design load with a return period of 200 years, see [51]. (The old level was 30-45 years.) In the Netherlands, after disastrous floods in 1953 with nearly 1900 deaths, the decision was taken to design the sea walls against return storms of 10 000 years.

Use of Eq. (2.11) gives the probability of catastrophical floods in the following $t = 50$ years, *i.e.* at least one flood, $P_t(A) \leq \frac{t}{T}$, which, in case of Ribe in Denmark, gives a considerable risk with likelihood $1/4$. Due to this risk, it is worth having some alarm system to warn the inhabitants of the possibility of a flood. Such systems are installed. In the Netherlands the chance is negligible if all computations and constructions have been done properly.

However, aspects not known at the time of the analysis have obviously not been taken into account. Wave climate in the Atlantic Sea may change, and knowledge about the impact of ice melting at the poles is uncertain. Besides this "model-type" uncertainties we need to acknowledge that we have also statistical uncertainty due to the fact that one wishes to find properties of storms that are very rare. Consequently our estimates will be very uncertain. For example, the storm that we consider as a 10 000-year storm may have a return period of 1000 years or less, (see Section 10.3.4 for further discussion).

However, the gathered information over many years can be used to *update* the value of return periods. This is of importance in computations of reliability of existing systems. (See Section 5.4.3, where examples of this type of problems are discussed).

Finally, note that we have not said how to find the size of the sea walls, which will sustain storms with return periods of 50 or 10 000 years, or equivalently how to find the level u^{crt}. This will be discussed in the last chapter.

□

2.6.2 Non-stationary streams

Computations are often done for stationary situations; however, most real phenomenon are non-stationary: simple environmental conditions vary with time, new safety technologies or regulations are introduced, systems deteriorate with time. Since introducing non-stationarity complicates mathematical modelling of uncertainties one often neglects it. However, there are situations when the non-stationary character of a problem is essential when safety of a system is evaluated.

Definition 2.4 (Intensity, non-stationary case). *Let s be a fixed time point and $\mathsf{P}_{st}(A) = \mathsf{P}(N_A(s,t) > 0)$ the probability that at least one event A occurs in the interval $[s,\, s + t]$, then the limiting value (if it exists)*

$$\lambda_A(s) = \lim_{t \to 0} \frac{\mathsf{P}_{st}(A)}{t}, \tag{2.15}$$

*will be called a **non-stationary intensity** of the stream of A.*

Obviously if $\mathsf{P}_{st}(A)$ does not depend on s, as in the stationary case, then $\lambda_A(s) = \lambda_A$. We now introduce the non-stationary Poisson stream.

Theorem 2.6 (Poisson stream of events). *Consider a stream of events A. Under some regularity assumptions (which are always satisfied in the problems studied in this book), if the conditions (I-III) are satisfied then*

$$\mathsf{P}_{st}(A) = 1 - e^{-\int_s^{s+t} \lambda_A(x)\, dx}, \tag{2.16}$$

where $\lambda_A(x)$ is given in Definition 2.4. Furthermore, the number of events observed in the time interval $[s,\, s + t]$, $N_A(s,t) \in Po(m)$, where $m = \int_s^{s+t} \lambda_A(x)\, dx$.

Example 2.14. Consider an event A whose intensity $\lambda_A(s)$ varies seasonally, *i.e.* it is a periodical function with period one year. Assume that it can be constant in one month; Using historical records, monthly intensities can be estimated by means of a formula similar to Eq. (2.8),

$$\lambda_{A_i} = \lim_{T \to \infty} \frac{N_{A_i}(T)}{T/12} \quad [\text{year}]^{-1},$$

where A_i = "Event A occurs and is in month i" and $T/12$ is the fraction of total recording time that falls into an individual month.

As an example, let us use records of daily rain amount measured at an airport in Venezuela during 1961-1999. (The data will be considered again in Chapter 10.) Define A = "Daily rain exceeds 50 mm" as an initiation event for possible hazard of proper operations of the airport. Clearly we have $T = 39$ years while the 12 observed values of $N_{A_i}(T)$ are

4   0   3   4   3   2   3   3   3   2   7   10.

Consequently a simple model could be to assume different intensities for, on the one hand, the months January to October, on the other, November and December:

$$\lambda_{A_i} \approx \begin{cases} \frac{27}{39(10/12)} = 0.83, & i = 1, \ldots, 10, \\ \frac{17}{39(2/12)} = 2.62, & i = 11, 12. \end{cases}$$

(The sign $\approx$ is used since these are only estimates of the intensities, $T = 39$ is not infinity.) Now $\lambda_A(s) = \lambda_{A_i}$ if s, having units years, falls in month i. It seems reasonable to assume that assumptions (I-III) in page 40 are satisfied and hence the stream of extreme rains is Poisson with intensity $\lambda_A(s)$.

Let N_1, N_2 be the number of huge rains in the first and second six months during next year, respectively. By Theorem 2.6 we know that $N_1 \in \text{Po}(m_1)$, while $N_2 \in \text{Po}(m_2)$ where

$$m_1 = \int_0^{1/2} \lambda_A(x)\,dx \approx 0.83 \cdot \frac{1}{2} = 0.415,$$

$$m_2 = \int_{1/2}^1 \lambda_A(x)\,dx \approx 0.83 \cdot \frac{4}{12} + 2.62 \cdot \frac{2}{12} = 0.713.$$

Now the probability that there will be more than two rains in the periods is given by

$$P(N_i > 2) = 1 - P(N_i = 0) - P(N_i = 1) - P(N_i = 2)$$
$$= 1 - e^{-m_i}\left(1 + m_i + \frac{m_i^2}{2}\right),$$

giving numerical values $P(N_1 > 2) \approx 0.009$, while $P(N_2 > 2) \approx 0.036$, which is four times higher. $\square$

Problems

2.1. Let X be the number of death casualties on a shipyard in a decade. It is assumed that $X \in \text{Po}(3)$. Calculate

(a) $P(X \leq 2)$,
(b) $P(0 \leq X \leq 1)$,
(c) $P(X > 0)$,
(d) $P(5 \leq X \leq 7 \,|\, X \geq 3)$.

2.2. In a highway, it is noticed that the probability of at least one accident in a given month involving lorries transporting hazardous materials is roughly 0.08.

(a) Calculate the probability of exactly 6 consecutive months without such an accident (an accident will thus happen in the seventh month).
(b) Estimate the intensity of accidents and compute the return period. (*Hint.* Use Definition 2.3.)

2.3. Suppose $P(A) = 0.20$, $P(B|A) = 0.75$, and $P(B) = 0.45$. Calculate $P(A|B)$.

2.4. The buildings in a district can roughly be characterized as either housing area or industrial zone. We study here emergency calls due to alarm. The probability for the brigade to turn out to a housing area is 0.45, the corresponding probability for industrial zone is 0.55. From available statistics, one assumes that the probability of a true fire at the arrival to a housing area is 0.90 while the probability of a true fire at the arrival to an industrial zone is 0.05.

The fire brigade returns to the station after a mission, they have put out a fire. Calculate the probability that they have returned from a mission to an industrial zone.

2.5. When coded messages are sent, errors in transmission sometimes occur. Consider Morse code, where "dots" and "dashes" are used. It is known that the odds for dot sent versus dash sent is 3:4.

Suppose there is interference on the transmission line: with probability 1/10 a dot is mistakenly received as a dash and vice versa. Calculate the probability of correctly receiving a dot.

2.6. This problem is based on a question posed by Stewart [75]:

(a) "Suppose Mr. and Mrs. Smith tell you they have two children, one of whom is a girl. What is the probability that the other is a girl?"
(b) Compute the probability if you know that Mr. and Mrs. Smith have two children and you see them walking with a girl (their girl).

2.7. Suppose a certain disease has a frequency 1 per 10 000. One can test whether a person is infected or not. Suppose the test has accuracy 99%, i.e. out of 100 infected persons on average 99 will be tested positive. The risk of "false alarms" is 0.1%, i.e. out of 1000 not infected persons, on average one will yield a positive test result[6]. Assume now that a person has been tested positively for the disease. Use Bayes' formula to compute the probability that the person is really infected.

2.8. Recall Problem 1.10, leakage of containers. Suppose that leakages in the deposit form a stationary stream with intensities

$$\lambda_{Corr} = 0.18, \quad \lambda_{Therm} = 0.45, \quad \lambda_{Other} = 0.002 \quad year^{-1}$$

The prior odds for the three conditions water flow, chemical interactions, and others are 4:1:95. Assume that the conditions are mutually excluding, one and only one of them is true.

The cost differs depending on scenario, hence it is of interest to update the probability distributions based on available information. Suppose that in 5 years, 3 leakages have been observed. Update the priors and give a comment on the result.

2.9. Oil pipelines are inspected by submarines in order to detect imperfections. A non-destructive (NDT) device is used to detect the location of cracks. Cracks may exist in various shapes and sizes, hence the probability that a crack will be detected

---

[6]Using terminology from medical science, the *sensitivity* is 0.99 in this problem while the *specificity* is 0.999.

by the NDT device is 0.8. Assume that the events of each crack being detected are statistically independent and that the NDT does not give false alarms.

(a) Suppose that along a fixed distance examined (say 5 m), there are two cracks in the pipeline. What is the probability that none of them would be detected?
(b) The actual number of cracks N along the distance examined is not known. However, a prior distribution is given as $P(N = 0) = 0.3$, $P(N = 1) = 0.6$, $P(N = 2) = 0.1$. Find the probability that the NDT device will detect 0 cracks in the pipeline.
(c) If the device detects 0 cracks, what is the probability that there is no crack at all?

2.10. A man walks across three main streets every morning on his way to work. In the afternoon, he walks across the same three streets when he returns home. Every time he walks across one of the main streets, he is subject to a risk of being hit by a car, which is *roughly* $5 \cdot 10^{-8}$. He goes to work approximately 200 days every year.

(a) Estimate the probability of being hit by a car at least once during 20 years.
(b) Determine roughly the return period (in years) for the event "being hit by a car".

2.11. In a factory, 5 accidents have been observed in 10 years.

(a) Estimate the intensity of accidents.
(b) Estimate the return period.
(c) Give an estimate of the probability that no accidents will occur in one month.

2.12. Consider the model for the intensity of fire ignition,

$$\lambda_A = t \, \exp(\beta_0 + \beta_1 \ln a), \quad [\text{year}^{-1}]$$

where $t = 1$ year, a is total floor area (m^2) and

$$A = \text{"Fire ignition at a hospital"}.$$

For hospitals in Great Britain, $\beta_0 \approx -7.1$, $\beta_1 \approx 0.75$.

In a county, there are two hospitals with total areas $6\,000 \text{ m}^2$ and $7\,500 \text{ m}^2$, respectively. Suppose the streams of fire ignitions are Poisson and fires start in both hospitals independently. Calculate the probability that there will be fire ignitions in both hospitals in one month.

2.13. Consider traffic accidents in the Swedish province of Dalecarlia (*Dalarna*). As reported to SRSA (Swedish Rescue Services Agency), the total number of accidents with trucks and the number of accidents involving tank trucks with dangerous goods signs were as follows for the years 2002-2004:

Year	2002	2003	2004
All trucks	48	26	44
Tank trucks	2	0	2

Assume a Poisson stream of events.

(a) Estimate the intensity of accidents involving trucks.

(b) Calculate the probability of at least one accident with a tank truck during one
month in Dalecarlia. Employ data for the whole of Sweden, see table below, for
estimation of the probability $P(B)$ that a truck accident involves a tank truck.

Year	2002	2003	2004
All trucks	1108	1089	1192
Tank trucks	37	41	39

2.14. A consultant in fire engineering investigates the risk for fire in a town. Obviously, the number of fires varies from year to year. However, from experience it is
assumed that fires occur according to a Poisson process with unknown intensity Λ
(year^{-1}).

The intensity of fires starting may depend on many factors. The consultant limits
himself to fires that start in dwellings or schools. In the literature, it is suggested
that the intensity of fires starting in these types of buildings is equal to the floor
area times a factor α, say, taking values between 10^{-6} and $4 \cdot 10^{-6}$ ($[\text{year}^{-1}\text{m}^{-2}]$).
Suppose that the total floor area in the town investigated is $2.5 \cdot 10^{6}$ m^{2}.

(a) As values of α, the consultant chooses 10^{-6}, $2 \cdot 10^{-6}$, $3 \cdot 10^{-6}$, and $4 \cdot 10^{-6}$.
Based on this choice, help her to estimate intensities and formulate suitable prior
odds for Λ.
(b) Suppose during the first two months, no fires were reported. Use this information
to update the prior odds.
(c) Use the updated odds to compute the probability of no fire in the following
month.

3

Distributions and Random Variables

Often in engineering or the natural sciences, outcomes of random experiments are numbers associated with some physical quantities. Obviously there are random experiments with outcomes that are *not* numerical, for example flipping a coin. However, the results in such experiments can also be identified numerically by artificially assigning numbers to each of the possible outcomes. For example, to the outcomes "tails" and "heads", one can (arbitrarily) assign the values 0 and 1, respectively.

In this section we consider random experiments with numerical outcomes. Such experiments are denoted by capital letters, *e.g.*, U, X, Y, N, K. The set $\mathcal{S}$ of possible values of a random variable is a sample space, which can be all real numbers, all integer numbers, or subsets thereof. Statements about random variables have truth sets that are subsets of $\mathcal{S}$.

A statement of the type "$X \leq x$" for any fixed real value x, *e.g.* $x = -2.1$ or $x = 5.375$, plays an important role in computation of probabilities for statements on random variables. More precisely, we introduce

$$F_X(x) = \mathsf{P}(X \leq x), \quad x \in \mathbb{R},$$

and call the function $F_X(x)$ the *probability distribution, cumulative distribution function* (cdf), or for short, the *distribution function*.

Example 3.1 (Exponential distribution). As presented later in this chapter, some distribution functions have their own names. One important example is the so-called exponential distribution

$$F_X(x) = \begin{cases} 1 - \mathrm{e}^{-x}, & \text{if} \quad x \geq 0, \\ 0, & \text{if} \quad x < 0, \end{cases}$$

often used to describe variability of life-length data for units under constant risk for accident.

Note that this function is increasing, starting at zero and approaching one as x tends to infinity. $\qquad\square$

The importance of the probability distribution function lies in the following fact:

> **Theorem 3.1.** *The probability of any statement about the random variable X is computable (at least in theory) when the distribution function $F_X(x)$ is known.*

Recall that the probability function P is defined on events (Section 1.1). Since, usually, the number of different events is higher than the number of all real numbers we cannot write the function P explicitly but can only give an algorithm how to compute the probabilities $P(A)$ for any event A. Theorem 3.1 says that the algorithm to compute $P(A)$ can be given if the distribution function $F_X(x)$ is known or estimated from data.

Examples of events and statements

Some simple and useful statements about X are "X exceeds a limit b" and "X is between two limits, $a < X \le b$". It is easy to show that $P(X > b) = 1 - F_X(b)$ and the fundamental relation

$$P(a < X \le b) = F_X(b) - F_X(a). \tag{3.1}$$

The slightly more complicated statement "$e^X \le b$" can also be computed, since it is equivalent to "$X \le \ln b$" and hence

$$P(e^X \le b) = P(X \le \ln b) = F_X(\ln b).$$

We turn now to another important example, considering the event "$X = b$" whose probability is given by

$$P(X = b) = \lim_{n \to \infty} P(b - 1/n < X \le b) = \lim_{n \to \infty} (F_X(b) - F_X(b - 1/n))$$

(cf. Eq. (3.1)). If the distribution function $F_X(x)$ is a continuous function then for any fixed b, $P(X = b) = 0$, *i.e.* it is impossible to guess the future value of X. The random variables with continuous distribution function are called *continuous random variables*.

Conclusion

We defined a random variable as a random experiment with numerical outcomes. To each random variable a distribution function F_X is assigned. We demonstrated that the distribution can be used to compute probabilities of different statements about X. We have, however, not specified methods how to find the distribution function $F_X(x)$ so that the computed probabilities can be used in taking rational decisions in risk analysis. This will be done in the next chapter.

3.1 Random Numbers

It is easy to see that the distribution function $F_X(x)$ is increasing in x, $F_X(-\infty) = 0$ while $F_X(+\infty) = 1$. Actually any function $F(x)$ satisfying the three properties is a distribution of some random variable. In this section we show how one constructs a random variable X, called random number, such that $P(X \leq x) = F(x)$.

3.1.1 Uniformly distributed random numbers

In Chapter 1, simple properties of probabilities were exemplified by random experiments having a finite number of possible outcomes. A random variable was defined as a number associated with the outcome of the experiment. In the same chapter, we introduced experiments with an infinite, but countable, number of possible outcomes (the geometric and the Poisson probability-mass functions). We now go further and use a series of coin-flipping experiments to create a random variable that can take any value between 0 and 1 with equal probability; hence, called *uniformly distributed* random number. The uniformly distributed random numbers will form the basis for construction of any non-uniformly distributed random numbers.

Binary representation of numbers

The procedure is based on a binary representation of a number u, $0 \leq u \leq 1$, *i.e.* as a sequence of zeros and ones; for example:

$$u = (011\,011\,01\ldots) : \quad \frac{0}{2} + \frac{1}{4} + \frac{1}{8} + \frac{0}{16} + \frac{1}{32} + \frac{1}{64} + \frac{0}{128} + \frac{1}{256} + \cdots \quad (3.2)$$

Let us use the binary representation in Eq. (3.2) to transform the result of two independent coin flips $00, 01, 10, 11$ into the following four numbers $0, 1/4, 1/2, 3/4$. We denote this transformation of the result of two flips of a coin to real numbers, which obviously is a random variable, by $U^{(2)}$. Clearly the probability $P(U^{(2)} = u) = 1/4$ for any of the four possible values of u.

In the same way, using Eq. (3.2), we can transform a result of 20 flips of a coin into a number u and denote it by $U^{(20)}$. Obviously there are now 2^{20} (more than one million) distinct values u that $U^{(20)}$ can take. By independence of individual flips, each of these values can occur with equal probability 2^{-20}. What is important is that all possible u-values are uniformly spread over the interval $[0, 1]$. Similarly, let $U^{(n)}$ be a number, which is a result of evaluating (3.2) on a result of n flips of a fair coin. Again, all possible resulting numbers u have equal probability (very small) of occurrence 2^{-n} that tends to zero as n goes to infinity and are uniformly spread over $[0, 1]$.

Definition 3.1. *A limit value of the random experiments* $U^{(n)}$ *as n tends to infinity will be called a* **uniformly** *on* $[0, 1]$ *distributed random number and denoted by* U, *i.e.*

$$U = \lim_{n \to \infty} U^{(n)}$$

Obviously, the variable U cannot be realized in practice — nobody can flip the coin infinitely many times. However, already $U^{(100)}$ has more than

1 000 000 000 000 000 000 000 000 000 000

possible outcomes and can be used as a practical version of the mathematically constructed variable U.

Distribution function

It is not too difficult to be convinced that the distribution function of U has the following form

$$F_U(u) = \mathsf{P}(U \le u) = \begin{cases} 0, & \text{if} \quad u < 0, \\ u, & \text{if} \quad 0 \le u \le 1, \\ 1, & \text{if} \quad u > 1, \end{cases}$$

with derivative (called probability-density function, see Section 3.2)

$$f_U(u) = \frac{\mathrm{d}}{\mathrm{d}u} F_U(u) = \begin{cases} 0, & \text{if} \quad u < 0, \\ 1, & \text{if} \quad 0 \le u \le 1, \\ 0, & \text{if} \quad u > 1. \end{cases}$$

3.1.2 Non-uniformly distributed random numbers

From science and technology we are familiar with deterministic scale transformations, *e.g.*

$$u = ax + b, \qquad u = \log x,$$

where u and x could be temperature in Celsius and Fahrenheit; amplification in real numbers and decibels, respectively. It can be shown that starting from a uniformly distributed random variable we can compute any existing random number by a suitable change of scales. We can view a uniformly distributed random number as a "dimensionless" standard number.

Theorem 3.2. *For any strictly increasing continuous function $F(x)$, x is a real number, taking values in the interval $[0,1]$, such that $F(-\infty) = 0$ and $F(+\infty) = 1$, the random variable X defined by*

$$U = F(X), \quad X = F^{-1}(U) \tag{3.3}$$

where U is a uniformly distributed random number, has probability distribution

$$F_X(x) = F(x).$$

The last equality simply follows from Eq. (3.3), *viz.*

$$\mathsf{P}(X \leq x) = \mathsf{P}(U \leq F(x)) = F(x),$$

since the statements "$X \leq x$" and "$U \leq F(x)$" are true for the same outcomes of the random experiment of infinitely many flips of a coin. Simply, in order to get a random number X that is smaller than a fixed number x, the uniformly distributed variable U has to be in the interval $(0, F(x))$.

However, there are distribution functions that are not strictly increasing or even have discontinuities; for example, see Figure 3.1. In such a case the solution to Eq. (3.3), $X = F^{-1}(U)$, may not be unique or defined. This is only a technical problem and one can define an (generalized) inverse function to $F(x)$, denoted by

$$x = F^-(y)$$

as follows. For any $y \in [0,1]$, let $F^-(y)$ be the maximal $\tilde{x}$ satisfying $F(\tilde{x}) \leq y$ (cf. Figure 3.1).

Remark 3.1. Any non-decreasing, right-continuous function $F(x)$ taking values in the interval $[0,1]$, such that $F(-\infty) = 0$ and $F(+\infty) = 1$, defines an inverse function $x = F^-(y)$ to be the maximum of all $\tilde{x}$ satisfying

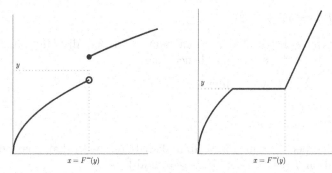

Fig. 3.1. Definition of the inverse $x = F^-(y)$, two situations. *Left:* Discontinuity of the distribution function. *Right:* Distribution function, not strictly increasing.

the inequality $F(\tilde{x}) \leq y$. The random number $X = F^{-}(U)$ has $F(x)$ as its distribution, *i.e.* $\mathsf{P}(X \leq x) = F(x)$. As before, U is a uniformly distributed random number. □

Equation (3.3) is fundamental since it provides a constructive way of defining random numbers as well as classifying them. More precisely, one can think of the statement "*a random variable with distribution $F(x)$*" means a procedure giving a random number X defined by Eq. (3.3).

Note that there are other methods to create and describe random numbers. Computer-generated random numbers are the basis of the so-called Monte Carlo algorithms, also called simulation methods. The random numbers created in a computer are called pseudo-random numbers, since these are created by means of deterministic algorithms. The pseudo-random numbers mimic properties of "true" random numbers created using random experiments similar to flipping a coin.

Remark 3.2. An important consequence of the definition of random numbers defined by means of Eq. (3.3) is the following observation. If we have two distributions $F_1(x)$ and $F_2(x)$ that are close to each other (in horizontal direction), then for a fixed value of U the random numbers X_1, X_2, which are solutions of the equations $U = F_1(X_1)$ and $U = F_2(X_2)$, are close. Practically speaking, random numbers with similar distribution can be used equivalently. Hence, of the distributions $F_1(x)$ and $F_2(x)$ the one is chosen that is easiest to handle. This continuity property explains why one considers classes of distributions that have nice explicit formulae but are flexible enough to be close to any particular distribution. □

As already mentioned, there are infinitely many different types of random numbers (variables) since there are infinitely many different scales. Some of them are simpler to handle in mathematical models, often fit real data well, have useful mathematical properties; hence, they are used more often and got specific names, see the following examples.

3.1.3 Examples of random numbers

Exponential distribution

An exponentially distributed random number X has distribution $F(x) = \mathsf{P}(X \leq x) = 1 - e^{-x}$, $x \geq 0$, and hence

$$U = 1 - e^{-X}, \qquad X = -\ln(1 - U).$$

Weibull distribution

A random number X, which is Weibull distributed, with shape parameter c, has distribution $F(x) = 1 - e^{-x^c}$, $x \geq 0$, and hence

$$U = 1 - e^{-X^c}, \qquad X = (-\ln(1 - U))^{1/c}.$$

Note that for $c = 1$ we have the exponential distribution, for $c = 2$ a Rayleigh distribution and $c = 3$ a Maxwell distribution.

Gumbel distribution

A random number X from the Gumbel distribution (also called double exponential distribution) has distribution $F(x) = \exp(-e^{-x})$, $-\infty < x < \infty$, and hence

$$U = e^{-e^{-X}}, \qquad X = -\ln(-\ln U).$$

Two-point distribution

A result of a flip of a coin, *i.e.* $X = 0$ if "Heads" showed up and $X = 1$ otherwise, has a distribution function satisfying $F(x) = 0$ for $x < 0$, $F(x) = 1/2$ for $0 \le x < 1$, and $F(x) = 1$ for $x \ge 1$, and hence

$$U = F(X), \qquad X = \begin{cases} 0, & \text{if } U \le 1/2, \\ 1, & \text{if } U > 1/2. \end{cases}$$

Obviously there are many other random numbers having special names that have already been presented: binomial-, Poisson-, geometric-distributed r.v. Others will be introduced later on in this book; some examples are normal (or Gauss), log-normal, Pareto, gamma, beta, Dirichlet, multinomial.

At this moment one may ask why we define X using the implicit equation $U = F(X)$ instead of just writing $X = g(U)$. Obviously, the variables

$$X = U^2, \quad X = \arctan U \quad \text{or} \quad X = U + e^U,$$

are random numbers by definition too. The problem is that the distribution of such defined variables $F_X(x) = \mathsf{P}(X \le x)$ may be hard to find. Many questions in safety analysis are often related to computations of distributions of explicitly defined functions of random variables. We return to these questions in Chapter 8, see also Section 3.3.1.

3.2 Some Properties of Distribution Functions

Hitherto we have shown how to create random numbers. We have used the uniformly distributed U and the distribution function $F(x) = \mathsf{P}(X \le x)$ to obtain new types of random numbers, denoted by X. Consequently, the distribution function completely characterizes a random number (variable). There are other concepts, usually functions of $F(x)$, that are also used to describe properties of random numbers. We now present three of these.

Probability-mass function

Let X take a finite or (countable) number of values (for simplicity, the values $0, 1, 2, \ldots$). One then speaks of *discrete* random variables and the distribution function $F(x)$ is a "stair" looking function that is constant except the possible jumps for $x = 0, 1, 2, \ldots$. The size of a jump at $x = k$, say, is equal to the probability $\mathsf{P}(X = k)$, denoted by p_k, which is called the *probability-mass function* (pmf). The function, or rather series, p_k defines uniquely the distribution since $F(x) = \sum_{k \leq x} p_k$. Consider for example a geometrically distributed r.v. K with pmf

$$p_k = 0.70^k \cdot 0.30, \quad k = 0, 1, 2, \ldots.$$

This distribution is shown in Figure 3.2 in the form of its distribution function (left panel) and pmf (right panel).

Probability-density function

For a uniformly distributed random variable $(X = U)$, the concept of probability mass does not have a sense since $\mathsf{P}(X = x) = 0$. However, one can write somewhat unprecisely but correctly that $\mathsf{P}(X \approx x) = \mathrm{d}x$ where $X \approx x$ means $x - 0.5\,\mathrm{d}x < X \leq x + 0.5\,\mathrm{d}x$, *i.e.* X has a value somewhere in an interval of length $\mathrm{d}x$ around x. We can interpret the relation as that the density of random numbers is constant and equal to one. We turn now to other random variables that are obtained by smooth scale changes, which gives non-constant intensities of random numbers. More precisely, if the distribution function $F(x)$ is differentiable, then the derivative

$$f(x) = \frac{\mathrm{d}}{\mathrm{d}x} F(x),$$

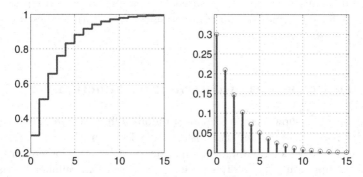

Fig. 3.2. Geometrical distribution with $p_k = 0.70^k \cdot 0.30$, for $k = 0, 1, 2, \ldots$. *Left:* Distribution function. *Right:* Probability-mass function.

called *probability-density function* (pdf), has the interpretation $P(X \approx x) = f(x)\,dx$. For random variables having a pdf, Eq. (3.1) can be written as

$$P(a < X \le b) = \int_a^b f(x)\,dx \tag{3.4}$$

and these are called *continuous* random variables. Consequently,

$$\int_{-\infty}^{\infty} f(x)\,dx = 1.$$

By direct differentiation we have that an exponentially distributed r.v. X has the density

$$f(x) = e^{-x}, \quad x \ge 0 \quad \text{and zero otherwise.}$$

Another example is the Weibull density, given by

$$f(x) = c\,x^{c-1}e^{-x^c}, \quad x \ge 0.$$

Standard normal distribution

The probability density $f(x)$ can be used to define a distribution function, since any non-negative function that integrates to one is a density of some distribution. Actually, the distribution of a standard *normal* (or standard Gaussian) random variable is defined by means of its density function. The density of a standard normal variable has its own symbol $\phi(x)$ and is given by

$$\phi(x) = \frac{1}{\sqrt{2\pi}}\,e^{-x^2/2}, \quad -\infty < x < \infty. \tag{3.5}$$

The r.v. X having this density is often denoted as $X \in \mathrm{N}(0,1)$. The distribution function of the variable, $F(x)$, has its own symbol $\Phi(x)$,

$$\Phi(x) = \int_{-\infty}^{x} \frac{1}{\sqrt{2\pi}}\,e^{-t^2/2}\,dt. \tag{3.6}$$

For an illustration of $\Phi(x)$ and $\phi(x)$, see Figure 3.3.

There is no analytical expression for $\Phi(x)$ and numerically computed values are often tabulated, see appendix. There are also very accurate polynomial approximations for $\Phi(x)$ that are basis for computer evaluations of its values.

Quantiles

The *median* $x_{0.5}$ of a random variable X is a value such that the probability that the outcome of X is not exceeding $x_{0.5}$ is equal to 0.5, *i.e.*

$$P(X \le x_{0.5}) = 0.5, \quad \text{and hence} \quad x_{0.5} = F_X^{-}(0.5).$$

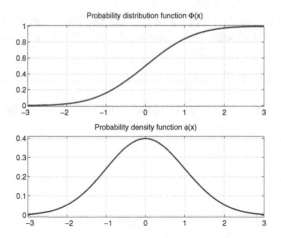

Fig. 3.3. Top: Distribution function $\Phi(x)$. Bottom: Density function $\phi(x)$

For an exponentially distributed variable X with $F_X(x) = 1 - e^{-x}$, we have that $x = F_X^-(y) = -\ln(1 - y)$, giving the median $x_{0.5} = -\ln(1 - 0.5) \approx 0.69$. Often income statistics is presented using median salary, which states that half of a population earns more than the median. The related concepts of *quartiles* denoted by $x_{0.75}, x_{0.25}$ are also often reported and mean the values of incomes that salaries of 75 %, 25 % of the population exceeds $x_{0.75}, x_{0.25}$, respectively. For the exponential variable X the quartiles $x_{0.75}, x_{0.25}$ are obtained by solving the equations

$$P(X \le x_{0.75}) = 1 - 0.75, \quad P(X \le x_{0.25}) = 1 - 0.25,$$

and are given by

$$x_{0.75} = F_X^-(0.25) = -\ln 0.75 \approx 0.29, \quad x_{0.25} = F_X^-(0.75) = -\ln 0.25 \approx 1.39,$$

respectively.

The α *quantile* x_α, $0 \le \alpha \le 1$, is a generalization of the concepts of median and quartiles and is defined as follows:

Definition 3.2. *The **quantile** x_α for a random variable X is defined by the following relations:*

$$P(X \le x_\alpha) = 1 - \alpha, \quad x_\alpha = F^-(1 - \alpha). \tag{3.7}$$

Remark 3.3. In some textbooks, quantiles are defined by the relation $P(X \le x_\alpha) = \alpha$; then the inverse function $F^-(y)$ could be called the "quantile function". $\square$

Table 3.1. Quantiles of the standard normal distribution.

α	0.10	0.05	0.025	0.01	0.005	0.001
λ_α	1.28	1.64	1.96	2.33	2.58	3.09

Remark 3.4. Obviously, knowing all quantiles x_α for a random variable X, we know the inverse function $x = F^-(y)$ and can easily construct the random-number generator for X. If U is a uniformly distributed random number, then $X = x_{1-U}$. $\quad\square$

In Chapter 4 where tools for statistical analysis of data are presented, we will make frequent use of quantiles for some common distributions:

Normal distribution. For a standard normal variable $X \in N(0, 1)$, the quantiles are denoted λ_α. Thus, $\Phi(\lambda_\alpha) = 1 - \alpha$. Values of λ_α are found in tables for standard choices of α and are also usually implemented in statistical software packages.

χ^2 **distribution.** The α quantiles of the so-called χ^2 distribution, to be presented in Section 3.3.1, are denoted as $\chi^2_\alpha(f)$, where f is an integer.

Quantiles of the standard normal distribution are given in Table 3.1 for some common choices of α.

Quantiles are important in statistics when constructing confidence intervals (see the next chapter). They are also of importance when focusing on applications to risk and safety, and are used to describe loads and strengths of components. We return to these issues in Chapters 8 and 9.

3.3 Scale and Location Parameters – Standard Distributions

As mentioned before, the somewhat artificial formula (3.3) is useful for construction of random variables with a desired distribution function $F(x)$. However, in practice we are often interested in distributions of functions of random variables. Maybe the simplest case is just a linear changing of scales. More precisely, for a fixed distribution $F_X(x)$ define a variable Y as follows

$$U = F_X(X), \qquad Y = aX + b,$$

where a and b are deterministic constants (may be unknown); a is called *scale parameter* and b is called *location parameter*. The distribution of Y is easy to compute:

$$F_Y(y) = \mathsf{P}(Y \leq y) = \mathsf{P}(aX + b \leq y) = \mathsf{P}(X \leq \frac{y-b}{a}) = F_X(\frac{y-b}{a}).$$

Definition 3.3. *If two variables X and Y have distributions satisfying the equation*

$$F_Y(y) = F_X\left(\frac{y-b}{a}\right)$$

*for some constants a and b, we shall say that the distributions F_Y and F_X belong to the same **class**.*

3.3.1 Some classes of distributions

Here we list some distributions of the continuous type that are focused on particularly in the sequel of this book. For an overview of relationships between some commonly used distributions, see the article by Leemis [48].

Exponential distribution

The class of exponentially distributed variables $Y = aX$ has the form

$$F_Y(y) = 1 - e^{-y/a}, \quad y \geq 0.$$

The density is

$$f_Y(y) = \frac{1}{a}e^{-y/a}, \quad y \geq 0 \tag{3.8}$$

while the quantile function, defined by Eq. (3.7), is given by

$$y_\alpha = -a\ln\alpha. \tag{3.9}$$

This class is often used in applications as a model for failure time, for example a machine breaking down or death caused by an accident.

Gamma distribution

A gamma distributed random variable Y has the probability density function

$$f_Y(y) = \frac{b^a}{\Gamma(a)}y^{a-1}e^{-by}, \quad y \geq 0 \tag{3.10}$$

where $a > 0$, $b > 0$ and $\Gamma(.)$ is the Gamma function[1]. Sometimes $Y \in$ Gamma(a,b) is used as a shorthand notation[2].

Several common distributions are obtained as special cases of the gamma distribution. For instance, Gamma$(n/2,1/2)$ leads to a chi-square distribution, notated as $\chi^2(n)$, and Gamma$(1,1/a)$ is the exponential distribution in the form presented in Eq. (3.8).

---

[1] $\Gamma(p) = \int_0^\infty t^{p-1}e^{-t}\,dt, \quad p > 0.$ For p an integer, $\Gamma(p) = (p-1)!$ When p is not an integer, the relation $p\Gamma(p) = \Gamma(p+1)$ is useful.

[2] Note that here $1/b$ is the scale parameter while a is a *form parameter*.

Weibull distribution

The general form of the three-parameter family of Weibull distributions $Y = aX + b$. With a shape parameter c,

$$F_Y(y) = 1 - e^{-((y-b)/a)^c}, \qquad y \geq b, \text{ and zero for } y < b.$$

(Observe that we usually assume that $b = 0$.) The density is

$$f_Y(y) = \frac{c}{a}\left(\frac{y-b}{a}\right)^{c-1} e^{-((y-b)/a)^c}, \qquad y \geq b, \text{ and zero otherwise}$$

while the quantile function, defined by Eq. (3.7), is given by $y_\alpha = b + a(-\ln\alpha)^{1/c}$. The Weibull distribution is commonly used as a model for strength of materials, obeying the "weakest link" principle: a chain will break when its weakest link breaks; cf. the original papers by Waloddi Weibull [80], [81].

Normal distribution

Let X be a standard normal variable denoted usually as $X \in N(0,1)$; then the variable $Y = \sigma X + m$ (for normal variables we customarily use m and σ instead of b, a, respectively) is also normally distributed, which we write as $Y \in N(m, \sigma^2)$. Note that the variable $-X$ has the same distribution as X and hence we need only to consider positive values of σ. The density of Y is

$$f_Y(y) = \frac{1}{\sigma\sqrt{2\pi}}e^{-(y-m)^2/2\sigma^2}, \qquad -\infty < y < \infty$$

and

$$F_Y(y) = P(Y \leq y) = P(\sigma X + m \leq y) = P(X \leq (y-m)/\sigma) = \Phi(\frac{y-m}{\sigma}),$$

where Φ (cf. Eq. (3.6)) is the distribution of an $N(0,1)$ variable. The quantile function y_α is given by

$$y_\alpha = m + \sigma\lambda_\alpha, \tag{3.11}$$

where λ_α is a quantile of X. The quantile λ_α is often used in statistical analysis and hence has been tabulated, see Table 3.1. It can also be found from a table for $\Phi(x)$ function by means of the inverse $\lambda_\alpha = \Phi^{-1}(1-\alpha)$.

The class of normal distributions is extremely versatile. From a theoretical point of view, it has many advantageous features; in addition, variability of measurements of quantities in science and technology are often well described by normal distributions.

Gumbel distribution

The family of Gumbel distributions has a form

$$F_Y(y) = \exp\left(-e^{-(y-b)/a}\right), \qquad -\infty < y < \infty.$$

The quantile function is $y_\alpha = b - a\ln(-\ln(1 - \alpha))$, while the density

$$f_Y(y) = \frac{1}{a} e^{-(y-b)/a} \exp(-e^{-(y-b)/a}), \qquad -\infty < y < \infty.$$

This class has proven to be useful in situations where the variable models the maximum load on a system. It is an important tool in design of engineering systems, *e.g.* in order to calculate design loads.

3.4 Independent Random Variables

The notion of independent events was introduced in Section 1.2. In the present section, we extend this notion and discuss independence for random variables. First, we introduce the concept of a sequence of independent identically distributed random variables.

Construction of iid variables

Let us consider a vector of k independent uniformly distributed variables[3] $U_1, U_2, \ldots, U_k$. Since the numbers are independent then, by solving k equations $U_i = F(X_i)$, we obtain k independent variables $X_1, X_2, \ldots, X_k$, each being $F(x)$ distributed. Such a vector is composed of the so-called *iid* (*independent, identically distributed*) variables.

Obviously, the construction easily extends to not identically distributed variables. Next we give a condition that has to be true in order to have independent variables.

Independent random variables

We now consider random variables having *different* distributions and start with the case of two distributions. In Chapter 1, we said that two events (statements) A_1, A_2 are independent if

$$P(A_1 \cap A_2) = P(A_1)P(A_2).$$

For random variables X_1 and X_2 with distribution functions $F_1(x)$, $F_2(x)$, respectively, we state that if any statement about X_1 is independent of a statement about X_2, then they are independent. Let A_1 be a statement about X_1, for example $A_1 = $"$X_1 \leq 5$", and $A_2 = $"$X_2 \leq -1$". Then

$$P(A_1 \cap A_2) = P(X_1 \leq 5 \text{ and } X_2 \leq -1) = P(X_1 \leq 5)P(X_2 \leq -1)$$
$$= F_1(5)F_2(-1).$$

---

[3]This can be interpreted as a result of k persons flipping independently, each of them say 100 times, a fair coin, rendering k uniformly distributed random numbers.

For random variables there is a convention that the word "and" relating the events is replaced by a comma and hence

$$P(X_1 \leq 5 \text{ and } X_2 \leq -1) = P(X_1 \leq 5, \ X_2 \leq -1).$$

It would be very hard to check whether *all* statements about X_1 and *all* statements about X_2 are independent and it is also not necessary. Again, the statements "$X_1 \leq x_1$" and "$X_2 \leq x_2$" will play an important role in defining independence between two variables X_1 and X_2, see the following definition.

> **Definition 3.4 (Independent random variables).** *The variables X_1 and X_2 with distributions $F_1(x)$ and $F_2(x)$, respectively, are **independent** if for all values x_1 and x_2*
>
> $$P(X_1 \leq x_1, X_2 \leq x_2) = F_1(x_1) \cdot F_2(x_2).$$

The function

$$F_{X_1, X_2}(x_1, x_2) = P(X_1 \leq x_1 \text{ and } X_2 \leq x_2) \tag{3.12}$$

is called the distribution function for a pair of random variables. The probability of any statement about X_1, X_2 can be computed (at least in theory) if the distribution function $F_{X_1, X_2}(x_1, x_2)$ is known (for example by means of Eq. (5.6), to be presented in Chapter 5). The distribution of a vector of n random variables is defined in a similar way. In the following chapters we shall mostly deal with independent random variables. In such a case their distribution is given by

$$\begin{aligned} F_{X_1,\ldots,X_n}(x_1,\ldots,x_n) &= P(X_1 \leq x_1,\ldots,X_n \leq x_n) \\ &= F_1(x_1) \cdot F_2(x_2) \cdot \ldots \cdot F_n(x_n). \end{aligned} \tag{3.13}$$

3.5 Averages – Law of Large Numbers

For a random variable X, the probability of any statement A about X can be computed when the distribution $F_X(x)$ is known. In Section 3.1.1 we introduced a procedure giving numbers as an output (numbers whose values cannot be known in advance). The procedure was called a random-number generator and was a way to construct a random variable with distribution $F(x)$. The proof that X had distribution $F(x)$ was based on the fact that X was a transformation of a random variable U, $X = F^-(U)$, while U had the property that any outcome was equally probable. Consequently, $P(X \leq x) = F(x)$ is the only possible probability that would satisfy Definition 1.2.

Now, let us consider an r.v. X, which is the unknown numerical output of a real-world experiment. If X is the result of rolling a die, then similar arguments, assuming the die is fair, would give the distribution of X. However, for many r.v. X used to model quantities in real-word experiments, the distributions $F(x)$ cannot be derived from the assumption that the outcomes of the experiment are equally probable. Hence another approach is needed.

The possible solution, in many cases, is based on the assumption that the experiment can be repeated in an independent manner, resulting in a vector $X_1, \ldots, X_n$ of r.v. all having the distribution $F(x)$. If the assumption of independence can be motivated, then the distribution $F(x)$ can be found using the following, fundamental result from probability theory: the Law of Large Numbers (LLN).

We return to the problem of finding $F(x)$ in the next chapter, where we discuss the classical inference theory, also called frequentistic approach.

Theorem 3.3. Law of large numbers: *Let $X_1, \ldots, X_k$ be a sequence of iid (independent identically distributed) variables all having the distribution $F_X(x)$. Denote by $\bar{X}$ the average of X_i, i.e.*

$$\bar{X} = \frac{1}{k}(X_1 + X_2 + \cdots + X_k).$$ \hfill (3.14)

(Obviously $\bar{X}$ is a random variable itself.) Let us also introduce a constant called the expected value of X, defined by

$$\mathsf{E}[X] = \int_{-\infty}^{+\infty} x f_X(x)\,\mathrm{d}x,$$

if the density $f_X(x) = \frac{\mathrm{d}}{\mathrm{d}x} F_X(x)$ exists, or

$$\mathsf{E}[X] = \sum_x x\,\mathsf{P}(X = x),$$

where summation is over those x for which $\mathsf{P}(X = x) > 0$. If the expected value of X exists and is finite then, as k increases (we are averaging more and more variables), $\bar{X} \approx \mathsf{E}[X]$ with equality when k approaches infinity.

Remark 3.5. Note that for random variables X_i such that $X_i = 1$ if A is true and zero otherwise, $\mathsf{E}[X] = \mathsf{P}(A)$. $\qquad\qquad\square$

For the most common distributions, the expectations have been calculated and can be found in tables. As illustration, we study two examples: one for a r.v. of discrete type, the other for a r.v. of continuous type.

Example 3.2. Recall the binomial distribution and let $X \in \mathrm{Bin}(n, p)$.

$$\mathsf{E}[X] = \sum_{k=0}^{n} k\,\mathsf{P}(X = k) = \sum_{k=0}^{n} k \binom{n}{k} p^k (1-p)^{n-k} = np.$$

(We have omitted the mathematical details when calculating the sum.) Known values of n and p immediately gives us the expectation; *e.g.* $n = 7$, $p = 0.15$ as in Example 1.7 yields $E[X] = 1.05$. ☐

Example 3.3. Let X be exponentially distributed with density function

$$f_X(x) = \frac{1}{a} e^{-x/a}, \quad x \geq 0.$$

Then $E[X]$ is given by

$$E[X] = \int_0^\infty x f_X(x)\, dx = \int_0^\infty \frac{x}{a} e^{-x/a}\, dx = \left[-x e^{-x/a} \right]_0^\infty + \int_0^\infty e^{-x/a}\, dx$$
$$= a,$$

where we used integration by parts. ☐

3.5.1 Expectations of functions of random variables

From LLN it follows that even the average of functions $Z_i = G(X_i)$, say, must converge to a constant that we denote by $E[Z] = E[G(X)]$, *i.e.*

$$\frac{1}{k}(G(X_1) + G(X_2) + \cdots + G(X_k)) \to E[G(X)] \quad \text{as} \quad k \to \infty, \tag{3.15}$$

if

$$E[G(X)] = \int_{-\infty}^{+\infty} G(x) f_X(x)\, dx, \quad \text{or} \quad E[G(X)] = \sum_x G(x) P(X = x), \tag{3.16}$$

exists.

Linear functions

A simple example is a linear function, that is, $G(x) = ax + b$. From Eq. (3.16) it then follows that

$$E[G(x)] = E[aX + b] = \int_{-\infty}^{+\infty} (ax + b) f_X(x)\, dx \tag{3.17}$$

$$= a \int_{-\infty}^{+\infty} x f_X(x)\, dx + b \int_{-\infty}^{+\infty} f_X(x)\, dx = a\, E[X] + b.$$

This linearity property is important, and we will make use of it in the next chapter in a generalized form for random variables $X_1, \ldots, X_n$ and coefficients $c_1, \ldots, c_n$

$$E[c_1 X_1 + \cdots + c_n X_n] = c_1 E[X_1] + \cdots + c_n E[X_n]. \tag{3.18}$$

Power functions. Variance

Especially important functions $G(x)$ are powers, *i.e.* $G(x) = x^k$. For $k = 2$, we obtain the so-called second moment of X, *i.e.* $\mathsf{E}[X^2]$. Somewhat more often used is the so-called *variance*

$$\mathsf{V}[X] = \mathsf{E}[(X - \mathsf{E}[X])^2] = \mathsf{E}[X^2] - \mathsf{E}[X]^2, \tag{3.19}$$

which measures the average squared distance between the random variable and its expected value. Variance is a measure of variability for random numbers (higher variance is related to higher variability).

One can show that for the variance,

$$\mathsf{V}[aX + b] = a^2 \mathsf{V}[X] \tag{3.20}$$

and that for a sequence of *independent* random variables $X_1, \ldots, X_n$ and coefficients $c_1, \ldots, c_n$

$$\mathsf{V}[c_1 X_1 + \cdots + c_n X_n] = c_1^2 \mathsf{V}[X_1] + \cdots + c_n^2 \mathsf{V}[X_n]. \tag{3.21}$$

This is an important result that will be used in Section 4.4; see also the following example.

Example 3.4. Consider the random variable

$$\bar{X} = \frac{1}{n}(X_1 + X_2 + \cdots + X_n)$$

where $X_1, X_2, \ldots, X_n$ are iid with $\mathsf{E}[X_i] = m$ and $\mathsf{V}[X_i] = \sigma^2$. By using Eqs. (3.18) and (3.21), one finds

$$\mathsf{E}[\bar{X}] = m, \qquad \mathsf{V}[\bar{X}] = \frac{\sigma^2}{n}. \tag{3.22}$$

$\square$

Standard deviation and coefficient of variation

Related concepts are the *standard deviation* $\mathsf{D}[X] = \sqrt{\mathsf{V}[X]}$, and for X with positive expectation (*i.e.* $\mathsf{E}[X] > 0$) the *coefficient of variation* defined as

$$\mathsf{R}[X] = \frac{\mathsf{D}[X]}{\mathsf{E}[X]}, \tag{3.23}$$

which measures "pure" variability of X: the influence of units in which X is measured is removed. Observe that if $\mathsf{D}[X] = 0$ then the variable is a deterministic constant. If $\mathsf{D}[X] \approx 0$ we may think that X is almost constant but it may be only a consequence that one is using wrong units. For example, let X be the length of a randomly chosen person measured in microns[4]; then the variance will be astronomically large. On the contrary, if we use kilometres as the scale of X, the variance will be close to zero and hence X almost constant. However, the coefficient of variation $\mathsf{R}[X]$ would be the same in both cases. It is also called the relative uncertainty. Consequently, if $\mathsf{R}[X] \approx 0$ then X is almost a constant independently of units used.

For the classes of standard distributions, it is not necessary to compute integrals to find values of $\mathsf{E}[X]$ or $\mathsf{V}[X]$. There are tables where these quantities are presented as a functions of parameters for different classes of distributions, see appendix.

Problems

3.1. The time intervals T (in hours) between emergency calls at a fire station are exponentially distributed as

$$F_T(t) = 1 - e^{-0.2t}, \quad t \geq 0.$$

(a) Find the probability for the time between emergencies to be longer than 3 hours.
(b) Find the expected value of T. (*Hint.* Use the table on page 252.)

3.2. Which of the following functions are probability density functions?

(i) $f(x) = \frac{1}{2}, \quad -1 \leq x \leq 1$
(ii) $f(x) = e^{-x}, \quad 0 \leq x \leq 1$
(iii) $f(x) = \pi^2 x e^{-\pi x}, \quad 0 \leq x < \infty$
(iv) $f(x) = \sin x, \quad 0 \leq x \leq \frac{3\pi}{2}$

3.3. *Specific load-bearing capacity* is defined as the 95 % quantile of the real load-bearing capacity. In other words, the probability that the real load-bearing capacity will exceed the specific one is 0.95. Calculate the specific load-bearing capacity if the real capacity is assumed to be Weibull distributed with distribution function

$$F(x) = 1 - e^{-(x/a)^k}, \quad x > 0.$$

The parameters are $a = 10$, $k = 5$.

3.4. The random variable X is Gumbel distributed. Give the distribution for $Y = e^X$.

3.5. A random variable Y for which

$$P(Y > y) = \begin{cases} 1, & y \leq 0, \\ e^{-y^2/a^2}, & y > 0, \end{cases}$$

is said to belong to a Rayleigh distribution ($a > 0$).

(a) Find the distribution function $F_Y(y)$.
(b) Give the density function $f_Y(y)$.

---

[4] 1 micron $= 10^{-6}$ m.

3.6. Show by partial integration that for a non-negative continuous random variable T with existing $\mathsf{E}[T]$, the expected value $\mathsf{E}[T]$ can be calculated as

$$\mathsf{E}[T] = \int_0^\infty 1 - F_T(t)\,dt.$$

3.7. Use the result in Problem 3.6 to calculate the expected value for a Rayleigh distributed random variable (see Problem 3.5). *Hint:* Use that $\int_{-\infty}^\infty e^{-u^2}\,du = \sqrt{\pi}$.

3.8. A random variable X with the density function

$$f(x) = \frac{1}{\pi(1+x^2)}, \quad -\infty < x < \infty$$

is said to belong to a Cauchy distribution.

(a) Calculate the median.
(b) Show that the expected value $\mathsf{E}[|X|] = \infty$ and hence $\mathsf{E}[X]$ does not exist.

3.9. Consider the high-water volume rate (m^3/s) in a certain river. Suppose the maximal rate during one year, X, is Gumbel distributed

$$F_X(x) = \exp(-e^{-(x-b)/a}), \quad -\infty < x < \infty,$$

where $a = 7.0$ m^3/s and $b = 35$ m^3/s. The 0.01 quantile $x_{0.01}$ of X is called the 100-year flow. Find the value of the 100-year flow for this river.

3.10. Let $X \in N(0,1)$. Find the quantiles $x_{0.01}$, $x_{0.025}$, and $x_{0.95}$.

3.11. Let $Z \in \chi^2(5)$. Find the quantiles $\chi^2_\alpha(5)$ for $\alpha = 0.001$, 0.01, 0.95.

3.12. Suppose that the height of a man in a certain population is normally distributed with mean 180 (cm) and standard deviation 7.5 (cm).

(a) Calculate the probability that a man is taller than 2 metres.
   (A practical interpretation of this result is that we have a population of men and choose one person at random (each person has the same chance to be chosen). If X is the length of the person, then $P(X > 200)$ is the fraction of the population with this property.)
(b) Calculate the quantile $x_{0.01}$ when $X \in N(180, 7.5^2)$. Interpretation?

3.13. Let $X \in \mathrm{Gamma}(10, 2)$. Define $Y = 3X - 5$ and calculate $\mathsf{E}[Y]$ and $\mathsf{V}[Y]$.

3.14. Let X be an exponentially distributed random variable with expectation $\mathsf{E}[X] = m$. Find the coefficient of variation $R[X]$.

4

Fitting Distributions to Data – Classical Inference

In Chapter 3, computations of probabilities assigned to statements about numerical outcomes of random experiments were discussed. The results of such experiments were denoted by X and identified as random variables[1].

Uncertainty in the values of X was described by a cumulative distribution function (cdf) $F_X(x) = P(X \leq x)$, since the probability of any statement about future values of X can be computed if $F_X(x)$ is known. Furthermore, knowing the distribution, random-number generators can be constructed, which give numerical outputs having the uncertainty described by $F_X(x)$. Random-number generators are very useful tools in risk assessments of complicated systems whose behaviours have to be simulated. Then the uncertain initial values (unknown constants, future measured quantities, etc.) can be generated using random-number generators with suitable distributions describing the uncertainty of the parameters or variability of not yet observed quantities.

In practice the distributions are seldom known and have to be determined in a coherent way. This chapter is devoted to the problem of finding (estimating) a function $F(x)$ that can be used as an approximation of the unknown distribution function $F_X(x)$. Rephrasing, one wishes to assign probabilities to statements about X, which well describe the uncertainties whether statements are true for future values of X.

The estimation of $F_X(x)$ is based on the assumption that *the random experiment is repeatable in an independent manner* (see discussion of different uses of probabilities in the Introduction of Chapter 2 and in Section 3.5). This assumption allows us to interpret the probabilities of a statement A, say, about X, as the relative frequency of times A is true as the number n of repetitions of a random experiment is increasing to infinity. This interpretation of probability $P(A)$ was introduced in Section 2.4 and motivated by the Law of Large Numbers (LLN). This law was given in Theorem 3.3, see also the following remark for more details on how the law is used in the present context.

---

[1]Formally, random variables are functions of outcomes, here identities $X(x) = x$ and hence identified with random experiments.

Remark 4.1. The Law of Large Numbers (LLN) states that, under some conditions, the average value of independent observations of random experiment converges to a constant, called the expected value. Now, let X_i be a sequence of iid (independent identically distributed) variables having the same distribution as X. (One can see X_i as the ith outcome of random experiment X.) Let

$$Z_i = \begin{cases} 1 & \text{if } A \text{ is true for } X_i, \\ 0 & \text{otherwise,} \end{cases}$$

then, by LLN, $\bar{Z} = \frac{1}{n} \sum_{i=1}^{n} Z_i$ converges to $\mathsf{E}[Z_i]$. Now

$$\mathsf{E}[Z_i] = 1 \cdot \mathsf{P}(Z_i = 1) + 0 \cdot \mathsf{P}(Z_i = 0) = \mathsf{P}(X \in A).$$

$\square$

Suppose that the experiment has been performed in an independent manner a number of times n, say, giving a sequence of values of the random variable X, $x_1, \ldots, x_n$. (In practice n is always finite.) The values x_i will be called *data* or *observations* and the distribution function will reflect variability in observed data. More precisely, let $\mathsf{P}_n(A)$ be the fraction of x_i for which A was true, *i.e.* $\mathsf{P}_n(A) = \frac{1}{n} \sum_{i=1}^{n} z_i$ with z_i as defined in Remark 4.1. By LLN, if n is large, $\mathsf{P}_n(A) \approx \mathsf{P}(A)$. Now $\mathsf{P}_n(\cdot)$ is a well-defined probability, satisfies the axiom given in Definition 1.2, and can be computed for any A. Since the probability describes the observed variability of data it is called an *empirical probability*. Similarly as in Chapter 3, statements $A =$"$X \leq x$" are particularly important and $\mathsf{P}_n(X \leq x) = F_n(x)$, see the following definition.

Definition 4.1. *Let $x_1, \ldots, x_n$ be a sequence of measurements (taking values in an unpredictable manner), then the fraction $F_n(x)$ of the observations satisfying the condition "$x_i \leq x$"*

$$F_n(x) = \frac{\text{number of } x_i \leq x, \ i = 1, \ldots, n}{n}$$

*is called the **empirical cumulative distribution function** (ecdf).*

Example 4.1 (Life times of ball bearings). In an experiment, the life times of ball bearings were recorded (million revolutions), see [34]. Consider the following 22 observations (sorted in order):

17.88, 28.92, 33.00, 41.52, 42.12, 45.60, 48.48, 51.84,
51.96, 54.12, 55.56, 67.80, 68.64, 68.88, 84.12, 93.12,
98.64, 105.12, 105.84, 127.92, 128.04, 173.40.

The ecdf obtained by use of the definition is shown in Figure 4.1. $\square$

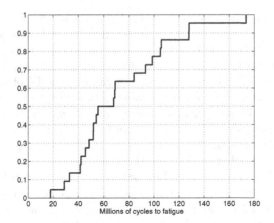

Fig. 4.1. Empirical distribution function, 22 observations of life times for ball bearings.

Obviously $F_n(x)$ is a non-decreasing function, further, $F_n(-\infty) = 0$ and $F_n(+\infty) = 1$ and hence it is a distribution. We now construct a random number that has distribution $F_n(x)$.

Remark 4.2 (Resampling algorithm). Let $x_1, \ldots, x_n$ be a sequence of measurements and $F_n(x)$ be the empirical cumulative distribution function. Let $\widetilde{X}$ be a random number having distribution $F_n(x)$. Independent observations of $\widetilde{X}$ can be generated according to the following algorithm:

Write the observed x_i on separate pieces of papers called lots. Put lots into an urn, mix them well, and draw one from the urn. The number written on the chosen lot, denoted by $\widetilde{x}_1$, is the observation of a r.v. $\widetilde{X}_1$ having distribution $F_n(x)$. Finally, put the lot back into the urn and mix well.

If one wishes to have a sequence of k observations of independent variables $\widetilde{X}_i$, $i = 1, \ldots, k$ with common distribution $F_n(x)$ just repeat the procedure k times.   □

Properties of $F_n(x)$ for large number of observations

The LLN tells us that if x_i are outcomes of iid random variables X_i with a common distribution $F_X(x)$ then, as n grows to infinity, the empirical distribution $F_n(x)$ converges to a distribution function $F_X(x)$. Even more can be said: the Glivenko–Cantelli Theorem, see *e.g.* [82], states that even the maximal distance between $F_n(x)$ and $F_X(x)$ tends to zero when n increases without bounds, *viz.* $\max_x |F_X(x) - F_n(x)| \to 0$ as $n \to \infty$ with probability one. However, n is always finite and moreover, in many problems encountered in safety or risk analysis, n is small. Using F_n as a model for F_X, *i.e.* assuming that $F_X(x) = F_n(x)$, means that the uncertainty in the future (yet unknown) value of the observation of X is the same as the uncertainty of drawing lots from an urn, where lots contain only the previously observed values x_i of X, see Remark 4.2. In many cases, such a random model of the variability of an observed sequence x_i can be sufficient. However, there are statements about an outcome of an experiment, which are false for all x_i that have been observed up to now. Consider again Example 4.1 and let $A =$ "Lifetime of a ball bearing is longer than 190 million revolutions". Using $F_X(x) = F_n(x)$ we find from Figure 4.1 that $P(A) = 0$. However, we are quite sure that if we wait long enough and test further ball bearings, $x_i > 190$ will happen.

Simply, the empirical distribution contains no information about possible extreme values that have not been observed in a finite sequence. If we wish to make some predictions about the chances of receiving extreme values without observing them, hypotheses are needed about the values of F_X outside the region of observations. One way of solving this problem is to assume that F_X belongs to a family of distributions, *e.g.* normal, exponential, Weibull, etc., which limits the possible shapes of the distributions. The problem boils down to *estimation* of scale- or location parameters (cf. Chapter 3.3) in the actual distribution. That problem is discussed in more detail in Section 4.3. Based on observations, one of the possible "shapes" are chosen, for example the one that is (in some sense) closest to the empirical distribution. Methods for this are presented in the following section.

4.1 Estimates of F_X

The previously defined empirical distribution function can be used as an approximation of the unknown distribution for a given data set. However, the convergence of $F_n(x)$ to $F_X(x)$ is slow, hundreds of observations are needed (see discussion in Chapter 2.4) to get acceptably small relative errors. Often in practical situations n can be small, especially when experiments are expensive to perform or seldom observed. For example, when estimating strength of material it is not rare to have less than 10 observations x_i. In such situations one wishes to use another estimate of $F_X(x)$ than the empirical distribution $F_n(x)$.

Example 4.2 (Periods between earthquakes). We return to Example 1.1. Periods between serious earthquakes are modelled by an r.v. X. By experience (see also Chapter 7 for theoretical motivations) we expect X to have an exponential cdf $F(x; a) = 1 - \exp(-x/a)$, $x \geq 0$, where a is an unknown parameter that has to be found. For example, one could choose a value a^* such that the empirical distribution $F_n(x)$ and $F^*(x) = F(x; a^*)$ are close to each other. Since the expected value of an exponentially distributed variable is just a and the mean $\bar{x}$ converges to the expectation (by LLN) as n tends to infinity, let us choose $a^* = \bar{x} = 437.2$ hours.

In Figure 4.2, left panel, we can see both $F_n(x)$ (stairwise function) and $F^*(x) = 1 - \exp(-x/437.2)$ (solid line). The curves seem to well follow each other. In order to motivate this opinion we perform a Monte Carlo experiment. We simulate 62 random numbers using an exponential random-number generator with mean 437.2 and based on this, compute the ecdf $\widetilde{F}_n(x)$. Now we know that the difference between $\widetilde{F}_n(x)$ and $F_X(x) = 1 - \exp(-x/437.2)$ reflects estimation error only due to the limited number of observations. The empirical distribution is presented in Figure 4.2, right panel. Conclusion: 62 observations is not much and the ecdf $\widetilde{F}_n(x)$ can differ quite a lot from the true distribution $F_X(x)$.   □

The discussion in the last example contained three main steps: choice of a model, finding the parameters, and analysis of error (in other words, checking if the model does not contradict the observations). These three steps are the core of a *parametric* estimation procedure to model the distribution $F_X(x)$. In the following we describe the three steps, introducing a more concise framework:

I **Modelling.** Choose a model, which means one of the standard distributions $F(x)$, for example normal, exponential, Weibull, Poisson, etc. Next postulate that

$$F_X(x) = F\left(\frac{x-b}{a}\right),$$

where a and b are unknown scale and location parameters. There are families of distributions that in addition have a shape parameter c. Examples encountered in this book are Weibull, GEV (generalized extreme value), and GPD (generalized Pareto distribution). For notational convenience, denote the vector of parameters by θ, *i.e.* $\theta = (a, b, c)$, and the model by $F(x; \theta)$.

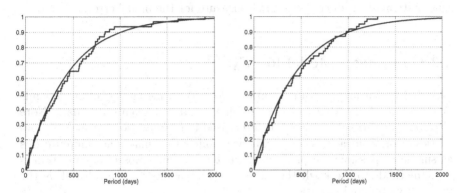

Fig. 4.2. *Left:* Original data, ecdf; *Right:* Simulated data, ecdf.

II **Estimation.** On the basis of the observations $\mathbf{x} = x_1, x_2, \ldots, x_n$ select a value of the parameter θ. Since the chosen value depends on data it will be denoted by

$$\theta^*(\mathbf{x}) = \big(a^*(\mathbf{x}),\, b^*(\mathbf{x}),\, c^*(\mathbf{x})\big).$$

The functions in $\theta^*(\mathbf{x})$ are called *estimates* of the unknown parameters in θ.

III **Error analysis.** The estimation error $e = \theta - \theta^*$ is in reality unknown and the best we can do is to study its variability. In order to do it we introduce the concept of an *estimator*, which consists of gathering of data and computation of estimates. More precisely, the values of $\mathbf{x}$ (which are unknown in advance) are treated as outcomes of a random vector $\mathbf{X} := (X_1, X_2, \ldots, X_n)$. Then the estimator

$$\Theta^* = \big(a^*(\mathbf{X}),\, b^*(\mathbf{X}),\, c^*(\mathbf{X})\big)$$

is a random variable modelling the uncertainty of the value of an estimate due to the variability of data. (Sometimes we also write $\Theta^* = (A^*, B^*, C^*)$.) Now the error $e = \theta - \theta^*$ is an outcome of the random variable $\mathcal{E} = \theta - \Theta^*$. The variability of the error can be described by finding the probability distributions of

$$\mathcal{E} = (\mathcal{E}_1, \mathcal{E}_2, \mathcal{E}_3) = (a - A^*, b - B^*, c - C^*).$$

If the chosen model contains $F_X(x)$, usually the error $F_X(x) - F(x; \theta^*)$ is much smaller than the error $F_X(x) - F_n(x)$. Hence our estimates of probabilities calculated from the distribution can be quite accurate even if the number of observations is limited. (It requires a lesser number of observations to get useful estimates of the probabilities of interest.) However, we face a problem of model error: simply, the distribution $F_X(x)$ looked for does not belong to the chosen class of distributions. Thus it is always recommended to make a sensitivity analysis of computed risk measure for the model error.

4.2 Choosing a Model for F_X

The choice of a family of distributions $F(x; \theta)$ to model F_X often depends on experience from studies of similar experiments or by analysis of data. Note also that the estimate $F(x; \theta^*)$ is often used in computations of some probabilities or other measures of risks, hence models that make the computations as simple as possible are preferable. In the following subsections we discuss some methods to check whether the chosen model is not contradicted by the variability observed in data.

4.2.1 A graphical method: probability paper

Let $F(x; \theta^*)$ be the cdf chosen to approximate the unknown probability distribution $F_X(x)$. A natural question is whether we can verify that $F(x; \theta^*)$ is a good model. How to perform this task is not obvious since the truth is unknown. Suppose now the observations are independent and hence the ecdf $F_n(x)$ is close to $F_X(x)$ at least when n is large. Consequently, the simplest check of the correctness of the model is to compare $F(x; \theta^*)$ with $F_n(x)$. Here it is important that the horizontal distance between $F(x; \theta^*)$ and $F_n(x)$ is small, which means that the quantiles are close. (Recall the definition of a quantile from Section 3.2.)

The visual estimation of the horizontal distance between $F(x; \theta^*)$ and $F_n(x)$ is not simple since for high and low values of x, both $F(x; \theta^*)$ and $F_n(x)$ are almost parallel to the abscissa. In order to avoid this nuisance, one uses the so-called *probability papers*. For historical reasons, one speaks of papers, in spite of the fact that computer programs with graphics facilities are used today. The graphical method is suitable for many of the distributions encountered in risk analysis: Weibull, Gumbel, exponential. Also the normal and lognormal distributions can be handled with this approach. In this book we use the papers only for the purpose of model validation. The idea is simple; change the scales so that the curve $(x, F(x; \theta))$ becomes a straight line for all values of the parameter θ. For simplicity, the probability scale, shown in Figure 4.3 right panel, is often omitted.

Suppose that $\theta = (a, b)$, i.e. $F(x; \theta) = F((x - b)/a)$, where a and b are scale and location parameters, respectively, while $F(.)$ is a known cdf. Let us assume that

$$F_X(x) = F\left(\frac{x - b}{a}\right). \tag{4.1}$$

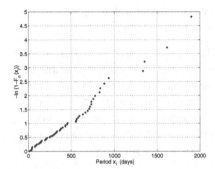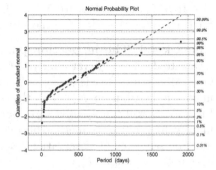

Fig. 4.3. *Left:* Exponential distribution investigated; observations plotted as $\bigl(x_i, -\ln(1 - F_n(x_i))\bigr)$. *Right:* Normal distribution investigated; observations plotted in normal probability paper.

Now let us solve for $(x - b)/a$ in Eq. (4.1); then one obtains

$$F^-\big(F_X(x)\big) = \frac{x - b}{a},$$

and hence dots with coordinates $(x, F^-(F_X(x)))$ lie on a straight line. Here $F^-(.)$ is the inverse function to $F(.)$ defined in Chapter 3.

The idea is simple but a practical problem is that $F_X(x)$ is unknown. Thus in practice, it is replaced by the ecdf $F_n(x)$. Now dots with coordinates $(x, F^-(F_n(x)))$ should be close to a straight line if n is large, since $F_n(x) \approx F_X(x)$. Consequently, if the curve $(x, F^-(F_n(x)))$ is not close to the straight line it gives strong indication that Eq. (4.1) cannot be true, *i.e.* our model is wrong.

Example 4.3 (Exponential distribution). Suppose one believes that the studied variable X is exponentially distributed. This means that $F_X(x) = 1 - e^{-x/a}$, *i.e.* $\theta = a$. The solution of Eq. (4.1) is $x/a = -\ln(1 - F_X(x))$. Since $F_X(x)$ is unknown it is replaced by the ecdf. Dots with coordinates $\big(x_i, -\ln(1 - F_n(x_i))\big)$ are plotted on the figure and they should be close to a straight line if the exponential distribution is a good model.

Periods between earthquakes. We turn again to studies of periods between earthquakes, cf. Example 4.2. As will be shown in Chapter 7, a more complex study of the variability of times of occurrences of earthquakes simplifies if the periods between earthquakes are independent and exponentially distributed. Consequently this is our first choice for the model and we will check whether the available data do not contradict it.

In Figure 4.3, we see that the points $\big(x_i, -\ln(1 - F_n(x_i))\big)$ follow a straight line and hence we will keep the model. Note that the same data plotted on normal probability paper clearly form a bent curve; hence normal distribution would not be appropriate to model variability of these data. □

In the approach with probability paper, one heavily makes use of an assumption that $F_n(x)$ is close to $F_X(x)$ if n is large. In the following remark we further discuss that this does not always have to be correct.

Remark 4.3 (The inspection paradox). The empirical distribution is a means to describe the variability of the observations of a random variable X. One can ask if $F_n(x)$ is always a "good" (although irregular) estimate of the unknown distribution $F_X(x)$. If the collected values are independent then by LLN $F_n(x)$ will converge to $F_X(x)$ as the number of observations n tends to infinity. In practice the method of collection of data may introduce dependence into the sequence of observations causing bias, which manifests that $F_n(x)$ does not converge to $F_X(x)$.

Suppose we wish to know the distribution of X, the average income per person in a family chosen at random. Data will be gathered according to the following scheme: Select at random a person (using a personal identification code) and then ask for an estimate of the average income in his/her family. The ecdf of the so collected data can be biased and not converge to F_X — the problem is that large families have higher probability to be selected than singles do.

The discussed sampling problem is a version of the so-called "inspection paradox", which states that the interval between accidents that contains a fixed time t, e.g. the period between earthquakes that contains the time Jan. 1, 2010, tends to be larger than an ordinary interval. Intuitively, larger intervals have a greater chance to contain the fixed time t. The inspection paradox, if overlooked, can lead to serious errors in practical situations ($F_n(x)$ may differ considerably from $F_X(x)$); see [67] for more detailed discussion.   □

4.2.2 Introduction to χ^2-method for goodness-of-fit tests

The method to be presented offers ways to test whether data do not contradict the model. (It does not imply that the model is correct, one just checks if it is not obviously wrong.) First, let us consider a simple version of the method, useful in situations where data are collected in classes as illustrated by the following example.

Example 4.4 (Rolling a die 20 000 times). In 1882, R. Wolf rolled a die $n = 20\,000$ times and recorded the number of eyes shown ([84], the data set is found in [34]). The result is given in the table below:

Number of eyes i	1	2	3	4	5	6
Frequency n_i	3407	3631	3176	2916	3448	3422

If the die were fair, then we have the probabilities

$$p_i = \mathsf{P}(\text{"The die shows number } i\text{"}) = \frac{1}{6}.$$

The estimated probabilities $p_i^* = n_i/n$ are equal to

$$p_i^* = 0.1704, \quad 0.1815, \quad 0.1588, \quad 0.1458, \quad 0.1724, \quad 0.1711.$$

Our problem is, on the basis of this data set, to decide whether we still can believe that the die is fair, and hence $p_i = 1/6$ in the next roll, or not. Can the difference $p_i - p_i^*$ be explained by the estimation error as n is finite? Or are errors $p_i - p_i^*$ too large and hence caused by model error?   □

The χ^2 test

The following method, called χ^2 test, has been developed by Karl Pearson (1857-1936). The quantity Q in Eq. (4.2) is sometimes called Pearson statistic. The interpretation of the test procedure is as follows. Denote by α the probability of rejecting a true hypothesis. This number is called the significance level and α is often chosen to be 0.05 or 0.01. Rejecting H_0 with a lower α indicates stronger evidence against H_0. Not rejecting the hypothesis does not mean that there is strong evidence that H_0 is true. It is recommendable to use the terminology "reject hypothesis H_0" or "not reject hypothesis H_0" but not to say "accept H_0".

Consider an experiment that can result in r different ways (classes). Let n_i, $i = 1, \ldots, r$, denote the number of experiments resulting in outcome i, while the total number of experiments $n = n_1 + \cdots + n_r$. Suppose that p_i, denoting the probability that any trial results in outcome i, are known.

χ^2 **test.** Consider the hypothesis

$$H_0 : \; \mathsf{P}(\text{"Experiment results in outcome } i\text{"}) = p_i, \quad i = 1, \ldots, r.$$

The test procedure is as follows:
1. Calculate

$$Q = \sum_{i=1}^{r} \frac{(n_i - np_i)^2}{np_i} \tag{4.2}$$

where p_i are the probabilities we are testing for.
2. Reject H_0 if $Q > \chi_\alpha^2(f)$, where $f = r - 1$.
Further, in order to use the test, as a rule of thumb one should check that $np_i > 5$ for all i (see [13], [85] and references therein).

If n is large, this test has approximatively significance α, *i.e.* the probability of rejecting true hypothesis is α.

Example 4.5 (Wolf's data). Using Eq. (4.2), we get

$$Q = 1.6280 + 26.5816 + 7.4261 + 52.2501 + 3.9445 + 2.3585 = 94.2$$

Since $f = r - 1 = 5$ and the quantile $\chi_{0.05}^2(f) = 11.1$, we have $Q > \chi_{0.05}^2(5)$, which leads to rejection of the hypothesis of a fair dice. $\qquad\square$

Goodness-of-fit tests

The χ^2 test can also be adapted to the situation when one observes results of experiments that are modelled by a continuous random variable X. Suppose that one wishes to check whether data do not contradict the model $F_X(x) = F(x, \theta)$. Our problem is that the χ^2 test is constructed for a discrete r.v., *i.e.* experiments have a *finite* number of possible results.

A way to go around this difficulty is to represent data by its histogram. More precisely let us introduce a partition

$$-\infty = c_0 < c_1 < c_2 < \ldots < c_{r-1} < c_r = +\infty$$

The observations in $\mathbf{x}$ are then classified into r groups by checking which of the conditions $c_{i-1} < x_j \leq c_i$ that are satisfied. Then n_i is the number of observations in $\mathbf{x}$ that falls in the interval $(c_{i-1}, c_i]$. Now if $F_X(x) = F(x, \theta)$, the probability of getting observations in the class i is

$$p_i(\theta) = F(c_i, \theta) - F(c_{i-1}, \theta), \quad i = 1, \ldots, r.$$

If the parameter θ were known, one could compute the χ^2 test as described before. This is a rare situation in risk or safety analysis and one would prefer to test whether data do not indicate the presence of model error, *i.e.* that the estimated model $F(x; \theta^*)$ does not fit the unknown $F_X(x)$. This can be done as follows:

For the partition $c_0, \ldots, c_r$, let n_i, $i = 1, \ldots, r$, denote the number of observations x_j satisfying $c_{i-1} < x_j \leq c_i$, while the total number of observations $n = n_1 + \cdots + n_r$. Let θ^* be an estimate of the parameter θ having k elements. (If $\theta = (a, b, c)$ then $k = 3$.) Next, let

$$p_i^* = p_i(\theta^*) = F(c_i, \theta^*) - F(c_{i-1}, \theta^*).$$

Goodness-of-fit test. Consider the hypothesis $H_0 : F_X(x) = F(x, \theta^*)$. The test procedure is as follows:
1. Calculate

$$Q = \sum_{i=1}^{r} \frac{(n_i - np_i^*)^2}{np_i^*}. \tag{4.3}$$

2. Reject H_0 if $Q > \chi_\alpha^2(f)$, where $f = r - k - 1$ and k is the number of estimated parameters.
(Again, as a rule of thumb one should check that $np_i > 5$ for all i.)

If n is large, at least around 100, this test has approximatively significance α, *i.e.* the probability of rejecting a true hypothesis is α. In the following example the number of observations will be too low in order to claim the significance of the test to be α.

Example 4.6 (Testing for exponential distribution). Consider the data set with 62 recorded periods between serious earthquakes (days):

840	157	145	44	33	121	150	280	434	736
584	887	263	1901	695	294	562	721	76	710
46	402	194	759	319	460	40	1336	335	1354
454	36	667	40	556	99	304	375	567	139
780	203	436	30	384	129	9	209	599	83
832	328	246	1617	638	937	735	38	365	92
82	220								

In Example 4.3, we discussed as a model for this situation an exponential distribution $F(x; \theta) = 1 - \exp(-x/\theta)$ with $\theta^* = 437.2$. We now perform a hypothesis test of this model.

Let us describe variability of data by means of the histogram and introduce $c_0 = 0$, $c_1 = 100$, $c_2 = 200$, $c_3 = 400$, $c_4 = 700$, $c_5 = 1000$, and $c_6 = \infty$. Consequently, $r = 6$ and n_i the number of observed periods between earthquakes x_j satisfying condition $c_{i-1} < x_j \leq c_i$. For example, n_1 is the number of observations not exceeding 100 and thus $n_1 = 14$. The remaining values of n_i are $n_2 = 7$, $n_3 = 14$, $n_4 = 13$, $n_5 = 10$, and $n_6 = 4$.

Returning to the exponential distribution, we now find

$$p_1^* = 1 - e^{-100/437.2} = 0.2045,$$

$$p_2^* = e^{-100/437.2} - e^{-200/437.2} = 0.1627,$$

and in a similar manner $p_3^* = 0.2323$, $p_4^* = 0.1989$, $p_5^* = 0.1001$, and $p_6^* = 0.1015$. The Pearson statistic is

$$Q = 0.1376 + 0.9449 + 0.0113 + 0.0362 + 2.3191 + 0.8355 = 4.285.$$

Now $f = 6 - 1 - 1$ and with $\alpha = 0.05$, the quantile $\chi^2_{0.05}(4) = 9.49$. Hence $Q < \chi^2_{0.05}(4)$, which leads to the conclusion that the exponential model can not be rejected. □

We end this subsection by a brief remark that there also exist other test procedures to test for continuous distributions, for instance, the Kolmogorov–Smirnov test, which measures the distance in a certain sense between the ecdf and the distribution given in the hypothesis. We refer to any textbook in statistics, e.g. [70], Chapter 8.5 or [3], Chapter 6.3.

4.3 Maximum Likelihood Estimates

4.3.1 Introductory example

The so-called Maximum Likelihood (ML) method is fundamental in finding estimates θ^* in a model $F(x; \theta)$ (recall Section 4.1 for an introductory discussion of the estimation problem). The theory of ML estimates has deep consequences for many fields in statistics; see Pawitan [60]. The statistical properties of the ML estimate are also useful, as demonstrated in Sections 4.4 and 4.5. Before we give details of the ML algorithm, we start with an example where X is of the discrete type.

Example 4.7 (Poisson distribution, accidents). The number of accidents in one year K, say, is unknown and may vary from year to year. Obviously K is a discrete r.v. and we wish to find the probability-mass function $p(k) = P(K = k)$.

Probabilistic model: Suppose that we can assume that the mechanism generating accidents is stationary and that the number of accidents in disjoint time periods are independent. Then we know that K is Poisson distributed, *i.e.*

$$p(k) = p(k; \theta) = \frac{\theta^k}{k!} e^{-\theta}, \quad k = 0, 1, \ldots,$$

where $\theta > 0$ is unknown.

Estimation: Suppose $k_1 = 2$ accidents were recorded during the first year. What is a reasonable estimate θ^* of θ on basis of this information? The ML method proposes to choose θ^* so that the probability that two accidents happen during one year is as high as possible. This is accomplished for θ^* such that

$$\mathsf{P}(K = 2) = p(2; \theta) = \frac{\theta^2}{2} e^{-\theta}$$

attains its maximal value for $\theta = \theta^*$. It is easy to check that $p(2; \theta)$ attains its maximal value for $\theta^* = 2$. Consequently the ML estimate $\theta^*(k_1) = k_1$.

Suppose that in the second year $K_2 = 0$ accidents were counted. By our assumptions K_1 and K_2 are independent, hence

$$\mathsf{P}(K_1 = 2, K_2 = 0) = \mathsf{P}(K_1 = 2)\mathsf{P}(K_2 = 0) = \frac{\theta^2}{2} e^{-\theta} \cdot e^{-\theta}. \tag{4.4}$$

Again the ML estimate θ^* is the value of parameter θ that makes the observed number of accidents most likely, which is $\theta^*(2, 0) = (2 + 0)/2 = 1$. $\quad\square$

Remark 4.4. The idea of the ML method presented in Example 4.7 is closely related to the issues discussed in Section 2.1. As in the previous example, let $K \in \mathrm{Po}(\theta)$, $\theta = \mathsf{E}[K]$ unknown. For instance, let A_θ be the alternative statement that "$\mathsf{E}[K] = \theta$". Then, given that $B_1 = $ "Two accidents first year" and $B_2 = $ "Zero accidents second year", the likelihood function $L(A_\theta) = L(\theta)$, say, is given by $L(\theta) = (\theta^2/2) \exp(-\theta) \exp(-\theta)$, *i.e.* the same as in Eq. (4.4).

Suppose now we have no information about possible value of $\mathsf{E}[K]$, *i.e.* odds for A_θ are 1 for all θ. The posterior odds given B_1 and B_2 are just $L(\theta)$ and the ML method proposes to choose θ^* as the alternative, which has the highest posterior odds.

We will return to this type of reasoning in Chapter 6. $\quad\square$

Suppose we have a random experiment (real or a random-number generator) that generates numbers with unknown distribution, having the density $f(x)$ or probability-mass function $p(x)$, say. We shall model the experiment by assuming that $f(x) = f(x; \theta)$ (or $p(x) = p(x; \theta)$), for some value of the parameter θ. The parameter θ can be a vector $\theta = (a, b, c, \ldots)$. Assume that we have n independent observations $x_1, \ldots, x_n$, outcomes of the random experiment. Our goal is to choose a value of the parameter θ.

Maximum Likelihood Method. Consider n independent observations $x_1, \ldots, x_n$ and study the *likelihood function* $L(\theta)$, defined as

$$L(\theta) = \begin{cases} f(x_1; \theta) \cdot f(x_2; \theta) \cdot \ldots \cdot f(x_n; \theta) & \text{(continuous r.v.)} \\ p(x_1; \theta) \cdot p(x_2; \theta) \cdot \ldots \cdot p(x_n; \theta) & \text{(discrete r.v.)} \end{cases} \tag{4.5}$$

where $f(x; \theta)$, $p(x; \theta)$ is probability density and probability-mass function, respectively.

The value of θ that maximizes $L(\theta)$ is denoted by θ^* and called the ML estimate.

Thus, to find the optimal value of the parameter θ in the sense of the ML method, one needs to find maximum of a function. For the standard distributions, explicit expressions for ML estimates of parameters have been derived. We now outline in several examples the main techniques of such derivations. However, calculations of this kind are not always possible and numerical algorithms to find maximum have to be used.

4.3.2 Derivation of ML estimates for some common models

Example 4.8 (ML estimation for Poisson distribution). Assume our model is the Poisson distribution with probability-mass function

$$p(x; \theta) = \frac{\theta^x}{x!} e^{-\theta}, \quad x = 0, 1, 2, \ldots$$

where θ is unknown, and that we have independent observations $x_1, \ldots, x_n$. The likelihood function is

$$L(\theta) = \prod_{i=1}^{n} \frac{\theta^{x_i}}{x_i!} e^{-\theta} = \frac{\theta^{\sum x_i}}{\prod x_i!} e^{-n\theta}$$

where $\prod_{i=1}^{n} a_i = a_1 \cdot a_2 \cdot \ldots \cdot a_n$. A common trick when deriving ML estimates is to study the logarithm of the likelihood function; this leads to easier expressions and the θ maximizing $L(\theta)$ does also maximize $l(\theta) = \ln L(\theta)$ given by

$$l(\theta) = \ln L(\theta) = \sum x_i \ln \theta - \ln(\prod x_i!) - n\theta.$$

Differentiating, we get

$$\dot{l}(\theta) = \frac{\mathrm{d}}{\mathrm{d}\theta} l(\theta) = \frac{\sum x_i}{\theta} - n, \qquad \ddot{l}(\theta) = \frac{\mathrm{d}^2}{\mathrm{d}\theta^2} l(\theta) = -\frac{\sum x_i}{\theta^2}. \tag{4.6}$$

and we now find extremum

$$\dot{l}(\theta) = 0 \quad \Longleftrightarrow \quad \theta = \frac{1}{n}\sum x_i = \bar{x}.$$

If the extremum at $\bar{x}$ is local maximum then ML estimate $\theta^* = \bar{x}$. Therefore, we should check that the second derivative of $l(\theta)$ at $\theta = \bar{x}$ is negative. Employing Eq. (4.6) we get that $\ddot{l}(\theta^*) = -\frac{n^2}{\sum x_i} < 0$ and hence $\theta^* = \bar{x}$.   □

Example 4.9 (Deaths from horse kicks). In this example we analyse some real data. In 1898, von Bortkiewicz published a dissertation about a law of low numbers where he proposed to use the Poisson probability-mass function in studying accidents [5] [2]. A part of his famous data is the number of soldiers killed by horse kicks in 1875–1894 in corps of the Prussian army, presented in [34], see also [62]. Here the data from corps II are presented:

$$0 \ \ 0 \ \ 0 \ \ 2 \ \ 0 \ \ 2 \ \ 0 \ \ 0 \ \ 1 \ \ 1 \ \ 0 \ \ 0 \ \ 2 \ \ 1 \ \ 1 \ \ 0 \ \ 0 \ \ 2 \ \ 0 \ \ 0$$

Clearly the ML estimate of θ is $\theta^* = 12/20$.   □

Example 4.10 (ML estimation for exponential distribution). Assume that our model is the exponential distribution with density

$$f(x) = \frac{1}{\theta}e^{-x/\theta}, \quad x \geq 0,$$

θ unknown, and that we have independent observations $x_1, \ldots, x_n$. The likelihood function is

$$L(\theta) = \prod_{i=1}^{n} \frac{1}{\theta}e^{-x_i/\theta} = \frac{1}{\theta^n}e^{-\sum x_i/\theta}.$$

The log-likelihood function $l(\theta) = \ln L(\theta)$ is given by

$$l(\theta) = \ln L(\theta) = \ln\frac{1}{\theta^n} - \frac{1}{\theta}\sum x_i = -n\ln\theta - \frac{1}{\theta}\sum x_i.$$

Differentiating, we get

$$\dot{l}(\theta) = -\frac{n}{\theta} + \frac{1}{\theta^2}\sum x_i, \quad \ddot{l}(\theta) = \frac{n}{\theta^2} - \frac{2}{\theta^3}\sum x_i. \tag{4.7}$$

and we now find extremum

$$\dot{l}(\theta) = 0 \quad \Longleftrightarrow \quad \theta = \frac{1}{n}\sum x_i = \bar{x}.$$

---

[2] In [29], the author argues that the Poisson distribution could have been named the von Bortkiewicz distribution.

This is a local maximum since by Eq. (4.7)

$$\ddot{l}(\bar{\mathbf{x}}) = -\frac{n^3}{(\sum x_i)^2} < 0,$$

and hence the obtained ML estimate is $\theta^* = \bar{\mathbf{x}}$.

For the data of earthquakes, the arithmetic mean of the observations, $\theta^* = 437.2$, thus is the ML estimate of the parameter.   □

Example 4.11 (ML estimation for normal distribution). Consider a normal variable $X \in N(m, \sigma^2)$; hence $f(x; \theta) = \frac{1}{\sigma\sqrt{2\pi}}e^{-(x-m)^2/2\sigma^2}$. Suppose we have n independent observations $\mathbf{x} = (x_1, \ldots, x_n)$ of X. We derive the ML estimates of $\theta = (\theta_1, \theta_2) = (m, \sigma^2)$. The likelihood function and log-likelihood function are given by

$$L(\theta) = \frac{1}{(2\pi\theta_2)^{n/2}}e^{-\sum(x_i - \theta_1)^2/2\theta_2},$$

$$l(\theta) = -\frac{n}{2}\left(\ln(2\pi) + \ln\theta_2\right) - \frac{1}{2\theta_2}\sum(x_i - \theta_1)^2.$$

Differentiating $l(\theta)$ with respect to θ_1 and θ_2, we obtain

$$\frac{\partial l}{\partial \theta_1} = \frac{1}{\theta_2}\sum(x_i - \theta_1) = \frac{1}{\theta_2}\sum x_i - n\frac{\theta_1}{\theta_2},$$

$$\frac{\partial l}{\partial \theta_2} = -\frac{n}{2\theta_2} + \frac{1}{2\theta_2^2}\sum(x_i - \theta_1)^2.$$

Solving the system of equations $\frac{\partial l}{\partial \theta_1} = 0$ and $\frac{\partial l}{\partial \theta_2} = 0$ leads to the ML estimates

$$\theta_1^* = \frac{x_1 + \cdots + x_n}{n} = \bar{\mathbf{x}}, \tag{4.8}$$

$$\theta_2^* = \frac{1}{n}\sum_{i=1}^n (x_i - \bar{\mathbf{x}})^2 = s_n^2. \tag{4.9}$$

Actually, we also should check if the matrix of second derivatives of $l(\theta)$ is negative definite to be sure that extremes are really local maxima. We do it here for completeness:

$$[\ddot{l}(\theta)] = \begin{bmatrix} \dfrac{\partial^2 l}{\partial \theta_1^2} & \dfrac{\partial^2 l}{\partial \theta_1 \partial \theta_2} \\[2mm] \dfrac{\partial^2 l}{\partial \theta_2 \partial \theta_1} & \dfrac{\partial^2 l}{\partial \theta_2^2} \end{bmatrix} = \begin{bmatrix} -\dfrac{n}{\theta_2} & -\dfrac{n\bar{\mathbf{x}}}{\theta_2^2} + \dfrac{n\theta_1}{\theta_2^2} \\[2mm] -\dfrac{n\bar{\mathbf{x}}}{\theta_2^2} + \dfrac{n\theta_1}{\theta_2^2} & \dfrac{n}{2\theta_2^2} - \dfrac{\sum(x_i - \theta_1)^2}{\theta_2^3} \end{bmatrix}.$$

$$\tag{4.10}$$

Now, it is easy to check that

$$[\ddot{l}(\theta^*)] = \begin{bmatrix} -\dfrac{n}{s_n^2} & 0 \\ 0 & -\dfrac{n}{2(s_n^2)^2} \end{bmatrix}, \tag{4.11}$$

i.e. $[\ddot{l}(\theta^*)]$ is a diagonal matrix with negative elements on the diagonal and hence the extremum at θ^* is the local maximum.   □

4.4 Analysis of Estimation Error

Suppose that the results of an experiment are uncertain values modelled by a random variable X with an unknown distribution F_X. We assumed that a family of distributions $F(x; \theta)$ contains the unknown cdf, *i.e.* there is a value of θ such that $F_X(x) = F(x; \theta)$ for all x. By this we neglect the possibility of a model error. Using observations of X, gathered in an n dimensional vector $\mathbf{x}$ we presented, in the previous subsection, the ML method to derive the estimates $\theta^*(\mathbf{x})$ of θ. As long as n is finite, $\theta^*(\mathbf{x}) \neq \theta$; in other words, the error $\theta - \theta^* \neq 0$ for finite n. Obviously, practically it is important to know if the error $\theta - \theta^*$ tends to zero as the number of observations n increases to infinity. The so-called *consistent estimators* possess the property.

In previous subsections, we proved that the ML estimate θ^* of the parameter θ in Poisson and exponential distributions was the average $\bar{\mathbf{x}}$. Consequently, in these models the estimator $\Theta^* = \bar{\mathbf{X}}$, where $\mathbf{X} = (X_1, X_2, \ldots, X_n)$ are iid with common cdf $F_X(x)$. Now, by LLN, Θ^* converges to $\mathsf{E}[X]$. For Poisson and exponential distributions $\mathsf{E}[X] = \theta$ and hence the estimator Θ^* is consistent.

The error analysis can be performed for any estimator, even if we limit ourselves here to the ML case. The reason for it is that ML estimators possess many good properties. For example, it can be shown (see [49] or for a review [60]) that the ML method results in consistent estimators, see the following theorem:

Theorem 4.1. Consistency of ML estimators. *Assume that $f(x; \theta)$ (or $p(x; \theta)$) satisfy certain regularity conditions, which are valid in examples discussed in this text, and let $X_1, X_2, \ldots$ be independent variables each having distribution given by $f(x; \theta)$ (or $p(x; \theta)$). Then the ML estimator $\Theta^* = \theta^*(X_1, X_2, \ldots, X_n)$ is a consistent estimator of θ, i.e.*

$$\mathsf{P}(C) = 1$$

where C is the statement "Θ^* converges to θ, as $n \to \infty$".

Example 4.12. Let $X = \theta \cdot U$, where U is a uniformly distributed variable (Chapter 3.1.1). Then the probability density of X is

$$f(x; \theta) = \begin{cases} \dfrac{1}{\theta}, & 0 < x < \theta, \\ 0, & \text{otherwise.} \end{cases}$$

This density does not satisfy the "regularity conditions" assumed in Theorems 4.1, 4.2, and 4.3. □

As we have mentioned above the exact value of the estimation error is unknown, in other words, it is an uncertain value. The variability of the error can be studied using the following random variable,

$$\mathcal{E} = \theta - \Theta^*,$$

the *estimation error*. For consistent estimators, $\mathcal{E}$ tends to zero as n increases without bounds. In practice, the values of the r.v. $\mathcal{E}$ cannot be observed except in Monte Carlo experiments using random-number generators, since then θ is an input to the program. Nevertheless we can study the distribution of $\mathcal{E}$, $F_{\mathcal{E}}(e)$, which, for example, can be used to find intervals such that with high confidence we can claim that θ is in these intervals (see Section 4.5). However, we first present some simpler measures to describe variability of $\mathcal{E}$.

4.4.1 Mean and variance of the estimation error $\mathcal{E}$

Finding the exact cdf of the error $\mathcal{E}$ can be difficult; hence first mean and variance of $\mathcal{E}$ are studied.

Mean of the estimation error

First the *expected* error may be checked:

$$m_{\mathcal{E}} = \mathsf{E}[\theta - \Theta^*] = \theta - \mathsf{E}[\Theta^*].$$

If the expected error is zero, we call the estimator *unbiased*.

Example 4.13. In Examples 4.8, 4.10, and 4.11, we proved that the ML estimate $\theta^* = \bar{x}$. Is the estimator $\Theta^* = \bar{X}$ unbiased? The answer is given by the following calculation, cf. Eq (3.18) and Example 3.4:

$$\mathsf{E}[\Theta^*] = \mathsf{E}\Big[\frac{X_1 + \cdots + X_n}{n}\Big] = \frac{1}{n}\mathsf{E}[X_1 + \cdots + X_n]$$

$$= \frac{1}{n}(\mathsf{E}[X_1] + \cdots + \mathsf{E}[X_n]) = \frac{1}{n}(n \cdot \mathsf{E}[X]) = \mathsf{E}[X].$$

Since $\mathsf{E}[X] = \theta$ in these examples, the estimator is unbiased. In other words, the expected value of the error $\mathsf{E}[\mathcal{E}] = 0$. □

Example 4.14. The ML estimator $\theta_2^* = s_n^2$ of $\theta_2 = \sigma^2$ in Example 4.11 (see Eq. (4.9)) is actually biased. Slightly changing the estimate

$$(\sigma^2)^* = \frac{1}{n-1} \sum_{i=1}^{n} (x_i - \bar{x})^2 = s_{n-1}^2, \tag{4.12}$$

will give unbiased estimation. One can show that the estimator

$$S_{n-1}^2 = \frac{1}{n-1} \sum_{i=1}^{n} (X_i - \bar{X})^2,$$

is an unbiased estimator of $\theta = V[X]$ for a general r.v. X. Here we kept the traditional symbols for the estimators. □

Variance of the estimation error

The variance is an important measure of variability of the error, denote it by $\sigma_{\mathcal{E}}^2$. Since for any r.v. ξ and a constant c, $V[\xi + c] = V[\xi]$ one has that

$$\sigma_{\mathcal{E}}^2 = V[\theta - \Theta^*] = V[\Theta^*]. \tag{4.13}$$

For *unbiased* estimators $m_{\mathcal{E}} = 0$. Moreover, *efficient* estimators should have as small variance $\sigma_{\mathcal{E}}^2$ as possible. For two unbiased estimators of the same parameter the one with lower $\sigma_{\mathcal{E}}^2$ is considered more efficient. Computation of $V[\Theta^*](= \sigma_{\mathcal{E}}^2)$, by (4.13), is important in evaluation of uncertainty in the estimate θ^*. Since $\Theta^* = \theta^*(X_1, \ldots, X_n)$ the variance can be theoretically computed if $F_X(x)$ is known (see Chapters 5 and 8 for definitions and approximate methods for computation of expectations of functions of random variables).

   Since $F_X(x) = F(x; \theta)$, even the variance $V[\Theta^*]$ is a function of an unknown parameter θ, which we write $V[\Theta^*] = f(\theta)$. Hence, most often, a numerical value for the variance cannot be given. However, since for consistent estimators $\theta^* \to \theta$ as n increases to infinity, the approximation $V(\Theta^*) \approx f(\theta^*)$ is made if n is large.

Example 4.15. Suppose X is exponentially or Poisson distributed with unknown mean θ, *i.e.* $E[X] = \theta$. Let $\mathbf{x}$ denote the data. In both cases the ML estimate of θ is $\theta^* = \bar{x}$. We have already demonstrated that $\Theta^* = \bar{X}$ is an unbiased estimator. Its variance $\sigma_{\mathcal{E}}^2 = V[\Theta^*]$ follows from the calculation (cf. Eq. (3.21) and Example 3.4):

$$V[\Theta^*] = V\left[\frac{X_1 + \cdots + X_n}{n}\right] = \frac{1}{n^2} V[X_1 + \cdots + X_n]$$

$$= \frac{1}{n^2}(V[X_1] + \cdots + V[X_n]) = \frac{1}{n^2}(n \cdot V[X]) = \frac{V[X]}{n}. \tag{4.14}$$

Now for exponentially distributed X, $V[X] = \theta^2$, while for Poisson distributed r.v. X, $V[X] = \theta$. The approximation of the variance $V[\Theta^*]$ is obtained by replacing the unknown parameter θ in the formulae for $V[X]$ by the estimates $\theta^* = \bar{x}$. In the case when X was distance between earthquakes, see Example 4.10, $V[\Theta^*] \approx 437.2^2/62 = 3083$, while for X being the number of perished from horse kicks during one year (see Example 4.9), $V[\Theta^*] \approx 0.6/20 = 0.03$.

□

Obviously, if $\sigma_{\mathcal{E}}^2 = 0$ there is no estimation error present and the estimate is equal to the parameter. However, in general it is not possible to have $\sigma_{\mathcal{E}}^2 = 0$. Under the assumptions of Theorem 4.1, one can actually demonstrate that there is a lower bound for the efficiency of the unbiased estimators, $i.e.$ the variance $\sigma_{\mathcal{E}}^2$ for an unbiased estimator is bounded from below by a positive constant σ_{MVB}^2 (MVB — Minimum Variance Bound). The value of σ_{MVB}^2 depends on the model $F(x; \theta)$ and it is proportional to $1/n$ (the inverse of the number of observations).

Theorem 4.2. *Suppose that the assumptions of Theorem 4.1 holds. Then*

$$E[\mathcal{E}] \to 0 \quad and \quad V[\mathcal{E}] \to 0 \quad as \quad n \to \infty.$$

In addition $\lim_{n\to\infty}(\sigma_{\mathcal{E}}^2/\sigma_{MVB}^2) = 1.$

The last theorem states that for large values n, the ML estimator Θ^* is approximately unbiased ($i.e.$ $E[\mathcal{E}] \approx 0$) and the error $\mathcal{E}$ has its variance close to the lowest possible value $\sigma_{\mathcal{E}}^2 \approx \sigma_{MVB}^2$.

A very important property for ML estimators Θ^* is that when n is large, the variance $V[\Theta^*] = \sigma_{\mathcal{E}}^2$ can be approximated using the second-order derivatives of the log-likelihood function computed at its maximum. The method is presented next.

Approximation of variance of ML estimators

Consider first the case when the model $F(x; \theta)$ for the cdf of X depends only on one parameter θ; for instance X is a binomial, Poisson, exponentially, Rayleigh distributed variable. Then

$$V[\Theta^*] = \sigma_{\mathcal{E}}^2 \approx -\frac{1}{\ddot{l}(\theta^*)} = (\sigma_{\mathcal{E}}^2)^*. \tag{4.15}$$

Programs used to compute ML estimates often also give $(\sigma_{\mathcal{E}}^2)^*$ as an output.

Example 4.16. Consider Examples 4.8 and 4.9, where a Poisson distribution was studied. With the ML estimate $\theta^* = \bar{x}$, it follows from Eq. (4.6) that $\ddot{l}(\theta) = -\sum x_i/\theta^2$ and hence

$$(\sigma_{\mathcal{E}}^2)^* = -\frac{1}{\ddot{l}(\theta^*)} = \frac{(\theta^*)^2}{\sum x_i} = \frac{(\theta^*)^2}{n\theta^*} = \frac{\theta^*}{n},$$

where $\theta^* = \bar{x}$ and is the same as derived in Example 4.15.

□

Example 4.17. Consider now an exponentially distributed r.v. X with mean θ. Again the ML estimate $\theta^* = \bar{x}$, while

$$\ddot{l}(\bar{x}) = -\frac{n^3}{(\sum x_i)^2} = -\frac{n}{(\theta^*)^2}.$$

Consequently, one finds that

$$(\sigma_{\mathcal{E}}^*)^2 = -\frac{1}{\ddot{l}(\theta^*)} = \frac{(\theta^*)^2}{n}.$$

□

Next consider the case when parameter θ is a vector, e.g. $\theta = (\theta_1, \theta_2)$ for Weibull, Gumbel, normal distribution, or $\theta = (\theta_1, \theta_2, \theta_3)$ for the GEV distribution, which will be used in Chapter 10. For a vector-valued parameter θ, $\ddot{l}(\theta)$ is a matrix of second-order derivatives, which we write as $[\ddot{l}(\theta)]$, see *e.g.* (4.10) in Example 4.11. Now variances $V(\Theta_i^*) = \sigma_{\mathcal{E}_i}^2$ can be approximated by $(\sigma_{\mathcal{E}_i}^2)^*$ equal to the ith element on the diagonal of the inverse matrix $-[\ddot{l}(\theta^*)]^{-1}$.

Example 4.18. Consider a normal variable $X \in N(m, \sigma^2)$, and let $\theta = (\theta_1, \theta_2) = (m, \sigma^2)$. For the data $\mathbf{x} = (x_1, \ldots, x_n)$, by Eqs. (4.8-4.9), the ML estimates $\theta_1^* = \bar{x}$, while $\theta_2^* = s_n^2$. For the matrix $[\ddot{l}(\theta^*)]$ given in Eq. (4.11) the inverse

$$[\ddot{l}(\theta^*)]^{-1} = \begin{bmatrix} -\dfrac{s_n^2}{n} & 0 \\ 0 & -\dfrac{2(s_n^2)^2}{n} \end{bmatrix}, \tag{4.16}$$

and thus we find

$$(\sigma_{\mathcal{E}_1}^2)^* = \frac{s_n^2}{n}, \quad (\sigma_{\mathcal{E}_2}^2)^* = \frac{2(s_n^2)^2}{n}, \tag{4.17}$$

and hence from Eq. (4.15) $V[\Theta_1^*] \approx (\sigma_{\mathcal{E}_1}^2)^*$ and $V[\Theta_2^*] \approx (\sigma_{\mathcal{E}_2}^2)^*$.   □

4.4.2 Distribution of error, large number of observations

In the previous subsection we described the variability of the estimation error by means of the mean $m_{\mathcal{E}}$ and variance $\sigma_{\mathcal{E}}^2$. The complete description of the variability of the estimation error $\mathcal{E}$ is first and foremost given by the cdf $F_{\mathcal{E}}(e) = P(\mathcal{E} \le e)$. If the cumulative distribution is known, the probability of making error larger than a specified threshold could be computed.

Finding the distribution is in general a difficult problem. Here we give two methods to approximate $F_{\mathcal{E}}(e)$. The methods are accurate when the number of observations n is large.

Asymptotic normality of the error distribution

The result presented in the following theorem is important when assessing the uncertainty of estimates. Based on this theorem, large-sample confidence intervals is considered later in this chapter. Further applications will be given in Chapters 7, 8, and 10. The theorem is valid for ML estimators for the so-called "regular" families of distributions; see Section 6.5 in Lehmann and Casella [49] where the exact assumptions are given:

Theorem 4.3. Asymptotic normality of ML estimators. *Assume that $f(x; \theta)$ (or $p(x; \theta)$) satisfies certain regularity conditions, which are satisfied in examples discussed in this text. Then*

$$P(\mathcal{E}/\sigma_{\mathcal{E}}^* \le e) \longrightarrow \Phi(e) \quad as \quad n \to \infty$$

where

$$\sigma_{\mathcal{E}}^* = 1/\sqrt{-\ddot{l}(\theta^*)}. \tag{4.18}$$

We shall also say that $\mathcal{E}$ is asymptotically normal distributed and write $\mathcal{E} \in AsN(0, (\sigma_{\mathcal{E}}^2)^*)$.

Asymptotic normality means that for large n, $P(\mathcal{E} \le e) \approx \Phi(e/\sigma_{\mathcal{E}}^*)$. In the following example, we summarize the variances $(\sigma_{\mathcal{E}}^2)^*$ for some common distributions.

Example 4.19. Consider again the three distributions encountered earlier in this section: Poisson, exponential, normal (see Examples 4.16-4.18 and for ML estimates of binomial distribution, see Problem 4.3).

Distribution	ML estimates	$(\sigma_{\mathcal{E}}^2)^*$
$X \in \mathrm{Po}(\theta)$	$\theta^* = \bar{x}$	$\dfrac{\theta^*}{n}$
$K \in \mathrm{Bin}(n, p)$	$\theta^* = \dfrac{k}{n}$	$\dfrac{\theta^*(1 - \theta^*)}{n}$
$X \in \mathrm{Exp}(\theta)$	$\theta^* = \bar{x}$	$\dfrac{(\theta^*)^2}{n}$
$X \in \mathrm{N}(m, \sigma^2)$	$\theta^* = (\bar{x}, s_n^2)$	$\left(\dfrac{s_n^2}{n}, \dfrac{2(s_n^2)^2}{n}\right)$

□

Theorem 4.3 is a generalization of a fundamental[3] result from probability theory, the Central Limit Theorem (CLT).

---

[3]Casella and Berger [10] describe this as "one of the most startling theorems in statistics".

Theorem 4.4. Central Limit Theorem. *Let* $\mathbf{X} = (X_1, \ldots, X_n)$ *be iid (independent identically distributed) variables all having the distribution* $F_X(x)$. *Assume that the expected value* $\mathsf{E}[X] = m$ *and variance* $\mathsf{V}[X] = \sigma^2$ *are finite. Then* $\bar{\mathbf{X}} \in AsN(m, \sigma^2/n)$.

The CLT tells us that for large n

$$\mathsf{P}\left(\frac{\bar{\mathbf{X}} - m}{\sigma/\sqrt{n}} \leq x\right) \approx \Phi(x), \quad \text{or} \quad \mathsf{P}(\bar{\mathbf{X}} \leq x) \approx \Phi\left(\frac{x - m}{\sigma/\sqrt{n}}\right).$$

How large n should be in order to be able to use the last approximation depends on the distribution $F_X(x)$. However, always valid is that

$$\text{if} \quad X \in N(m, \sigma^2) \quad \text{then} \quad \bar{\mathbf{X}} \in N(m, \sigma^2/n) \tag{4.19}$$

for any value of n.

Using Bootstrap to estimate the error distribution

In the past decades, the use of bootstrap techniques has attracted a lot of interest, from scientists in different fields handling data, as well as researchers in statistical theory. Roughly speaking, bootstrap techniques combine notions from classical inference with computer-intensive methods. Much literature exists; for an introduction, we refer the interested reader to [23], [36]. Bradley Efron is honoured to have invented the bootstrap method, and he gives an overview in [19].

We here only point out some of the basic ideas and demonstrate how to use bootstrap to derive the distribution of the estimation error $\mathcal{E}$. Bootstrap methods are most useful for complicated statistical problems, *e.g.* when the parameter θ is a large vector, and when an analytical approach is not possible or adequate.

Parametric bootstrap

Let us neglect the possibility of a model error, *i.e.* we assume that there is a value of a parameter θ such that $F_X(x) = F(x; \theta)$. Assume that we have n independent observations $\mathbf{x} = (x_1, \ldots, x_n)$ with X having distribution $F_X(x)$. The parameter θ is estimated using an estimate $\theta^*(\mathbf{x})$. Since, as before, usually $\theta^* \neq \theta$, we wish to find the distribution of the error $\mathcal{E} = \theta - \Theta^*$. This can be done numerically by parametric bootstrap.

For bootstrap methods, a computer program for Monte Carlo simulation is necessary. If the parameter θ, equivalently, the distribution $F_X(x)$ is known, such a program can simulate independent samples $\mathbf{x}_i = (x_1, \ldots, x_n)$, $i = 1, \ldots, N_B$, where N_B is some large integer. All these samples have the same random properties as our initial sample $\mathbf{x}$ and from each sample are calculated estimates $\theta_i^* = \theta^*(\mathbf{x}_i)$ and the errors

$$e_i = \theta - \theta_i^*, \quad i = 1, \ldots, N_B. \tag{4.20}$$

The error distribution $F_{\mathcal{E}}(e)$ can be approximated by means of the empirical distribution of $(e_1, e_2, \ldots, e_{N_B})$, with increasing accuracy as N_B goes to infinity. However, in most cases the distribution $F_X(x)$ is unknown. Still we can use the same simulation principle as outlined above if there is strong evidence that our model is correct, *i.e.* there is a θ such that $F_X(x) = F(x; \theta)$. Simply, replace the unknown distribution $F_X(x)$ by the closest we can get, $F(x; \theta^*)$, and the parameter θ in (4.20) by θ^*.

This is the so-called *parametric bootstrap*: simulate N_B times a sample with n independent random numbers having distribution $F(x; \theta^*)$, $\mathbf{x}_i = (x_1, \ldots, x_n)$, $i = 1, \ldots, N_B$. From each sample estimates $\theta_i^B = \theta^*(\mathbf{x}_i)$ and the errors

$$e_i^B = \theta^* - \theta_i^B, \quad i = 1, \ldots, N_B.$$

are calculated

Let $F_{\mathcal{E}}^B(e)$ be the empirical distribution describing the variability of the sequence e_i^B. (Note that the empirical distribution depends both on the number n of observations in our original data set and the number N_B of bootstrap simulations.) Usually N_B is much larger than n since it is only limited by the computer time we wish to spend for the simulations. Finally, one can prove that, under suitable conditions, with $N_B > n$,

$$F_{\mathcal{E}}^B \to F_{\mathcal{E}}(e) \quad \text{as} \quad n \to \infty. \tag{4.21}$$

Using the last result, if n is large we have an approximation of the error distribution $\mathcal{E}$.

4.5 Confidence Intervals

In this section, we present the idea of *confidence intervals*. Such intervals summarize the information on the estimation error. We study how an interval that covers the true value of the parameter with high probability can be constructed.

As pointed out in Section 4.1, page 73, the parameter θ can be a vector. Hence also the error can be a vector, $e = \theta - \theta^*$. For simplicity we consider each component of the error vector separately. For example, in the case of a normal model when $\theta = (m, \sigma^2)$ and $\theta^* = (\bar{\mathbf{x}}, s_n^2)$ we have two errors $m - \bar{\mathbf{x}}$ and $\sigma^2 - s_n^2$ and here we only consider the first one, *i.e.* $e = m - \bar{\mathbf{x}}$.

4.5.1 Introduction. Calculation of bounds

The error distribution $F_{\mathcal{E}}(e)$ describes the size of error as well as the frequency with which that occurs. For instance, $F_{\mathcal{E}}(e_U) - F_{\mathcal{E}}(e_L)$ is the probability (frequency) that the error will be between a lower bound e_L and an upper bound e_U, *i.e.* $e_L \le \mathcal{E} \le e_U$. This probability can be computed for any pair $e_L < e_U$. However, in most situations it is enough to choose only one interval $[e_L, e_U]$ and give the probability that the error will fall in the interval as a rough characterization of the variability of the estimation error. Actually, one usually starts by first choosing the probability, α, for some low value α and then looks for suitable bounds such that

$$P(e_L \le \mathcal{E} \le e_U) = 1 - \alpha.$$

Typical values of α are 0.01, 0.05, or 0.1; then e_L and e_U are chosen to be the quantiles $e_{1-\alpha/2}$, $e_{\alpha/2}$, respectively. The quantiles are solutions to the following equations, see also Chapter 3.2,

$$P(\mathcal{E} \le e_{1-\alpha/2}) = P(\mathcal{E} \ge e_{\alpha/2}) = \alpha/2. \tag{4.22}$$

Obviously we have that $P(e_{1-\alpha/2} \le \mathcal{E} \le e_{\alpha/2}) = 1 - \alpha$. An equivalent way, and more often used in practice, of presenting the bounds is derived from the defining equation $\mathcal{E} = \theta - \Theta^*$:

$$1 - \alpha = P(e_{1-\alpha/2} \le \mathcal{E} \le e_{\alpha/2}) = P\left(\Theta^* + e_{1-\alpha/2} \le \theta \le \Theta^* + e_{\alpha/2}\right)$$
$$= P\left(\theta \in [\Theta^* + e_{1-\alpha/2}, \; \Theta^* + e_{\alpha/2}]\right).$$

Before we continue, let us return to the interpretation of the concept of probability. Suppose we have an experiment with numerical outcomes, *i.e.* a random variable X, and let A be a statement about properties of an outcome of the experiment. Then $P(A)$ measures chances that for a yet unknown outcome x, the statement A will be true. Obviously when the outcome x is available then one usually, but not always, knows if A is true or not.

Let $X = \Theta^*$ while A is the statement $\theta \in [\Theta^* + e_{1-\alpha/2}, \; \Theta^* + e_{\alpha/2}]$. Since α is small, the probability that A will be true is high (0.9, 0.95, or 0.99). The outcome of our experiment is now the estimate θ^*, *i.e.* $x = \theta^*$. Now the problem starts: the statement A is of such nature that one cannot tell whether A is true or not for $\Theta^* = \theta^*$. In order to measure this lack of knowledge, one uses the probability $P(A) = 1 - \alpha$ but call this *confidence* instead. Thus we say that with confidence $1 - \alpha$,

$$\theta \in [\theta^* + e_{1-\alpha/2}, \; \theta^* + e_{\alpha/2}]. \tag{4.23}$$

Remark 4.5 (One-sided intervals). In some applications it can be more important to find *one-sided* confidence intervals. In the case when positive errors are "beneficial" (for instance, when estimating θ, the average volume of milk in a one-litre package) positive errors mean that on average consumers get more milk than the estimated value. Then one finds the $1 - \alpha$ quantile of error distribution $P(\mathcal{E} \ge e_{1-\alpha}) = 1 - \alpha$, leading to

$$P(\Theta^* + e_{1-\alpha} \le \theta) = 1 - \alpha, \qquad \theta^* + e_{1-\alpha} \le \theta$$

with confidence $1 - \alpha$. Similarly, when negative errors are beneficial, *e.g.* θ being the average concentration of pollutant, then the α quantile of error distribution $P(\mathcal{E} \leq e_\alpha) = 1 - \alpha$, leading to

$$P(\theta \leq \Theta^* + e_\alpha) = 1 - \alpha, \qquad \theta \leq \theta^* + e_\alpha,$$

with confidence $1 - \alpha$. □

4.5.2 Asymptotic intervals

Theorem 4.3 tells us that for large values of n, the error of the ML estimator $\mathcal{E} = \theta - \Theta^*$, based on iid observations $x_1, \ldots, x_n$, is approximately normally distributed $\mathcal{E} \in \text{AsN}(0, (\sigma_{\mathcal{E}}^2)^*)$, which means that for large values of n

$$P(\mathcal{E} \leq e) \approx \Phi(e/\sigma_{\mathcal{E}}^*).$$

Consequently the quantiles

$$e_{1-\alpha/2} \approx \lambda_{1-\alpha/2} \, \sigma_{\mathcal{E}}^* = -\lambda_{\alpha/2} \sigma_{\mathcal{E}}^*, \qquad e_{\alpha/2} = \lambda_{\alpha/2} \, \sigma_{\mathcal{E}}^*$$

and hence, for large n, with approximately $1 - \alpha$ confidence,

$$\theta \in [\theta^* - \lambda_{\alpha/2} \sigma_{\mathcal{E}}^*, \ \theta^* + \lambda_{\alpha/2} \sigma_{\mathcal{E}}^*], \tag{4.24}$$

where $\sigma_{\mathcal{E}}^*$ is given in Theorem 4.3, see Eq. (4.18). The number of observations n needs to be quite large in order to be sure that the true confidence level of the interval is close to $1 - \alpha$.

Remark 4.6. Suppose we have independent observations $x_1, \ldots, x_n$ from $N(m, \sigma^2)$, σ *unknown*, and we want to construct a confidence interval for m. If the number of observations is not large enough, use of the interval in Eq. (4.24) is not justified. However, with σ estimated as

$$(\sigma^2)^* = \frac{1}{n-1} \sum_{i=1}^{n} (x_i - \bar{x})^2 = s_{n-1}^2,$$

(see Example 4.14), one can construct an *exact* interval. Without going into details, the exact confidence interval for m is given by

$$\left[\bar{x} - t_{\alpha/2}(n-1) \frac{s_{n-1}}{\sqrt{n}}, \ \bar{x} + t_{\alpha/2}(n-1) \frac{s_{n-1}}{\sqrt{n}} \right] \tag{4.25}$$

where $t_{\alpha/2}(f)$ are quantiles of the so-called *Student's t distribution* with $f = n - 1$ degrees of freedom. This could be compared with Eq. (4.24)

$$\left[\bar{x} - \lambda_{\alpha/2} \frac{s_n}{\sqrt{n}}, \ \bar{x} + \lambda_{\alpha/2} \frac{s_n}{\sqrt{n}} \right].$$

Consider $\alpha = 0.05$. Then $\lambda_{\alpha/2} = 1.96$ and for $n = 10$, one has $t_{\alpha/2}(9) = 2.26$ while for $n = 25$, $t_{\alpha/2}(24) = 2.06$, which is closer to $\lambda_{\alpha/2} = 1.96$. □

4.5.3 Bootstrap confidence intervals

Using resampling techniques one can approximate the error distribution (see Eq. (4.21)) and hence for large n we have that $F_{\mathcal{E}}^{\mathrm{B}}(e) \approx F_{\mathcal{E}}(e)$. Consequently, the bootstrap quantiles defined by

$$F_{\mathcal{E}}^{\mathrm{B}}(e_{1-\alpha/2}^{\mathrm{B}}) = \alpha/2, \quad F_{\mathcal{E}}^{\mathrm{B}}(e_{\alpha/2}^{\mathrm{B}}) = 1 - \alpha/2,$$

are close to the quantiles $e_{1-\alpha/2}, e_{\alpha/2}$ given in Eq. (4.22). (The quantiles $e_{1-\alpha/2}^{\mathrm{B}}, e_{\alpha/2}^{\mathrm{B}}$ can be found graphically or by means of a suitable computer program.) Thus an interval, which with (approximately) $1 - \alpha$ confidence, covers the unknown parameter θ is given by

$$[\theta^* + e_{1-\alpha/2}^{\mathrm{B}}, \ \theta^* + e_{\alpha/2}^{\mathrm{B}}]. \tag{4.26}$$

Here we replaced the error distribution $F_{\mathcal{E}}$ by $F_{\mathcal{E}}^{\mathrm{B}}$, hence a so-called simple (or standard) bootstrap confidence interval was obtained. For other methods see [23], Ch. 12-14, or [36], Ch. 5.7.

4.5.4 Examples

Example 4.20. Return to the data set with periods between earthquakes. From the previous analysis, we concluded that a suitable model for $F_X(x)$ is the exponential distribution: $F_X(x) = 1 - \exp(-x/\theta)$. In Example 4.10, we found the ML estimate θ^* as the average observed over the $n = 62$ periods: $\theta^* = \bar{x} = 437.2$ days.

Asymptotic interval. For exponentially distributed X, the ML estimate $\theta^* = \bar{x} = 437.2$. Further,

$$\sigma_{\mathcal{E}}^* = \theta^*/\sqrt{n} = 437.2/\sqrt{62} = 55.5.$$

Using Eq. (4.24), an asymptotic 0.95-confidence interval for the parameter θ is

$$[437.2 - 1.96 \cdot 55.5, \ \ 437.2 + 1.96 \cdot 55.5] = [328, \ 546].$$

The interpretation of the interval is as follows: the risk of making error when claiming that θ (the average period between serious earthquakes) is some number between 328 and 546 days is approximately 0.05, *i.e.* similar as getting four heads in four flips of a fair coin.

Exact interval, exponential distribution. Next one can ask if the number of observations $n = 62$ is large enough to allow us to use the asymptotic normal approximation for the error distribution. Now if the true model for distribution of X is exponential and the observed values of X are independent then the distribution of relative error $\mathcal{E} = \Theta^*/\theta$ can be found and exact confidence intervals for the parameter θ can be derived. Without going into details, we just give formulas: with confidence $1 - \alpha$

$$\theta \in \left[\frac{2n\theta^*}{\chi^2_{\alpha/2}(2n)}, \ \frac{2n\theta^*}{\chi^2_{1-\alpha/2}(2n)} \right], \tag{4.27}$$

where $\chi^2_\alpha(f)$ is the α quantile of the $\chi^2(f)$ distribution. These quantiles are tabulated or for large n computed using Eq. (4.28).

For $\alpha = 0.05$ and $n = 62$, we find by Eq. (4.28) $\chi^2_{1-\alpha/2}(2n) = 95.07$ and $\chi^2_{\alpha/2}(2n) = 156.71$ and hence Eq. (4.27) gives that

$$\theta \in [346, \ 570]$$

with confidence 0.95. We can see that the confidence interval based on asymptotic normality of the errors is, for practical use, sufficiently close to the exact confidence interval. For higher n values the intervals become closer.

Bootstrap interval. In order to make the comparison more complete we also use the bootstrap methodology to estimate the error distribution and derive the confidence intervals. The distribution $F^B_{\mathcal{E}}(e)$ has been derived with $n = 62$, the number of observed periods between major earthquakes, and the number of bootstrap simulations $N_B = 5000$. The distribution is shown in Figure 4.4 (right) where the quantiles $e^B_{1-\alpha/2}$, $e^B_{\alpha/2}$, $\alpha = 0.05$ are marked as stars. We obtain the following bootstrap confidence interval for θ (the unknown return period between the earthquakes) with approximately 0.95 confidence:

$$\left[\theta^* + e^B_{1-\alpha/2}, \ \ \theta^* + e^B_{\alpha/2} \right] = [437.2 + (-107.9), \ \ 437.2 + 97.7] = [329, \ 535].$$

The interval is very similar to the one obtained using the normal approximation of the error distribution. It is important to note that, although both

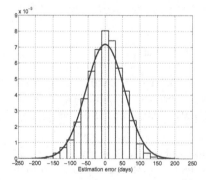 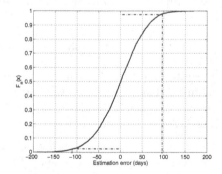

Fig. 4.4. Illustration of the distribution of bootstrap errors. *Left:* A histogram for the bootstrap errors compared with the pdf of normally distributed errors. *Right:* The empirical distribution $F^B_{\mathcal{E}}(e)$ with quantiles $e^B_{1-\alpha/2}$, $e^B_{\alpha/2}$, marked as stars.

methods are derived under the assumption that n is large[4], they have different theoretical motivations.

$\square$

Remark 4.7 (Accurate approximations). Computation of quantiles of a $\chi^2(f)$ distribution might be problematic when f is large. By the central limit theorem, $X \in \chi^2(f)$ can be approximated by an $N(f, 2f)$ distribution and hence the following approximation is valid:

$$\chi^2_\alpha(f) \approx f + \lambda_\alpha \sqrt{2f}.$$

However, this approximation is not particularly accurate unless f is rather large. Better approximations are for instance the Wilson-Hilferty approximation,

$$\chi^2_\alpha(f) \approx f \left(\sqrt{\frac{2}{9f}} \lambda_\alpha + 1 - \frac{2}{9f} \right)^3, \tag{4.28}$$

originally given by Wilson and Hilferty [83] and discussed in [41], Section 18.5.

$\square$

Example 4.21. In this example we turn to the data set with the number of killed people due to horse kicks, cf. Example 4.9. For the intensity of accidents, we give approximate confidence intervals as well as exact.

We assumed a Poisson distribution and found the ML estimate $\theta^* = \bar{x} = 0.6$. The total number of victims is 12 (in 20 years, $n = 20$), which we consider sufficiently large to apply asymptotic normality.

Approximate interval. For a Poisson variable, $(\sigma^2_{\mathcal{E}})^* = \theta^*/n$, hence $\sigma^*_{\mathcal{E}} = \sqrt{\theta^*/20} = 0.173$. Now, by Eq. (4.24), with approximate confidence 0.95, the true intensity of deaths due to horse kicks

$$\theta \in \left[0.6 - 1.96 \cdot 0.173, \ 0.6 + 1.96 \cdot 0.173 \right] = [0.26, \ 0.94].$$

Exact interval, Poisson distribution. Similarly as in the exponential model, one can even here propose confidence intervals with exactly $1 - \alpha$ coverage. Again without going into details, if n is the number of years accidents are observed, $\mathbf{x} = x_1, x_2, \ldots, x_n$ are the observed numbers of accidents during the successive years. With confidence $1 - \alpha$, the expected number of accidents θ during one year is in

$$\theta \in \left[\frac{\chi^2_{1-\alpha/2}(2n\theta^*)}{2n}, \ \frac{\chi^2_{\alpha/2}(2n\theta^* + 2)}{2n} \right], \tag{4.29}$$

where $\chi^2_\alpha(f)$ is the α quantile of the $\chi^2(f)$ distribution. This interval was first derived by Garwood in 1936 [26]; see also [10], page 434 for the derivation.

---

[4]The interval is based on $n = 62$ observed periods.

Now with $\theta^* = 0.6$ we get

$$\theta \in [0.32, \ 1.05]$$

since $\chi^2_{1-\alpha/2}(2n\theta^*) = \chi^2_{0.975}(24) = 12.40$ while $\chi^2_{\alpha/2}(2n\theta^*+2) = \chi^2_{0.025}(26) = 41.92$.

Again the difference between the confidence intervals would be smaller if the number of degrees of freedom $f = 2n\theta^*$ were higher — in other words, if we had observed more accidents. This can be done by observing longer periods of time than 20 years, or by studying situations with higher intensities of accidents.

□

4.6 Uncertainties of Quantiles

In safety applications we are often interested in estimations of the following quantities used to measure risks:

(1) The probability that some measured quantity exceeds a critical level u^{crt},
   e.g. $p = \mathsf{P}(X > u^{\mathrm{crt}}) = 1 - F_X(u^{\mathrm{crt}})$.
(2) The α quantile, i.e. x_α such that $\mathsf{P}(X > x_\alpha) = \alpha$, i.e. $x_\alpha = F^-(1-\alpha)$.

The two quantities p and x_α can be seen as functions defined on the distribution F_X, which we write as $g(F_X)$.

Two types of estimates can be considered: non-parametric, when F_X is approximated by the empirical distribution F_n and parametric when F_X is approximated by $F(x;\theta^*)$. Here we assume that $F(x;\theta)$ is a family of distributions such that there is a value of θ for which $F_X(x) = F(x;\theta)$. The unknown value of parameter is estimated by θ^*, e.g. using the ML method. As we mentioned earlier, using the non-parametric method model error of the type that $F_X(x) \neq F(x;\theta)$ for all θ is avoided, but the price to be paid is that the estimates have usually larger errors.

In this section we present means to find the distribution of errors $e = x_\alpha - x_\alpha^*$, $e = p - p^*$, where x_α^* and p^* are parametrically estimated. First, the asymptotic normality of the ML estimate θ^* is employed to describe the statistical error only. Second, we use the so-called statistical bootstrap to estimate the distribution $F_\mathcal{E}$. The method is designed so that the error e contains two possible sources of error: the *statistical error* due to finiteness of the sample size n; and *model error*, i.e. F_X does not belong to the family of distributions $F(x;\theta)$.

4.6.1 Asymptotic normality

As we have seen in the previous subsection, the estimate of θ^* itself serves only as an intermediate step to compute probabilities of type $p = \mathsf{P}(X > x)$

$(p^* = 1 - F(x; \theta^*))$ or quantiles x_α $(x_\alpha^* = F^-(1 - \alpha; \theta^*))$. Generally, we are interested in the estimates of functions of the parameter, $g(\theta)$ say, by means of $g(\theta^*)$. One can ask the question whether $g(\Theta^*)$ is a consistent estimator of $g(\theta)$, and further, what is the distribution of the error $g(\theta) - g(\Theta^*)$. (Here we neglect the possibility of model error.)

Let $g(r)$ possess a continuous derivative $\dot{g}(r)$. If the assumptions of Theorem 4.3 are satisfied then

$$g(\theta) - g(\Theta^*) \in \mathrm{AsN}\left(0, \dot{g}(\theta^*)^2 (\sigma_{\mathcal{E}}^2)^*\right), \tag{4.30}$$

where $(\sigma_{\mathcal{E}}^2)^*$ is an estimate of the variance of $\mathcal{E} = \theta - \Theta^*$. The result is an application of Taylor's formula

$$g(\theta) - g(\Theta^*) \approx \dot{g}(\theta)(\theta - \Theta^*) = \dot{g}(\theta)\mathcal{E}$$

and shows us that $g(\Theta^*)$ is a consistent estimate of $g(\theta)$ and for large n the estimation error $g(\theta) - g(\Theta^*)$ is approximately normally distributed. Note, however, that usually a higher number of observations n are needed than in Theorem 4.3, especially if g is a strongly non-linear function of θ. Hence, the approximation should be used with caution. Further discussion is found in Chapter 8 on the so-called Gauss approximation and the delta method.

Example 4.22 (Earthquake data). This is a continuation of Example 4.23 where the objective was to estimate the probability

$$p = \mathrm{P}(X > 1500) = e^{-1500/\theta}.$$

Let $g(\theta) = e^{-1500/\theta}$, then $p^* = g(\theta^*)$. Based on the ML estimate $\theta^* = 437.2$ found earlier in this chapter, we find

$$p^* = g(\theta^*) = 0.032.$$

We want to have an idea of the uncertainty of this estimate. In order to use Eq. (4.30) we need to compute $\dot{g}(\theta^*) = (1500/(\theta^*)^2)p^*$ and $(\sigma_{\mathcal{E}}^2)^* = (\theta^*)^2/n$ (see table in Example 4.19) and hence

$$\left[\dot{g}(\theta^*)\sigma_{\mathcal{E}}^*\right]^2 = \frac{(1500\,p^*)^2}{(\theta^*)^4} \cdot \frac{(\theta^*)^2}{n} = \frac{(1500 \cdot 0.032)^2}{62 \cdot 437.2^2} = 1.944 \cdot 10^{-4}.$$

In the right panel of Figure 4.5, the solid curve is the asymptotic normal pdf of the estimation error $g(\theta) - g(\Theta^*) \in \mathrm{AsN}(0, 1.944 \cdot 10^{-4})$. By comparing to the corresponding bootstrap-estimation error (the normalized histogram) we can see that the two approaches give in general similar results, although the normal asymptotic approximation seems to be a somewhat more crude approach, giving symmetrical errors.

Finally, we give an approximate 0.95-confidence interval, based on the asymptotic normal approximation of the error distribution:

$$[0.032 - 1.96 \cdot 0.014,\ 0.032 + 1.96 \cdot 0.014] = [0.005,\ 0.06].$$

$\square$

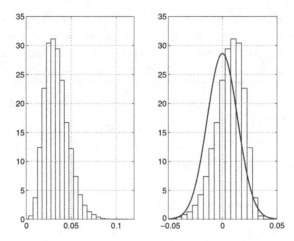

Fig. 4.5. *Left:* Histogram, bootstrap estimate of $p = \mathrm{P}(X > 1500)$; *Right:* Histogram, bootstrap-estimation error.

4.6.2 Statistical bootstrap

Suppose we are interested in the estimation error of the quantity $g(F_X)$, where g is a real-valued functional like $p = 1 - F_X(u^{\mathrm{crt}})$ or x_α. Recall the resampling algorithm of generating independent observations described in Remark 4.2, page 71. Now, a *resampling* technique is used as follows. Simulate N_B times from the empirical distribution a sample of n observations. This results in the bootstrap estimates θ_i^B, $i = 1, \ldots, N_B$. Bootstrap-error estimates are given by

$$e_i^B = g(F_n) - g(F(x; \theta_i^B)), \quad i = 1, \ldots, N_B. \tag{4.31}$$

The distribution of the error $\mathcal{E}$ is then estimated by means of empirical distribution of e_i^B. Note that here both errors are incorporated: estimation error and modelling error.

Example 4.23 (Bootstrap study: earthquake data). In this example we return to earthquake data. Our objective is to get an opinion about the probability of a period of more than 1500 days between serious earthquakes. This will be summarized in histograms of the probability itself and the error given in Eq. (4.31).

The parametric model is exponential, *i.e.*

$$F(x; \theta) = 1 - \mathrm{e}^{-x/\theta}, \quad x > 0,$$

and we are interested in the quantity

$$p = \mathrm{P}(X > 1500) = 1 - F(1500; \theta) = \mathrm{e}^{-1500/\theta}.$$

Using the previously derived ML estimate $\theta^* = 437.2$, we find $p^* = \exp(-1500/437.2) = 0.032$.

We want to have an idea of the uncertainty of the value p^* and wish to employ statistical bootstrap. Let

$$g(F_n) = 1 - F_n(1500)$$

while

$$g(F(x;\theta)) = 1 - F(1500;\theta) = e^{-1500/\theta} = g(\theta),$$

say.

We turn to the bootstrap algorithm to derive the distribution of estimation error $e = p - p^*$. From the original data set of 62 observations, we find that $1 - F_n(1354) = 3/62$ while $1 - F_n(1617) = 2/62$. Hence, by linear interpolation,

$$g(F_n) = 1 - F_n(1500) = 0.04.$$

We now resample $N_B = 10\ 000$ bootstrap samples from the original data. In each sample, the parameter θ is estimated (by taking the average of the bootstrap sample) and plugged in, yielding

$$g(\theta_i^B) = e^{-1500/\theta_i^B}, \quad i = 1, \ldots, N_B. \tag{4.32}$$

A histogram of the resulting estimates $p_i^* = \exp(-1500/\theta_i^B)$ from Eq. (4.32) is shown in Figure 4.5, left panel. From the histogram, we get an idea of the variability of p^*; note that the distribution is skewed to the right. In the right panel, a histogram of the bootstrap-estimation error $e_i^B = g(F_n) - g(\theta_i^B)$, see Eq. (4.31), is shown. This can be used to find quantiles $e_{1-\alpha/2}^B$, $e_{\alpha/2}^B$ and hence a bootstrap interval follows from Eq. (4.26) as $[0.01, 0.06]$, since $e_{0.975}^B = -0.022$ and $e_{0.025}^B = 0.028$. This could be compared to the result found in Example 4.22: $[0.005, 0.06]$. Practically, the intervals are equivalent. □

Problems

4.1. Assume that $x_1, \ldots, x_4$ are independent observations from a distribution with $E[X] = m$ and $V[X] = \sigma^2$. Consider the following estimators for m:

$$M_1^* = \frac{1}{2}(X_1 + X_4), \quad M_2^* = \frac{1}{2}(X_1 + 2X_4), \quad M_3^* = \bar{X} = \frac{1}{4}(X_1 + \cdots + X_4)$$

(a) Check which of the proposed estimators are unbiased.
(b) Calculate variances of the proposed estimators. Which one has the smallest variance?

4.2. The annual maximum of the water level at a place by the sea was observed for 70 years. The numerical values $x_1, \ldots, x_{70}$ of the registered maximum levels are observations of independent, identically distributed random variables $X_1, \ldots, X_{70}$, which are log-normally distributed with parameters m and σ, that is to say that $\ln X_i \in N(m, \sigma^2)$.

(a) Find unbiased estimates of the parameters m and σ when

$$\sum_{i=1}^{70} \ln x_i = 69.3, \qquad \sum_{i=1}^{70} (\ln x_i)^2 = 74.8.$$

Hint: Use that, with $\bar{y} = \frac{1}{n} \sum y_i$, $\sum (y_i - \bar{y})^2 = \sum y_i^2 - n\bar{y}^2$.

(b) Find an estimate of the constant h_{1000}, called 1000-year sea level, defined by the equation $P(X > h_{1000}) = 1/1000$.

4.3. Consider an r.v. $K \in \text{Bin}(n, p)$ with probability-mass function

$$P(K = k) = \binom{n}{k} p^k (1-p)^{n-k}, \quad k = 0, 1, 2, \ldots, n$$

(a) Derive the ML estimate p^*.

(b) Find the variance of the estimation error $(\sigma_\varepsilon^2)^*$.

4.4. Assume that the strength of a wire of length 10 cm, i.e. the maximal load it can carry, R is Rayleigh distributed with density function

$$f(x) = \frac{2x}{a^2} e^{-x^2/a^2}, \quad x > 0,$$

where $a > 0$ is a scale parameter. For eight tested wires, the following observations were found (unknown unit):

$$2.5 \quad 2.9 \quad 1.8 \quad 0.9 \quad 1.7 \quad 2.1 \quad 2.2 \quad 2.8.$$

(a) Give the maximum likelihood estimation of a on the assumption that the observations above are independent. *Hint:* Study the logarithm of the likelihood function.

(b) Compute $(\sigma_\varepsilon^2)^*$ and give an asymptotic 0.9 confidence interval for a.

(c) It can be shown that $R^2 \in \text{Exp}(a^2)$ and then Eq. (4.27) gives an exact confidence interval as

$$a \in \left[a^* \sqrt{2n/\chi_{\alpha/2}^2(2n)}, \ a^* \sqrt{2n/\chi_{1-\alpha/2}^2(2n)} \right]$$

Compute an exact confidence interval for a using this information. (Use the quantiles $\chi_{\alpha/2}^2(16) = 26.30$, $\chi_{1-\alpha/2}^2(16) = 7.962$.)

4.5. A sample of nine iron bars were tested for tensile strength, and the sample mean was 20 kN. Assume normally distributed strengths.

(a) Give a 90 % confidence interval for the expectation, if the standard deviation is assumed to be equal to 3 kN.

(b) How many more bars would have had to be tested to keep (at least not increase) the *width* of the interval but increase the *confidence level* to 95%?

4.6. One of the most important researchers in the history of mathematical statistics was Karl Pearson (1857-1936). For instance, he is considered "father" of the χ^2 test. Around 1900, Pearson tossed a coin 24 000 times and received 12 012 heads. Test the hypothesis "Coin is fair" at the significance level 0.05.

4.7. Consider a traditional deck of cards and the following simple game. A person draws a card, checks whether it was a spade, and puts the card back. The deck is shuffled, and the person draws again. This is repeated one more time, i.e. a person has drawn 3 cards in total.

(a) Suggest a probability model for $X =$"Total number of spades in three trials".
    Hint: Binomial distribution.
(b) One has noticed the outcomes of 4096 people playing this game; the independent observations are given in the table below:

Value	0	1	2	3
Observations	1764	1692	552	88

Test at the significance level 1 % that data follow the distribution suggested in (a).

4.8. Suppose Θ_1^* and Θ_2^* are each unbiased estimators of θ. Further, $V(\Theta_1^*) = \sigma_1^2$ and $V(\Theta_2^*) = \sigma_2^2$. A new unbiased estimator for θ is constructed as

$$\Theta_3^* = a\Theta_1^* + (1 - a)\Theta_2^*$$

where $0 \le a \le 1$. Assuming that Θ_1^* and Θ_2^* are independent, how should a be chosen so that $V(\Theta_3^*)$ is minimized?

4.9. The following data set gives the number of motorcycle riders killed in Sweden in 1990–1999:

39   30   28   38   27   29   38   33   33   36.

Assume that the number of killed drivers per year is modelled as a random variable $N \in Po(m)$.

(a) Give the ML estimate of m.
(b) Calculate an approximate 95%-confidence interval.
(c) Calculate an exact 95%-confidence interval (use Eqs. (4.28-4.29)).

4.10. The Environmental Protection Agency has collected data on the LC50 measurements for certain chemicals, likely to be found in freshwater rivers and lakes. With LC50 is meant the concentration killing 50% of the tested animals in a specified time interval. For a certain species of fish, the LC50 measurements (in ppm) for DDT in 12 experiments resulted in the following data set:

16   5   21   19   10   5   8   2   7   2   4   9

(cf. [70], Chapter 7.3). Assume that these measurements are approximately normally distributed with mean $m =$LC50 (unbiased estimates).

(a) Which is of interest in this application, to find a lower or upper confidence bound?

(b) Estimate m^* and calculate a one-sided 95% confidence bound for the mean, following the suggestion from (a).

4.11. Two researchers A and B at an institute analysed the same data set. Both assumed that data originated from $N(m, \sigma^2)$, σ known and wanted to compute a confidence interval for m. However, they did use different confidence level $1 - \alpha$. The intervals are as follow – A: [2.41, 4.59]; B: [2.19, 4.81]. Deduce from these results which researcher used $\alpha = 0.10$ and $\alpha = 0.05$, respectively.

4.12. Below are given the total numbers of yearly hurricanes for the North Atlantic basin[5] for the years 1950 – 2004.

```
11  9  4   4  5  6  9  7  8  11  8
 8  4  8   7  6  6  7  4  4   9  9
 6  3  3   6  3  5  2  3  4   3  4
 6  7  7   5  4  5  3  5  4  10  7
 8  7  6  12  4  6  5  7  3   8  9
```

(a) Let N_i be the number of hurricanes in the Northern Atlantic during year i. Assume that N_i are independent and Poisson distributed with mean m. Test if data do not contradict the assumed distribution. *Hint.* Use a χ^2 test, divide into classes $< 3, 3, 4, \ldots, 9, > 9$.
(b) Give an approximate 0.95-confidence upper bound for m.

---

[5]Data are found at http://weather.unisys.com/hurricane/atlantic/ and are courtesy of Tropical Prediction Center.

5

Conditional Distributions with Applications

In this chapter, some more notions from probability theory are provided like correlation, conditional distributions, and densities. Some results are extended and generalized, for instance, we present in terms of distributions the law of total probability and Bayes' formula. Some of the material will be needed in the following chapter to further develop Bayesian methods to analyse data.

5.1 Dependent Observations

When the outcome of an experiment is numerical we call it a random variable. Obviously, for one and the same experimental outcome many numerical properties can be measured. For instance, at a meteorological station the weather situation is measured in the form of temperature, pressure, wind speed, etc. Thus weather is described as a vector of random variables $X_1, \ldots, X_n$, say, defined on the same outcome.

Example 5.1 (Wave parameters). We study here measurements of wave data from the North Sea. Data were recorded on 24th December 1989 at the Gullfaks C platform. The so-called significant crest height for data is 3.4 m and the peak period is 10.5 s.

An observed wave can be considered as an outcome of a random experiment. Clearly, a huge number of waves are found in the actual data set. In ocean engineering a number of quantities and measures are used to characterize an individual wave, the so-called characteristic wave parameters. We consider two of them in this example: crest amplitude A_c and crest period T_c. A definition is given in Figure 5.1.

A computer program was applied to extract crest periods and related crest amplitudes from (a part of) data; the procedure resulted in 199 pairs (T_c, A_c), and these are illustrated in a *scatter plot* (see Figure 5.2, left panel). In the scatter plot, each outcome of a random experiment is represented as a dot in a Cartesian coordinate system. For each wave we have a pair (T_c, A_c), which is represented as a dot in the plane. Thus the variability of the wave characteristics is represented as 199 dots.

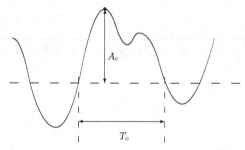

Fig. 5.1. Some characteristic wave parameters: A_c (crest amplitude) and T_c (crest period)

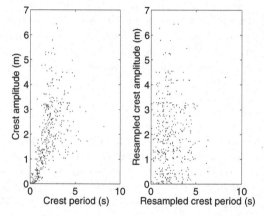

Fig. 5.2. *Left:* Scatter plot of crest period and crest amplitude. Real data. *Right:* Scatter plot of crest period and crest amplitude, resampled from original data.

We note that the measurements follow a certain pattern; high crest periods tend to have higher crest amplitudes, which also is reasonable from a physical point of view.

Obviously, variability is present in this problem and T_c and A_c can be modelled as random variables. The question of which distributions that might be suitable to describe T_c and A_c is subject for much research, and we do not tackle it here.

We know from previous chapters how to generate independent variables by means of a random-number generator. If we use the empirical distribution functions for T_c and A_c, respectively, random numbers can be produced by the resampling technique described in Chapter 4. We then obtain two samples of 199 independent observations each for T_c and A_c.

However, the scatter plot of the resampled observations, shown in Figure 5.2 (right), does not resemble the original scatter plot (left). Although the individual distributions for T_c and A_c before and after resampling are about the same, the simultaneous behaviour of T_c and A_c is lost. Clearly Figure 5.2 (left) does not represent independent observations. The concept of *dependent distributions* is therefore studied next. □

The analysis of data simplifies when observations can be assumed to be independent. However, as we have seen in Example 5.1, variables may both have different distributions $F_i(x)$ and be dependent. In Chapter 3.4, we defined the notion of independent random variables, and presented Eq. (3.13) that in the case of two random variables X_1 and X_2 can be written

$$F_{X_1, X_2}(x_1, x_2) = P(X_1 \leq x_1 \text{ and } X_2 \leq x_2) = P(X_1 \leq x_1) \cdot P(X_2 \leq x_2).$$

We now investigate this relation for our example with wave parameters.

Example 5.2 (Wave parameters). Are T_c and A_c independent? From the data available, we can calculate, for example,

$$
\begin{aligned}
F_{T_c, A_c}(1.0, 2.0) &= P(T_c \leq 1.0, A_c \leq 2.0) \\
&\approx \frac{\text{Number of waves with } T_c \leq 1.0 \text{ and } A_c \leq 2.0}{\text{Total number of waves}} \\
&= \frac{31}{199} = 0.156.
\end{aligned}
$$

Now, using values from the empirical distributions, we have that

$$F_{T_c}(1.0) \approx 0.161, \quad F_{A_c}(2.0) \approx 0.558.$$

Hence $F_{T_c}(1.0) \cdot F_{A_c}(2.0) \approx 0.161 \cdot 0.558 = 0.0898 \neq 0.156$. Thus we conclude that T_c and A_c are dependent. □

5.2 Some Properties of Two-dimensional Distributions

In this section we assume that we have only two random variables, $n = 2$, and, in order to simplify notation, we denote X_1, X_2 by X, Y. The distribution function $F_{X_1, X_2}(x_1, x_2)$ is also denoted by

$$F_{X,Y}(x, y) = P(X \leq x, Y \leq y),$$

which we often write simplified as $F(x, y)$. The distributions of the variables X and Y is denoted by $F(x) = P(X \leq x)$ and $F(y) = P(Y \leq y)$, respectively. From the definition of $F(x, y)$, it follows immediately that

$$F(x) = F(x, +\infty), \quad F(y) = F(+\infty, y).$$

Probability-mass function

If X and Y take only a finite or (countable) number of values (for simplicity only, let X, Y take values $0, 1, 2, \ldots$), then the distribution $F(x, y)$ is a "stair" looking function, that is constant except the possible jumps for $x, y = 0, 1, 2, \ldots$. The function

$$p_{jk} = \mathsf{P}(X = j, Y = k)$$

or rather matrix (can have infinitely many elements) is called *probability-mass function* (pmf) and defines the distribution

$$F(x, y) = \sum_{j \leq x, k \leq y} p_{jk}.$$

A pmf p_{jk} often used in applications is the *multinomial* pmf, which is a generalization of the binomial pmf (see Eq. (1.9)) to higher dimensions:

$$\mathsf{P}(X = j, Y = k) = \frac{n!}{j!\, k!\, (n - j - k)!}\, p_A^j p_B^k (1 - p_A - p_B)^{n-j-k} \qquad (5.1)$$

for $0 \leq j + k \leq n$ and zero otherwise, where n, $0 \leq p_A \leq 1$, and $0 \leq p_B \leq 1$ are parameters.

Obviously the variables X and Y are discrete and their probability-mass functions can be computed (using Eq. (1.3))

$$p_j = \mathsf{P}(X = j) = \sum_{k=0}^{\infty} p_{jk}, \quad p_k = \mathsf{P}(Y = k) = \sum_{j=0}^{\infty} p_{jk}. \qquad (5.2)$$

These are called the *marginal* probability-mass functions for X and Y, respectively. It is easy to show (Definition 3.4) that if X and Y are independent,

$$p_{jk} = p_j p_k.$$

Note that multinomially distributed variables X, Y are dependent and that the probabilities p_j, p_k are given by the binomial pmf, $X \in \mathrm{Bin}(n, p_A)$, $Y \in \mathrm{Bin}(n, p_B)$.

An application of the multinomial distribution is now presented.

Example 5.3 (Multinomial distribution – Chess players). Two persons, called A and B, play chess once a week. Let us assume that the results of their games are independent. Further, suppose that their capacities to win (knowledge of the game) are unchanged as time passes. Obviously a game of chess can end (result) in three ways: A wins, B wins, or neither A nor B win (draw).

Probabilistic model. Let X be the number of times A wins, while Y is the number of times B wins and let p_A and p_B be the corresponding probabilities in an individual game. Then for a fixed number n of games, X, Y have the multinomial pmf in Eq. (5.1).

Obviously the parameters p_A and p_B have to be estimated in some way. Suppose that after one year of playing the score is: A won 10 times while B won 20 times. Since there are 52 weeks in a year, our estimates of probabilities are $p_A^* = 10/52$ and $p_B^* = 20/52$. Obviously, the estimates are uncertain values. This is a frequentist approach where p_A, p_B are unknown constants — frequencies. Here we assume that capacities of victories for A and B, respectively (probabilities p_A, p_B), are unchanged for 52 games and that results are independent. However, this assumption of constant capacity for such a long time is quite unrealistic, hence the classical approach is questionable.

This example will be revisited in the next chapter, where we give a systematic account of a Bayesian solution to the problem.   □

Probability-density function

If the distribution $F(x, y)$ is differentiable with respect to x and y, the derivative

$$f(x, y) = \frac{\partial^2 F(x, y)}{\partial x \partial y}$$

is called the probability-density function (pdf) and

$$F(x, y) = \int_{-\infty}^{x} \int_{-\infty}^{y} f(\tilde{x}, \tilde{y}) \, d\tilde{x} \, d\tilde{y}.$$

Any non-negative function $f(x, y)$ that integrates to one

$$\int_{-\infty}^{+\infty} \int_{-\infty}^{+\infty} f(x, y) \, dx \, dy = 1 \tag{5.3}$$

is a density of some random variables (X, Y). Often one specifies the density and computes the distribution function by integration. The one-dimensional, *marginal* densities of X, Y can be computed from the joint density by means of the following integrals

$$f(x) = \int_{-\infty}^{+\infty} f(x, \tilde{y}) \, d\tilde{y}, \quad f(y) = \int_{-\infty}^{+\infty} f(\tilde{x}, y) \, d\tilde{x}.$$

It is easy to prove (Definition 3.4) that for independent X and Y

$$f(x, y) = f(x) f(y). \tag{5.4}$$

Two-dimensional normal distribution

Suppose that X and Y are normal r.v., with distributions $N(m_X, \sigma_X^2)$, $N(m_Y, \sigma_Y^2)$, respectively. This means that their probability-density functions are written

$$f(x) = \frac{1}{\sigma_X \sqrt{2\pi}} e^{-\frac{1}{2\sigma_X^2}(x-m_X)^2}, \quad f(y) = \frac{1}{\sigma_Y \sqrt{2\pi}} e^{-\frac{1}{2\sigma_Y^2}(y-m_Y)^2},$$

defined for $-\infty < x < \infty$, $-\infty < y < \infty$, respectively. If X and Y are independent, their joint probability density $f(x,y)$ is given by

$$f(x,y) = f(x)f(y) = \frac{1}{2\pi\sigma_X\sigma_Y} e^{-\frac{1}{2}\left\{\frac{(x-m_X)^2}{\sigma_X^2} + \frac{(y-m_Y)^2}{\sigma_Y^2}\right\}},$$

defined for $-\infty < x < \infty$, $-\infty < y < \infty$, respectively.

The variables X and Y can also be dependent. Then there is a parameter $-1 \le \rho \le 1$, called *correlation* (to be introduced later on), that measures the degree of dependence between X and Y. If $\rho = 0$ then X and Y are independent. Consequently, five parameters define the two-dimensional normal distribution. These are m_X, m_Y, σ_X^2, σ_Y^2, and ρ, and the statement that (X,Y) is normal,

$$(X,Y) \in N(m_X, m_Y, \sigma_X^2, \sigma_Y^2, \rho),$$

means that the joint density of (X,Y) is given by

$$f(x,y) = \frac{1}{2\pi\sigma_X\sigma_Y\sqrt{1-\rho^2}} e^{-\frac{1}{2(1-\rho^2)}\left\{\frac{(x-m_X)^2}{\sigma_X^2} + \frac{(y-m_Y)^2}{\sigma_Y^2} - 2\rho\frac{(x-m_X)}{\sigma_X}\frac{(y-m_Y)}{\sigma_Y}\right\}}.$$

$$(5.5)$$

Remark 5.1 (Simulation). The question is how to generate correlated normally distributed random variables. Suppose we want to create observations from

$$(X,Y) \in N(m_X, m_Y, \sigma_X^2, \sigma_Y^2, \rho).$$

We first consider the case with independent random variables, *i.e.* $\rho = 0$ (follows from Eqs. (5.4-5.5)). Let U_i be independent uniformly distributed random variables. Then Z_i defined by $U_i = \Phi(Z_i)$ are independent and $N(0,1)$ (see Section 3.1.2). Then

$$\begin{cases} X = m_X + \sigma_X Z_1 \\ Y = m_Y + \sigma_Y Z_2 \end{cases}$$

are $N(m_X, m_Y, \sigma_X, \sigma_Y, 0)$. In the case of dependent variables, the relation is given as

$$\begin{cases} X = m_X + \sigma_X Z_1 \\ Y = m_Y + \rho\sigma_Y Z_1 + \sigma_Y \sqrt{1 - \rho^2} Z_2 \end{cases}$$

and $(X, Y) \in N(m_X, m_Y, \sigma_X^2, \sigma_Y^2, \rho)$. (cf. Problem 5.8). □

Example 5.4 (Length and weight of children). In a medical study, length and weight of 725 newborn children were registered. In Figure 5.3, left panel, we show a histogram for the weights along with a fitted pdf for $N(m_W, \sigma_W^2)$, where $m_W^* = 3\,343$ [g], $\sigma_W^* = 528$ [g] (estimated from the sample). Ditto plot for length is shown in the right panel, the pdf for $N(m_L, \sigma_L^2)$ where the estimated parameter values are given as $m_L^* = 49.8$ [cm], $\sigma_L^* = 2.5$ [cm].

Now, study the joint distribution. An estimate of the correlation is $\rho^* = 0.75$ (see Eq. (5.14)). In Figure 5.4, contour lines of a two-dimensional normal density are shown as well as a scatter plot of the original data. Note that some observations are not well described by the distribution. Usually in scientific investigations, such not "normal" observations have to be examined closer. This simple example shows that attention has to be paid in modelling situations of real data. □

For an r.v. having a pdf, probabilities of statements can be obtained by integrating the density function (see Eq. (3.4)). In the case of a two-dimensional distribution, we have that for any events A and B,

$$P(X \in A \text{ and } Y \in B) = \int_A \int_B f(x, y) \, dx \, dy, \tag{5.6}$$

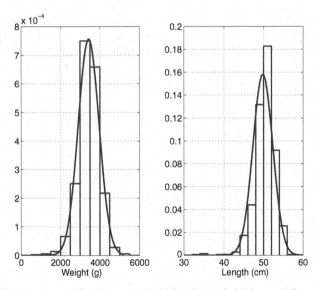

Fig. 5.3. Histogram and fitted normal pdf for data of children. *Left:* weight. *Right:* length.

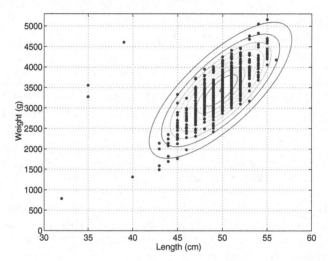

Fig. 5.4. Scatter plot of observations of length and weight and fitted two-dimensional normal density ($\rho^* = 0.75$). Note that some observations are not well described by the distribution.

that is, a double integral. Quite often, the last formula has to be computed numerically, even for simple sets A, B. For example, this is the case when X, Y are normally distributed.

Expected values of functions

Consider a function $z = h(x, y)$. Define a new random variable Z as $Z = h(X, Y)$. Simple examples are $Z = X + Y$ or $Z = XY$. We want to find the expected value of Z. Obviously if the distribution of Z is known, we can compute the expectation of Z directly by use of Eq. (3.16). However, it can also be done by means of the following formulae:

$$\mathsf{E}[Z] = \mathsf{E}[h(X, Y)] = \int_{-\infty}^{+\infty} \int_{-\infty}^{+\infty} h(x, y) f(x, y) \, \mathrm{d}x \, \mathrm{d}y \tag{5.7}$$

or

$$\mathsf{E}[Z] = \mathsf{E}[h(X, Y)] = \sum_{j=0}^{+\infty} \sum_{k=0}^{+\infty} h(j, k) \, p_{jk}.$$

In the special case of a linear combination $h(X, Y) = aX + bY$, it follows that

$$\mathsf{E}[aX + bY] = a \, \mathsf{E}[X] + b \, \mathsf{E}[Y],$$

for any constants a and b (cf. Eq. (3.18)).

5.2.1 Covariance and correlation

It is easy to check (Eqs. (5.4) and (5.7)) that for independent variables X and Y, the relation

$$E[X \cdot Y] = E[X] \cdot E[Y] \tag{5.8}$$

is always valid. Variables X and Y for which Eq. (5.8) holds are called *uncorrelated*. (All independent variables are uncorrelated but not conversely.) Now, if the equation does not hold, the difference between the terms is a measure of dependence between the variables X and Y. This measure is called *covariance* and is defined by

$$Cov[X, Y] = E[X \cdot Y] - E[X] \cdot E[Y]. \tag{5.9}$$

A glance at Eq. (3.19) convinces oneself that $Cov[X, X] = V[X]$ and obviously $Cov[Y, Y] = V[Y]$.

When one has two random variables, their variances and covariances are often represented in the form of a symmetric matrix

$$Cov[X, Y; \ X, Y] = \begin{bmatrix} V[X] & Cov[X, Y] \\ Cov[X, Y] & V[Y] \end{bmatrix}. \tag{5.10}$$

The variance of a sum of correlated variables will be needed for computation of variance in the following chapters. Starting from the definition of variance and covariance, the following general formula can be derived (do it as an exercise):

$$V[aX + bY + c] = a^2 V[X] + b^2 V[Y] + 2ab \, Cov[X, Y]. \tag{5.11}$$

The last formula easily generalizes to

$$V\left[\sum_{i=1}^{n} a_i X_i\right] = \sum_{i=1}^{n} a_i^2 V[X_i] + 2 \sum_{i=2}^{n} \sum_{j=1}^{i-1} a_i a_j Cov[X_i, X_j]. \tag{5.12}$$

The property $Cov[aX, bY] = ab \cdot Cov[X, Y]$ means that by changing the units in which variables X and Y are measured, the covariance can be made very close to zero. This could be misinterpreted as X and Y being only weakly dependent. Consequently, the covariance is often scaled so that it becomes independent of the units in which the variables are measured. Such a scaled covariance is called *correlation* and is defined as follows

$$\rho_{XY} = \frac{Cov[X, Y]}{D[X]D[Y]}, \tag{5.13}$$

where $D[X] = \sqrt{V[X]}$, $D[Y] = \sqrt{V[Y]}$. The correlation is always between one and minus one, see the following theorem (a proof is given in [10], Section 4.5).

> **Theorem 5.1.** *Let X and Y be two random variables such that $|\rho_{XY}| = 1$. Then there are constants a, b, and c (all not equal to zero) such that $aX + bY + c = 0$ with probability one.*

However, not all functionally dependent variables X and Y are perfectly correlated ($|\rho_{XY}| = 1$). For example, for $X \in \mathrm{N}(0,1)$, define $Y = X^3$. Obviously, if we know the outcome of the random experiment then $X = x$ while $Y = x^3$. However, the correlation is given by $\rho_{XY} = 3/\sqrt{15} < 1$.

Remark 5.2 (Estimation of correlation). Having observed (x_i, y_i), $i = 1, \ldots, n$, an estimate of ρ_{XY}, the correlation between the variables X and Y, is given by

$$\rho_{XY}^* = \frac{\Sigma(x_i - \bar{x})(y_i - \bar{y})}{\sqrt{\sum(x_i - \bar{x})^2 \sum(y_i - \bar{y})^2}}. \tag{5.14}$$

For instance, consider a bivariate normal distribution, with density function given by Eq. (5.5). In Example 5.4, length and weight were positively correlated with $\rho_{XY}^* = 0.75$ (see also Figure 5.4). □

We end this section with an important application of the two-dimensional normal density to approximate the error distribution when the estimated parameter θ has two components: $\theta = (\theta_1, \theta_2)$. For example, in Example 4.11, $\theta_1 = m$ and $\theta_2 = \sigma^2$ are mean and variance in a $\mathrm{N}(m, \sigma^2)$ distribution. Now the estimation error

$$\mathcal{E} = (\mathcal{E}_1, \mathcal{E}_2) = (\theta_1 - \Theta_1^*,\ \theta_2 - \Theta_2^*)$$

consists of two random variables. If (Θ_1^*, Θ_2^*) are ML estimators, then, for large values of n, $\mathsf{E}[\mathcal{E}_i] \approx 0$ and the covariance matrix

$$\mathsf{Cov}[\mathcal{E}_1, \mathcal{E}_2;\ \mathcal{E}_1, \mathcal{E}_2] \approx - \begin{bmatrix} \dfrac{\partial^2 l}{\partial \theta_1^2} & \dfrac{\partial^2 l}{\partial \theta_1 \partial \theta_2} \\ \dfrac{\partial^2 l}{\partial \theta_2 \partial \theta_1} & \dfrac{\partial^2 l}{\partial \theta_2^2} \end{bmatrix}^{-1} = - \left[\ddot{l}(\theta_1^*, \theta_2^*) \right]^{-1} \tag{5.15}$$

i.e. the partial derivatives are computed for θ_i equal to the ML estimates θ_i^*. (Here A^{-1} means the inverse of the matrix A). Furthermore, the errors are asymptotically normally distributed.

Example 5.5. As in Example 4.11, let $X \in \mathrm{N}(m, \sigma^2)$ and suppose we have n independent observations $\mathbf{x} = (x_1, \ldots, x_n)$ of X. The ML estimates of $\theta = (\theta_1, \theta_2) = (m, \sigma^2)$ are

$$\theta_1^* = \frac{x_1 + \cdots + x_n}{n} = \bar{\mathbf{x}},$$

$$\theta_2^* = \frac{1}{n} \sum_{i=1}^{n} (x_i - \bar{\mathbf{x}})^2 = s_n^2.$$

The errors $(\mathcal{E}_1, \mathcal{E}_2) = (m - \bar{X}, \sigma^2 - S_n^2)$ have for large n mean approximately equal to zero and covariance matrix

$$\mathsf{Cov}[\mathcal{E}_1, \mathcal{E}_2; \ \mathcal{E}_1, \mathcal{E}_2] \approx - \begin{bmatrix} -\dfrac{n}{s_n^2} & -\dfrac{n\bar{\mathbf{x}}}{(s_n^2)^2} + \dfrac{n\bar{\mathbf{x}}}{(s_n^2)^2} \\[2ex] -\dfrac{n\bar{\mathbf{x}}}{(s_n^2)^2} + \dfrac{n\bar{\mathbf{x}}}{(s_n^2)^2} & \dfrac{n}{2(s_n^2)^2} - \dfrac{n s_n^2}{(s_n^2)^3} \end{bmatrix}^{-1}$$

$$= \begin{bmatrix} \dfrac{s_n^2}{n} & 0 \\[2ex] 0 & \dfrac{2(s_n^2)^2}{n} \end{bmatrix}.$$

Consequently, $(\mathcal{E}_1, \mathcal{E}_2) \in \mathrm{AsN}(0, 0, s_n^2/n, 2(s_n^2)^2/n, 0)$ are asymptotically uncorrelated and normally distributed. Asymptotical confidence intervals can be constructed as presented in Chapter 4.

It is easy to prove, using Eqs. (5.4-5.5), that uncorrelated ($\rho = 0$) normal variables are actually independent. $\square$

5.3 Conditional Distributions and Densities

In Section 1.3 we introduced the concept of conditional probabilities. We give the definition again: Suppose we are told that the event A, such that $P(A) > 0$, has occurred. Then the probability that B occurs, given that A has occurred, is

$$P(B|A) = \frac{P(A \cap B)}{P(A)}.$$

This notion is now extended to random variables and distributions, which is needed in order to introduce the Bayesian analysis in Chapter 6.

5.3.1 Discrete random variables

For discrete random variables X, Y with pmf $p_{jk} = P(X = j, Y = k)$ the conditional probabilities

$$P(X = j|Y = k) = \frac{P(X = j, Y = k)}{P(Y = k)} = \frac{p_{jk}}{p_k} = p(j|k), \quad j = 0, 1, \ldots \quad (5.16)$$

are well defined for all k such that $p_k > 0$ (if $p_k = 0$ we can let $p(j|k) = 0$ too). The conditional probabilities $p(j|k) = P(X = j|Y = k)$ sum to one, since $\sum_{j=0}^{\infty} p_{jk} = p_k$. This means that $p(j|k)$, as a function of j, is a probability-mass function.

Suppose we observed the value of Y, e.g. we know that $Y = y$, but X is not observed yet. An important question is if the uncertainty about X

is affected by our knowledge that $Y = y$. Uncertainty is measured by the distribution function, which we denote by $F(x|y) = \mathsf{P}(X \leq x|Y = y)$. (If X and Y are independent then obviously $F(x|y) = F_X(x)$ and Y gives us no knowledge about X.) For discrete X, Y we have

$$F(x|k) = \mathsf{P}(X \leq x|Y = k) = \frac{\mathsf{P}(X \leq x, Y = k)}{\mathsf{P}(Y = k)} = \frac{\sum_{j \leq x} p_{jk}}{p_k} = \sum_{j \leq x} p(j|k),$$

i.e. $p(j|k)$ is the probability-mass function of the conditional distribution $F(x|y)$. That is why we call $p(j|k)$ a *conditional probability-mass function*.

5.3.2 Continuous random variables

Consider now variables X, Y having continuous distributions. We wish to find the conditional distribution $F(x|Y = y)$. However, we face a problem since for continuous variables Y, $\mathsf{P}(Y = y) = 0$ for all y. An easy solution to this problem can be found if X, Y has the density $f(x, y)$. In such a case we can follow the formula (5.16) and define

$$f(x|y) = \frac{f(x, y)}{f(y)}, \quad \text{if } f(y) > 0 \text{ and zero otherwise.} \tag{5.17}$$

Since for a fixed value y, $f(x|y)$ as a function of x integrates to one (Eq. 5.3), $f(x|y)$ is a probability density function. Let us denote by $F(x|y)$ a distribution having the density $f(x|y)$, *i.e.* for any x

$$F(x|y) = \int_{-\infty}^{x} f(\tilde{x}|y) \, d\tilde{x}. \tag{5.18}$$

Now a combination of Eqs. (5.6), (5.17), and (5.18) leads to the following important result

$$\mathsf{P}(X \leq x) = F(x) = \int_{-\infty}^{+\infty} F(x|y) f(y) \, dy. \tag{5.19}$$

The last equation is a special case of the law of total probability given in the following subsection and is the motivation why we call $F(x|y)$ the conditional distribution of X given that $Y = y$, which we also write

$$F(x|y) = \mathsf{P}(X \leq x|Y = y).$$

Consequently, we also call $f(x|y)$ the conditional density of X given $Y = y$. Note that since $f(x, y) = f(x|y) f(y)$ then

$$f(x) = \int_{-\infty}^{+\infty} f(x|y) f(y) \, dy, \tag{5.20}$$

which also could be used to demonstrate (5.19).

5.4 Application of Conditional Probabilities

In Chapters 1 and 2, the most important applications of the conditional probabilities were *the law of total probability* and *Bayes' formula*. We now give more general versions of these two results, starting with the law of total probability.

5.4.1 Law of total probability

In Chapter 1, we considered an event B and a partition of the sample space $\mathcal{S}$, *i.e.* a collection of n disjoint events A_i that sums to the whole space $\mathcal{S}$. Then the probability

$$P(B) = P(B|A_1)P(A_1) + \cdots + P(B|A_n)P(A_n).$$

Now, consider a partition generated by a random variable Y, say, that takes only n values $1, \ldots, n$. Obviously $A_i = \{Y = i\}$ is a well-defined partition and hence we can write

$$P(B) = P(B|Y = 1)P(Y = 1) + \cdots + P(B|Y = n)P(Y = n).$$

An extension to any discrete variable Y that can take infinitely many values is straightforward:

Theorem 5.2 (Law of total probability, discrete distributions). *Let B be an event and Y a random variable of discrete type. Then*

$$P(B) = \sum_{i=0}^{\infty} P(B|Y = i)P(Y = i). \tag{5.21}$$

Example 5.6 (Inspection of cracks). In this hypothetical example we consider an old tanker that has a large number of surface cracks. We model the total number of cracks by a Poisson distributed variable N, say, which means that

$$P(N = k) = \frac{m^k}{k!}e^{-m}, \quad k = 0, 1, \ldots \ .$$

The parameter m is the average number of cracks, to be derived next. Considering the age of the tanker, one assumes that the intensity of cracks is $\lambda = 0.01$ m^{-2}, while the total area of the surface of the tanker is 5000 m^2 giving on average $m = \lambda \cdot 5000 = 50$ cracks.

Assume that an automatic device is used to detect and repair cracks. From laboratory experiments it is known that the detection probability is extremely high and equal to 0.999. Further, failures in detection are assumed to be independent. We are interested in the probability that all cracks have been detected and repaired.

Let us introduce $B =$ "All cracks have been repaired", and $A_i =$ "There have been i cracks on the surface before inspection", i.e. $A_i = \{N = i\}$. Then

$$P(B) = \sum_{i=0}^{\infty} P(B|N = i)P(N = i) = \sum_{i=0}^{\infty} (0.999)^i \frac{m^i}{i!} e^{-m}$$

$$= e^{-m} \sum_{i=0}^{\infty} \frac{(0.999m)^i}{i!} = e^{-m} e^{0.999m} = e^{-0.001m} = e^{-0.05} \approx 0.95.$$

Since the surface of the tanker is large, the probability of missing some cracks is not negligible.   □

A generalization of Eq. (5.21) to the case of Y having a density function is given in the following theorem.

Theorem 5.3 (Law of total probability, continuous distributions).
Assume that a random experiment renders values of an r.v. Y and that we in addition are interested in any statement B, say, about the outcomes of the experiment. Then, for each y, there exists a probability $P(B|Y = y)$ such that

$$P(B) = \int_{-\infty}^{+\infty} P(B|Y = y) f_Y(y) \, dy. \tag{5.22}$$

If X and Y have joint density $f(x, y)$ and B is a statement about X, then

$$P(B \mid Y = y) = \int_B f(x|y) \, dx,$$

where $f(x|y)$ is the conditional probability density defined by Eq. (5.17).

The following formula, a simple consequence of the last theorem, is often used.

Remark 5.3. Suppose that X, Y are independent variables and that Y has a density $f_Y(y)$. Consider a statement $B =$ "$X \leq Y$". It is not difficult to prove $P(B|Y = y) = P(X \leq y)$, and hence using Eq. (5.22), we have that

$$P(B) = P(X \leq Y) = \int_{-\infty}^{\infty} P(X \leq y) f_Y(y) \, dy. \tag{5.23}$$

□

5.4.2 Bayes' formula

Suppose that for an outcome of a random experiment it is known that B is true. Let Y be an r.v. with pdf $f_Y(y)$ (or pmf $p(y)$). The conditional distribution of Y given that B is true is

$$F_{Y|B}(y) = P(Y \leq y \mid B).$$

The pdf (or pmf) of this distribution is given by Bayes' formula as

$$f_{Y|B}(y) = c \cdot P(B \,|\, Y = y)\, f_Y(y) \tag{5.24}$$

where $1/c = P(B)$ and $P(B)$ is given by Eq. (5.22). Again, $L(y) = P(B|Y = y)$ is called likelihood function (cf. Chapter 2, page 22).

5.4.3 Example: Reliability of a system

The following example demonstrates the use of Eqs. (5.23-5.24) to compute the probability of failure of a simple system. The features of this simple system are common in many applications to engineering design, where typically "loads" and "strengths" have to be modelled. In Chapters 9–10, we will study design questions in more detail, when discussing how to compute the so-called characteristic strengths and design loads.

Example 5.7 (Reliability of a wire). A wire to be used in a construction needs to be designed to support weights that varies from day to day in an unpredictable way, without breaking in a year of operation. We will now give probabilistic models for load and strength, respectively, and compute the probability of failure.

Modelling the load. Denote by X the maximal weight that will be carried by the wire for a period of 1 year. Obviously the weight, and hence X, is unknown and X will be modelled as a random variable. As will be shown in Chapter 10, X can have a Gumbel distribution,

$$P(X \le x) = e^{-e^{-(x-b)/a}}, \quad -\infty < x < \infty.$$

Suppose the load X has mean 1000 kg and standard deviation 200 kg. From the expressions for mean and variance of a Gumbel distributed variable, one finds that $a = 156$ and $b = 910$. We neglect the fact that the parameters are estimated and hence are uncertain values.

Modelling the strength. When ordering a wire, wires of different capacity of carrying loads can be chosen. The producer specifies the quality of his wires by giving the average strength of a wire and the coefficient of variation. From experience it is known that the strength of wires follows a Weibull distribution.

The last three sentences describe a random experiment of getting a product from a population of wires with variable strength. The variability is described using a Weibull distribution with specified mean and coefficient of variation. Let us ignore the fact that the parameters are estimated and hence uncertain, or even that the choice of a Weibull model might be wrong. We agree that the strength y (*i.e.* a capacity to carry a load) is unknown before we get the wire and hence is modelled by a random variable Y with density

$$f(y) = \frac{c}{\alpha} \left(\frac{y}{\alpha}\right)^{c-1} e^{-(y/\alpha)^c}, \quad y \ge 0.$$

Suppose that one decides to order wires with average strength 1000 kg and coefficient of variation $R[Y] = 0.2$. This implies (see appendix, Table 4) that the shape parameter in the Weibull distribution $c = 5.79$ while the scale parameter $\alpha = 1000/0.9259 \approx 1080$.

Reliability of the wire. Introduce the statement

$B =$ "Safe operation during 1 year".

Clearly $P(B) = P(X < Y)$ and we can use Eq. (5.23) to compute the probability. (Note that $P(X = Y) = 0$.)

Suppose that we know the value of the strength of the wire, $Y = y$. Then the conditional probability of safe operation given the strength $Y = y$ is given as

$$P(B \mid y) = P(X \leq y) = e^{-e^{-(y-b)/a}},$$

since the load X is independent of the value of the strength Y. Now Eq. (5.23) gives the probability of safe operation

$$P(B) = \int_0^\infty \exp(-e^{-(y-910)/156}) \frac{5.79}{1080} \left(\frac{y}{1080}\right)^{4.79} e^{-(y/1080)^{5.79}} dy$$

$$\approx 0.533. \tag{5.25}$$

The likelihood takes into account both the variability of the load and the variability of material properties and manufacturing, leading to an unknown value of the strength of the wire.

(The result in Eq. (5.25) is not particularly surprising since we took the average strength to be equal to the expected load. Thus we expect that the odds are approximately 1:1 that the load will exceed the strength.) $\quad\square$

Often decisions need to be made about safety of existing structures, for example, whether a bridge used for 40 years has to be renovated or be used for some additional period of time, *e.g.* 10 years. Money spent on renovation of a safe (with high probability) bridge could be used on other measures to increase the safety of traffic.

The following example illustrates some aspects of evaluation of safety of existing structures. Again, we study the simple system of a wire under variable load. A serious simplification is made by assuming that the wire is not getting weaker with age.

Example 5.8 (Safety of existing structures). Suppose the wire ordered in Example 5.7 survived one year of exploitation, which means that

$B =$ "The wire supported a load during one year"

is true. The probability for this was computed in Eq. (5.25), $P(B) = 0.533$. We have to make a decision whether to keep the wire for the next year or to replace it by a new one of higher quality with $E[Y] = 1200$ kg and $R[Y] = 0.2$. The decision is taken on the basis of computed reliability, *i.e.* $P(C)$ where

$C =$ "Safe operation during next year".

New wire. We begin with the reliability of a new wire with $E[Y] = 1200$ kg and $R[Y] = 0.2$, which means that we need to find the new scale parameter a for the Weibull strength. (Since $E[Y] = a\Gamma(1 + 1/c)$ we have that $a = 1200/0.9259 \approx 1300$.) Then we recompute the integral (5.25)

$$P(C) = \int_0^\infty \exp(-e^{-(y-910)/156}) \frac{5.79}{1300} \left(\frac{y}{1300}\right)^{4.79} e^{-(y/1300)^{5.79}} \, dy \approx 0.757.$$

One-year-old wire. Next we compute the reliability of a one-year-old wire. Obviously we wish to include in our analysis the information that B is true, *i.e.* we wish to modify the density $f(y)$ describing the strength of wire and compute the posterior density $f^{\text{post}}(y)$ using Bayes' formula. Since the likelihood function $P(B|y) = P(X \le y)$ and $P(B) = 0.533$, we can use Eq. (5.24) to compute the conditional density, *viz.*

$$f^{\text{post}}(y) = \frac{1}{P(B)} P(B|y) f(y)$$

$$= \frac{1}{0.533} \exp(-e^{-(y-910)/156}) \frac{5.79}{1080} \left(\frac{y}{1080}\right)^{4.79} e^{-(y/1080)^{5.79}}.$$

Consequently, with this posterior density, the probability of safe operations during the following year is found as

$$P(C) = \int_0^\infty P(C|y) f^{\text{post}}(y) \, dy = 0.705.$$

Clearly, the decision is not easy. The reliability of a used wire is slightly lower (compared to 0.757), but by keeping it one saves the cost of buying and installing a new one. Against the decision of keeping the used wire is the possibility of ageing, *i.e.* losing strength over time.

Conditional independence. We end this example with a warning about the possible erroneous analysis of failure during the time period investigated. Denote by X_1 and X_2 the maximal load during the first and second year, respectively. By our assumption, the variables X_1 and X_2 are independent Gumbel distributed. Further let $B_1 = \text{``}X_1 < Y\text{''}$ and $B_2 = \text{``}X_2 < Y\text{''}$ be the event that the strength is higher than the load during the first and second year, respectively. Since the load is independent of the strength and $Y = y$ is fixed, although unknown, one could think that B_1 and B_2 are independent, giving

$$P(\text{``The wire survives two years''}) = P(B_1 \cap B_2) = P(B_1) \cdot P(B_2) = 0.533^2.$$

However, this is not correct since we have only independency conditionally that we know the strength "$Y = y$",

$$P(B_1 \cap B_2 | Y = y) = P(B_1 | Y = y)^2$$

and the correct answer is $P(B_1 \cap B_2) = P(B_2|B_1)P(B_1) = 0.705 \cdot 0.533$.

Conditional independence will be further discussed in the next chapter.

$\square$

Problems

5.1. The random variables X and Y are independent and have probability mass functions

j	1	2	3
p_j	0.20	0.60	0.20

k	1	2	3	4
p_k	0.25	0.25	0.25	0.25

Calculate the following probabilities:

(a) $P(X = 2, Y = 3)$
(b) $P(X \leq 2, Y \leq 3)$

5.2. From the National Fire Incident Reporting Service (in the U.S.), we have that among residential fires, approximately 73% are in family homes, 20% are in apartments, and the remaining 7% are in other types of dwellings [70].

Consider five fires, independently reported during one week. Find the probability that three are in family homes, one is in an apartment, and one is in another type of dwelling.

5.3. A friendly tournament in football (association football, "soccer") between two football teams, A and B, is so arranged that there will be precisely two matches played. In a single match, the probability that A will win is $p_A = 0.35$. The probability that B will win in the same match is $p_B = 0.25$. So the probability of a draw is thus $1 - p_A - p_B = 0.40$. Let X_A be the number of matches won by A, and X_B the number of matches won by B.

(a) Give the joint probability-mass function p_{X_A, X_B} for X_A and X_B.
(b) Calculate $E[X_A]$, $V[X_A]$, $Cov[X_A, X_B]$, and $\rho[X_A, X_B]$.

5.4. Let X denote the number of interruptions in a newly installed computer network: 1, 2, or 3 times per week. Let Y denote the number of times an expert technician is called on an emergency call, related to interruptions during a week. A statistician has established the following probability-mass function $p_{jk} = P(X = j, Y = k)$:

		j	
	1	2	3
1	0.05	0.05	0.1
k 2	0.05	0.1	0.35
3	0	0.2	0.1

(a) Give the marginal distributions of X and Y.
(b) Calculate $P(Y = 3 | X = 2)$.
(c) Explain theoretically what the probability in (b) means.

5.5. A region by the sea with a square area with sides one length unit is frequently used by e.g. old tankers that might leak oil. Let X and Y denote the coordinates of a ship when a leakage starts. Suppose the position of the ship is uniformly located over the square area. Then a model for (X, Y) is given by

$$f_{X,Y}(x, y) = \begin{cases} 1, & 0 \leq x \leq 1, 0 \leq y \leq 1, \\ 0, & \text{elsewhere.} \end{cases}$$

Find the probability $P(X \leq 0.3, Y \leq 0.4)$.

5.6. Let $X \in \text{Gamma}(7, 2)$, $Y \in \text{Gamma}(6, 4)$. Calculate $E[2X + 3Y]$.

5.7. Let $N_1 \in \text{Bin}(20, 0.3)$, $N_2 \in \text{Bin}(10, 0.5)$. Further, the variables are correlated, $\text{Cov}[N_1, N_2] = 0.85$. Calculate $\text{V}[N_1 - N_2]$.

5.8. Let $X_1 \in \text{N}(0, 1)$ and $X_2 \in \text{N}(0, 1)$ be two independent random variables. Let Y_1 and Y_2 be two other random variables defined by

$$\begin{cases} Y_1 = X_1 \\ Y_2 = \varrho X_1 + \sqrt{1 - \varrho^2} X_2 \end{cases}$$

where ϱ is a real-valued constant such that $-1 \le \varrho \le 1$.

(a) Give $\text{E}[Y_1]$, $\text{E}[Y_2]$, $\text{V}[Y_1]$, $\text{V}[Y_2]$, $\text{Cov}[Y_1, Y_2]$, and the correlation coefficient $\rho[Y_1, Y_2]$.

(b) One can show that (Y_1, Y_2) has a bivariate Gaussian distribution. Write down the joint density function for Y_1 and Y_2.

5.9. (a) Let X be a random variable with distribution function F_X. Assume that we have observed that $X > 0$. Give, in terms of F_X, the conditional distribution function of X, given that information, i.e.

$$F_{X|X>0}(x) = \text{P}(X \le x | X > 0), \qquad x \in \mathbb{R}.$$

You may assume that $\text{P}(X > 0) \ne 0$.

(b) A property of the normal distribution is that it always produces negative numbers with a non-zero probability. If one wants to model something that is strictly non-negative (T, say) but still retain the bell-shaped curve of the density function, one can model T by means of a *truncated* normal distribution, i.e.

$$F_T(t) = \text{P}(X \le t | X > 0)$$

where $X \in \text{N}(m, \sigma^2)$. Use (a) to obtain $F_T(t)$.

Hint: $F_X(x) = \Phi\left(\frac{x-m}{\sigma}\right)$.

(c) Give the density function of the truncated normal distribution of T, obtained in (b).

5.10. Consider the independent random variables $X \in \text{Po}(m_1)$ and $Y \in \text{Po}(m_2)$. Show that the conditional probability-mass function for X, given $X + Y = n$, is binomial. Use that $X + Y \in \text{Po}(m_1 + m_2)$.

5.11. A classic example of a *hierarchical model* is as follows: An insect lays a large number of eggs, each surviving with probability p. On average, how many eggs will survive? A probabilistic framework is given below.

First, the large number of eggs laid is often modelled as a Poisson variable with expectation m, say. Further, if the survival of each egg is independent, we have a binomial model for the number of survivors. With $X =$ "Number of survivors" and $Y =$ "Number of eggs laid", we have

$$X \,|\, Y = y \in \text{Bin}(y, p), \quad Y \in \text{Po}(m).$$

Compute

$$\text{P}(X = x) = \sum_{y=0}^{\infty} \text{P}(X = x | Y = y) \text{P}(Y = y)$$

and identify the distribution for X. Then find the number asked for, the average number of survivors, as $\text{E}[X]$.

6

Introduction to Bayesian Inference

In this chapter we further develop Bayesian methods to analyse data and estimate probabilities for the different scenarios first discussed in Chapter 2. The probabilities, often used as measures for risks, depend on a mathematical model of the random experiment, observations (data) and experience from similar problems.

In Chapter 4, we presented some statistical methods to fit distributions to data. The methods were based on interpretation of probabilities as frequencies, *i.e.* if one has an infinite sequence of independent outcomes of the experiment, then by means of LLN one can compute the probability p, say, of any statement by finding the relative frequency of times that the statement is true. However, since we never have infinite series of observations, even in the frequentist framework the estimated probabilities are uncertain. Consequently the classical inference results in an estimate p^* of the probability and a random variable $\mathcal{E}$ that models the variability of the estimation error. Often confidence intervals are used to describe possible size of error.

In the Bayesian approach, probability densities (pdfs) are used instead of confidence intervals to describe uncertainty in the value of a risk (a probability of suitably chosen scenario) due to finite length of observed data. However, a more important difference is that even uncertainties originating from our "lack of knowledge" as well as experience can be included in a measure of a risk for a particular scenario. This is often used when probabilities of occurrence of non-repeatable scenarios have to be analysed, for example damage of the vital parts in a specific nuclear power plant or a collision of a ship with a bridge, etc. Even here probabilities are used to measure risks; however, those have no frequentistic interpretation (the Law of Large Numbers cannot be applied).

This chapter is a brief introduction to the Bayesian methodology to analyse data and compute probabilities. Only some of the methods in this theory are mentioned. For deeper studies, we refer to the book by Gelman *et al.* [28]. For a discussion of the philosophy of the Bayesian interpretation of probability and reasoning, much debated among statisticians over decades, see for instance [52], and for a review, Chapters 1.4-1.5 in [60].

6.1 Introductory Examples

Bayesian statistics is a general methodology to analyse and draw conclusions from data. Here we mainly focus on two problems of interest in risk analysis:

- The first one deals with the estimation of a probability $p_B = \mathsf{P}(B)$, say, of some event B, for example the probability of failure of some system.
- The second one is estimation of the probability that at least once an event A occurs in a time period of length t. The problem reduces itself to estimation of the intensity λ_A of A.

Both the continuous parameters p_B and λ_A are attributes of some physical system, e.g. if $B =$"A water sample passes tests" then $p_B = \mathsf{P}(B)$ is a measure of efficiency of a waste-water cleaning process. The intensity λ_A of accidents may characterize a particular road crossing. Obviously, the parameters p_B and λ_A are unknown.

For simplicity of presentation, let θ denote the unknown value of p_B, λ_A, or any other quantity. Similarly as in Section 2.3, let us introduce odds q_θ, which for any pair θ_1, θ_2 represents our belief of which θ_1 or θ_2 is more likely to be the unknown value of θ, i.e. $q_{\theta_1} : q_{\theta_2}$ are odds for the alternatives $A_1 =$ "$\theta = \theta_1$" against $A_2 =$ "$\theta = \theta_2$". Since there are here uncountable number of alternatives, we require that q_θ integrates to one and hence $f(\theta) = q_\theta$ is a probability-density function representing our belief about the value of θ. The random variable Θ having the pdf serves as a mathematical model for uncertainty in the value of θ. Let us turn to two illustrative examples.

Estimation of probability $\mathsf{P}(B)$

Suppose we are interested in the probability of an event B, for example $B =$"Victim of an traffic accident needs hospitalization", where outcomes of the random experiment are accidents on a specific road crossing. We are interested in the frequency of serious accidents in which hospitalization for one or more of involved people is needed. We assume that B, for different accidents, happen independently with the same probability $p_B = \mathsf{P}(B)$. In other words, if B_i denotes the event that B is true in the ith accident then for any $i \neq j$,

$$\mathsf{P}(B_i \cap B_j) = \mathsf{P}(B_i)\mathsf{P}(B_j) = p_B^2.$$

Consequently, if K is the number of accidents leading to hospitalization of any of the people involved in the accident, then $K \in \mathrm{Bin}(n, p_B)$, where n is the total number of accidents considered. The goal is to estimate the probability p_B.

Classical estimate of the probability p_B: The probability p_B is an unknown constant. A commonly used estimate of the frequency is $p_B^* = k/n$, where k is the number of times B were true in n trials. Since n is finite the estimate is an uncertain value, and very likely $p_B \neq p_B^*$. The uncertainty is quantified using a random variable $\mathcal{E}$ and measured by means of confidence interval.

Bayesian approach: If parameters (or constants) are unknown, the uncertainty of which value is true can be described using a pdf $f(p)$, say. As mentioned before, the ratio $f(p_1) : f(p_2)$ is our odds for p_1 against p_2. Methods to find $f(p)$ is the main subject of this chapter.

Suppose $f(p)$ has been selected and denote by P a random variable having pdf $f(p)$. A plot of $f(p)$ is an illustrative measure of how likely the different values of p_B are. If only one value of the probability is needed, the Bayesian methodology proposes to use the so-called *predictive probability*, which is simply the mean of P:

$$\mathsf{P}^{\mathrm{pred}}(B) = \mathsf{E}[P] = \int pf(p)\, \mathrm{d}p. \tag{6.1}$$

The predictive probability is a properly defined probability that measures the likelihood that B occurs in future. By the Law of Total Probability, the predictive probability combines two sources of uncertainty: the unpredictability whether B will be true in a future accident and the uncertainty in the value of probability p_B.

Example 6.1. Suppose we have no idea how harmful accidents are. We express this total "lack of knowledge" by choosing $f(p)$ to be a uniform density function equal to 1 for all $0 \le p \le 1$ and zero otherwise (see Figure 6.1 left panel, dashed line). Obviously, for any two probabilities p_1 and p_2 the ratio $f(p_1) : f(p_2)$ is 1:1, which means that p_1 is equally likely to be true as p_2. Finally, using Eq. (6.1), the predictive probability

$$\mathsf{P}^{\mathrm{pred}}(B) = \mathsf{E}[P] = \int_0^1 p\, \mathrm{d}p = \frac{1}{2}.$$

$\square$

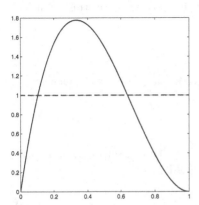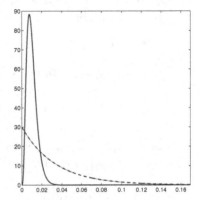

Fig. 6.1. *Left:* the pdf $f(p)$ in Examples 6.1 (dashed line) and 6.4 (solid line). *Right:* the pdf $f(p)$ in Examples 6.2 (dashed line) and 6.3 (solid line).

Estimation of probability $P_t(A)$

Suppose one is interested in the probability of occurrences of at least one accident at a specific road crossing. Let A be the event that an accident is recorded. Times instants when A occurs form a stream of A. Suppose that the stream is stationary and has intensity λ_A. The goal is to compute $P_t(A)$, *e.g.* $t = 1$ day, *i.e.* the probability of at least one accident in the period t. If it is reasonable to assume that the stream satisfies conditions (I-III) from Section 2.5.1, then the stream is Poisson and $P_t(A) = 1 - e^{-\lambda_A t}$. If $\lambda_A t$ is small, the probability $P_t(A)$ can be approximated as follows

$$P_t(A) = 1 - e^{-\lambda_A t} \approx \lambda_A t. \tag{6.2}$$

Actually, the approximation is also a bound for the probability since

$$P_t(A) \leq \lambda_A t \tag{6.3}$$

for any stationary stream of A (see Eq. (2.11), Theorem 2.5). Since we are mainly interested in the situations when $P_t(A)$ is small, in this chapter, we always estimate $P_t(A)$ by $\lambda_A t$, which is a *conservative* measure of a risk.

Classical estimate of the probability: Obviously the intensity λ_A is unknown and the commonly used estimate of the intensity is $\lambda_A^* = N_A(T)/T$, where $N_A(T)$ is the number of accidents that occurred in the period of time T. Consequently by Eq. (6.2) the estimate p^* of $P_t(A)$ is simply $p^* = \lambda_A^* t$. Since T is finite, the estimate is an uncertain value and one can analyse the size of the estimation error $\mathcal{E}$. The error can be expressed using a confidence interval, *e.g.* using Eq. (4.29).

Bayesian approach: Again, the Bayesian methodology models the uncertainty in the value of λ_A by means of a probability-density function $f_\Lambda(\lambda)$. The density $f_\Lambda(\lambda)$ describes our knowledge about possible values of λ_A. Denote by Λ a random variable having the pdf $f_\Lambda(\lambda)$. Recall that by Eqs. (6.2-6.3), $P_t(A) \approx \lambda_A t$. To describe the uncertainty of this quantity, a random variable

$$P = \Lambda t$$

is introduced. Since $P(P \leq p) = P(\Lambda \leq p/t)$, the pdf of P is given by

$$f(p) = \frac{\mathrm{d}}{\mathrm{d}p} P(P \leq p) = \frac{1}{t} f_\Lambda(p/t). \tag{6.4}$$

As before, if only one single value of the probability is needed, the Bayesian approach proposes to use the predictive probability

$$P_t^{\mathrm{pred}}(A) \approx E[P] = t E[\Lambda] = t \int \lambda f_\Lambda(\lambda) \, \mathrm{d}\lambda. \tag{6.5}$$

This is a measure of the risk that A occurs, combining two sources of uncertainty: the variability of the stream of A and the uncertainty in the intensity of accidents λ_A.

We see that the crucial point in the Bayesian methodology is the choice of the density $f_\Lambda(\lambda)$. The density reflects all our knowledge about the studied problem — we give a simple example.

Example 6.2. The exact value of λ_A is unknown and the pdf $f(\lambda)$ expresses our uncertainty in the value of λ_A. Suppose that our experience (belief) is that the intensity of accidents λ_A varies between road crossings. However, for the type of crossing considered, on average, the intensity is $1/30$ [day^{-1}]. In Section 6.5.1 we discuss methods to choose $f(\lambda)$ when not much is known about the intensity. It is shown that in the present situation a convenient choice is the exponential density. Since $\mathsf{E}[\Lambda] = 1/30$, $f(\lambda) = 30\,e^{-30\lambda}$, $\lambda \geq 0$ [day^{-1}]. The density is shown in Figure 6.1, right panel, dashed line. Let $t = 1$ day, then $\mathsf{P}_t(A)$ is approximated by $P = \Lambda$. From the plot one can see that the uncertainty in the value of P is quite high. Actually the probability can be any value between zero and $1/10$ with higher odds for small values of P. Finally, the Bayesian predictive probability is

$$\mathsf{P}_t^{\text{pred}}(A) \approx t\,\mathsf{E}[\Lambda] = \frac{t}{30}.$$

□

Again, let θ be the unknown parameter ($\theta = p_B$, $\theta = \lambda_A$ in Examples 6.1, and 6.2, respectively) while Θ denotes any of the variables P or Λ. Since θ is unknown, it is seen as a value taken by a random variable Θ with pdf $f(\theta)$[1]. In both examples we assumed that the probability densities $f(\theta)$ were somehow selected and we claimed that the densities represented our knowledge about the possible value of the parameter θ.

If $f(\theta)$ is chosen on basis of experience without including observations of outcomes of an experiment then the density $f(\theta)$ is called a *prior density* and denoted by $f^{\text{prior}}(\theta)$. However, as time passes, our knowledge may change, especially if we observe some outcomes of the experiment that can influence our opinions about the values of parameter θ reflecting in the new density $f(\theta)$. The modified density $f(\theta)$ will be called the *posterior density* and denoted by $f^{\text{post}}(\theta)$. The method to update $f(\theta)$ is discussed in detail in the following section. Selection of prior densities is discussed in Section 6.5. In Section 6.4, the so-called conjugated priors are introduced. These are the priors that are particularly convenient for recursive updating procedures, *i.e.* when new observations arrive at different time instants.

---

[1]For discrete distributions $F(\theta)$, *i.e.* when θ can take only a countable number of values θ_i, in all formulae integration on θ should be replaced by summation over all possible values of θ. Updation of odds was discussed in Chapter 2.

6.2 Compromising Between Data and Prior Knowledge

In the previous section we have introduced the prior density $f^{\mathrm{prior}}(\theta)$, which represents information about the random experiment before any observations (measurements) of the experiment were collected. Suppose now that the experiment resulted in that a statement C about the outcome is true. If the information is relevant it should influence our opinion about θ. The new density, called posterior and denoted as $f^{\mathrm{post}}(\theta)$, incorporates our *a priori* knowledge (experience) and the information that C is true. How to modify $f^{\mathrm{prior}}(\theta)$ to include this new piece of information is the subject of this section.

Suppose one can compute the probability of C when the unknown parameter has value θ, *i.e.* $\mathsf{P}(C|\Theta = \theta)$, for all values of θ, then Bayes' formula in Eq. (5.24) can be employed to update the prior density, *viz.*

$$f^{\mathrm{post}}(\theta) = c\mathsf{P}(C|\Theta = \theta)f^{\mathrm{prior}}(\theta), \qquad (6.6)$$

where the constant c is chosen so that $\int f^{\mathrm{post}}(\theta)\,\mathrm{d}\theta = 1$, since $f^{\mathrm{post}}(\theta)$ is a probability density function. Consequently

$$\frac{1}{c} = \int_{-\infty}^{+\infty} \mathsf{P}(C|\Theta = \theta)f^{\mathrm{prior}}(\theta)\,\mathrm{d}\theta.$$

Example 6.3. We continue the analysis from Example 6.2, where the prior pdf was chosen to be exponential with mean $1/30$ [days^{-1}], *i.e.* $f^{\mathrm{prior}}(\lambda) = 30\mathrm{e}^{-30\lambda}$, where λ is the intensity of accidents at a particular crossing.

Now suppose that after one year of monitoring the crossing three accidents have been recorded. Let us denote this information by C. The fact that C is true should affect our uncertainty about the intensity λ_A and also a measure of risk: the probability of at least one accident during one day. In order to use Eq. (6.6) to compute the posterior density $f^{\mathrm{post}}(\lambda)$ the likelihood function, *i.e.* the conditional probability $\mathsf{P}(C|\Lambda = \lambda)$, needs to be found. This is done next.

Let $N_A(T)$ be the number of accidents that has been recorded during a period of time T. Using Theorem 2.5, Eq. (2.12), it follows that

$$\mathsf{P}(C) = \mathsf{P}(N_A(T) = 3) = \frac{(\lambda_A T)^3}{3!}\mathrm{e}^{-\lambda_A T}$$

if the stream is Poissonian (the assumptions I–III can be motivated). Consequently $\mathsf{P}(C\,|\,\Lambda = \lambda) = \frac{(\lambda T)^3}{3!}\mathrm{e}^{-\lambda T}$ and Bayes' formula (6.6) gives

$$f^{\mathrm{post}}(\lambda) = c\,\lambda^3\mathrm{e}^{-\lambda T}\,f^{\mathrm{prior}}(\lambda) = c\,\lambda^3\,\mathrm{e}^{-\lambda T}\mathrm{e}^{-30\lambda} = c\,\lambda^3\,\mathrm{e}^{-(30+T)\lambda},$$

where c is a constant that needs to be determined and $T = 365$ [days]. The posterior density is recognized to be the pdf for the gamma distribution,

$\Theta \in \text{Gamma}(4, 395)$, (see Eq. (6.16) for definition and some simple properties of gamma distributed variables). The posterior density $f^{\text{post}}(\lambda)$ is given in Figure 6.1 right panel, solid line. We observe that the updated density $f(\lambda)$ is more concentrated around its peak, showing that the uncertainty for the probability $P_t(A) \approx t\Lambda$ decreased considerably after monitoring the crossing for a year.

Finally, as in Example 6.2, we compute the predictive probability for an accident during one day using Eq. (6.5) for the updated pdf $f^{\text{post}}(\lambda)$. Since for a random variable $\Lambda \in \text{Gamma}(a, b)$, the mean $\mathsf{E}[\Lambda] = a/b$, we have that

$$P_t^{\text{pred}}(A) \approx t\,\mathsf{E}[\Lambda] = t\frac{4}{395}$$

which for $t = 1$ gives the value $P_t^{\text{pred}}(A) = 0.01$. The computed risk is around three times smaller than computed in Example 6.1.

□

The following example shows another use of Bayes' formula.

Example 6.4. We continue the analysis from Example 6.1 and study the probability p_B for a serious accident (leading to hospitalization of one or more victims). Due to complete ignorance about the order of the probability, the prior pdf $f^{\text{pred}}(p)$ was chosen to be uniform $U(0, 1)$.

Now suppose that in the first year of monitoring this crossing three accidents were recorded and only one of these was serious. Denote this event by C. Using this information the posterior pdf is computed by means of Eq. (6.6).

Let N be the number of serious accidents among three accidents. If the probability p_B were known then $N \in \text{Bin}(3, p_B)$; consequently, with $C =$ "$N = 1$",

$$P(C \mid P = p) = 3\,p\,(1 - p)^2, \qquad 0 \le p \le 1,$$

and by (6.6)

$$f^{\text{post}}(p) = c\,p\,(1 - p)^2 \cdot 1, \qquad 0 \le p \le 1,$$

where c is a constant that needs to be determined. The posterior density is recognized to be the pdf for the beta distribution, $P \in \text{Beta}(2, 3)$, (see Eq. (6.10) for definition and some simple properties of beta distributed variables). The constant $c = 4!/2! = 12$. The prior and posterior pdf of P are presented in Figure 6.1 (right panel). We note that $f(p)$ is more concentrated, showing that the uncertainty for the probability $P(B)$ decreased slightly after monitoring the crossing for one year. More data are needed to be more certain about the size of the probability. Finally, using Eq. (6.1), the predictive probability $P^{\text{pred}}(B) = 2/5$. Note that the classical estimate $p_B^* = 1/3$; this is the value that maximizes the density f^{post}.

□

6.2.1 Bayesian credibility intervals

In the previous section a random variable P with posterior pdf $f^{\mathrm{post}}(p)$ was used to describe uncertainty in the values of the probabilities $\mathsf{P}(B)$ and $\mathsf{P}_t(A)$. The plot of $f^{\mathrm{post}}(p)$ visualizes the uncertainty while the predictive probability $\mathsf{E}[P]$ gives a single value as a measure of risk. By computing the predictive probability the information about the uncertainty in values of the probabilities $\mathsf{P}(B)$, $\mathsf{P}_t(A)$ is lost. A compromise between a complete description of uncertainty $f^{\mathrm{post}}(p)$ and averaging the possible values of p in the predictive probability $\mathsf{E}[P]$ is to use quantiles p_α of P. Definition and some applications of quantiles were given in Chapter 3. For convenience we give the defining equation here: p_α, where $0 < \alpha < 1$, is the α quantile of P if it satisfies the following equality: $\mathsf{P}(P > p_\alpha) = \alpha$.

Often instead of giving the posterior density, the uncertainty of P is described by means of a few quantiles, for example for $\alpha =0.975$, 0.9, 0.75, 0.5, 0.25, 0.1, and 0.025. In particular the interval $[p_{0.975},\ p_{0.025}]$ is called the 0.95-credibility interval since $\mathsf{P}(p_{0.975} \leq P \leq p_{0.025}) = 0.95$.

Example 6.5. Continuation of Example 6.4, where the posterior density for $f(p) \in \mathrm{Beta}(2,3)$. Quantiles are given below:

α	0.975	0.9	0.75	0.5	0.25	0.1	0.025
p_α	0.068	0.143	0.243	0.386	0.544	0.680	0.806

The 0.95-credibility interval is $[0.068,\ 0.806]$.

□

Quantiles and credibility intervals can also be used to describe the uncertainty of parameters, for example the intensity λ_A of a stream A. See the following example.

Example 6.6. In Example 6.3, an intensity λ_A was studied. As before, let $\varLambda$ be a r.v. having a pdf $f(\lambda)$; thus, $\varLambda \in \mathrm{Gamma}(4,395)$. Quantiles for $\varLambda$, denoted by q_α, are as follows:

α	0.975	0.9	0.75	0.5	0.25	0.1	0.025
q_α	0.003	0.004	0.006	0.009	0.013	0.017	0.022

Thus the 0.95-credibility interval is $[0.003, 0.022]$.

□

6.3 Bayesian Inference

In the previous section we considered two problems: the estimation of the probability that a statement B about an outcome of a random experiment will be true and the probability $\mathsf{P}_t(A)$ that at least one event A occurs in a period of time t. In both cases the uncertainty of some parameter θ, equal to p_B, λ_A, respectively, needed to be modelled. In this section we consider a more general situation, which is a parallel to the problem studied in Chapter 4, *i.e.* modelling observed variability in the data. The scope of problems that is discussed in this section is much narrower than in Chapter 4 and should only be regarded as a short introduction to some issues treated in Bayesian statistics.

6.3.1 Choice of a model for the data – conditional independence

Suppose we wish to model the variability of an experiment whose output is a value x of a random variable X. Let us choose a class of distributions $F(x; \theta)$. In order to keep things simple, we write $f(x; \theta)$ for the density, or probability-mass function, defining the distribution $F(x; \theta)$. As in Chapter 4.2, assume that there is a θ (seen as a property of an experiment), for which, if known, $F_X(x) = F(x; \theta)$. The important step in Bayesian modelling is to assume that $F(x; \theta)$ is actually the conditional distribution of X given that $\Theta = \theta$, *i.e.*

$$F(x|\theta) = \mathsf{P}(X \leq x|\Theta = \theta) = F(x; \theta). \tag{6.7}$$

We denote the density (or probability-mass function) of the conditional distribution $F(x|\theta)$ by $f(x|\theta)$.

Example 6.7. Consider an experiment of flipping a coin n times in an independent manner and let X be the number of "Heads" recorded. If the coin is fair, then $X \in \mathrm{Bin}(n, 1/2)$. Obviously, there exist no exactly fair coins and we let $\theta = p$ be the unknown probability of getting "Heads", a property of a coin. The natural choice of the model of $\mathsf{P}(X \leq x|\Theta = \theta) = F(x|\theta)$ is binomial, $\mathrm{Bin}(n, \theta)$. Here the parameter θ is equal to the probability p of getting "Heads" in a single flip of the coin. $\qquad\square$

Example 6.8 (Periods between earthquakes). Let us turn to the data set with periods between earthquakes (cf. Example 1.1). Denote by X the time between large earthquakes. Suppose that large earthquakes form a Poisson stream A with intensity λ_A; then, as will be shown in the following chapter, $\mathsf{P}(X \leq x|\Theta = \theta) = F(x|\theta)$ is exponential and hence θ is equal to the intensity of earthquakes λ_A. $\qquad\square$

Now we turn to the most important assumption of conditional independence of outcomes (observations) of the random variable X. As in the previous chapters we denote the not-yet observed values of X by $X_1, X_2, \ldots, X_n$. In Chapter 4, we assumed that $X_1, X_2, \ldots, X_n$ are iid (independent identically distributed) with the same distribution as X, *i.e.* $F(x; \theta_0)$, where θ_0 is the true parameter whose value is usually not known.

Here in the Bayesian set-up both X_i and the parameter value, represented by Θ, are random and hence the independence is assumed to be valid for any value θ of the parameter, *i.e.* when $\Theta = \theta$. This is written more formally as

$$\mathsf{P}(X_1 \leq x_1, \ldots, X_n \leq x_n \,|\, \Theta = \theta) = F(x_1|\theta) \cdot \ldots \cdot F(x_n|\theta) \tag{6.8}$$

for all θ and called *conditional independence*. This subject was already discussed; in Section 2.3 and Section 5.4.3. The assumption Eq. (6.8) allows for recursive updating of the priors.

6.3.2 Bayesian updating and likelihood functions

Suppose we have observed n values of X: $X_1 = x_1$, $X_2 = x_2, \ldots, X_n = x_n$. We assume that the conditional density (probability-mass function) $f(x|\theta)$ is known and that the observations of X_i are conditionally independent (see Eq. (6.8)), which means that the joint density (probability-mass function) satisfies

$$f(x_1, \ldots, x_n|\theta) = f(x_1|\theta) \cdot \ldots \cdot f(x_n|\theta) = L(\theta).$$

Here $L(\theta)$ is the likelihood function defined in Eq. (4.5), which was maximized to find the ML estimate θ^* of θ. Obviously, one should include these observations into the model for the variability of the parameter θ. The following version of Bayes' formula can be used to update the prior density $f^{\mathrm{prior}}(\theta)$ to the posterior density,

$$f^{\mathrm{post}}(\theta) = c\, L(\theta) f^{\mathrm{prior}}(\theta), \quad c^{-1} = \int_{-\infty}^{+\infty} L(\theta) f^{\mathrm{prior}}(\theta)\, d\theta. \tag{6.9}$$

The following example illustrates the updating procedure.

Example 6.9 (Prediction of earthquake tomorrow). Suppose that one is interested in the probability of occurrence of at least one major earthquake tomorrow anywhere in the world, *i.e.* $\mathsf{P}_t(A) \approx \theta\, t$, $t = 1$ day. Denote by X the time period between the major earthquakes. As seen in Example 6.8, $f_X(x|\theta) = \theta e^{-x\theta}$. Here $\theta = \lambda_A$ is the unknown intensity of earthquakes while observations x_i are time between ith and the next earthquake.

Choice of prior density. First we need to describe our experience of uncertainty in the possible value of θ by means of a prior pdf $f^{\mathrm{prior}}(\theta)$. For example, suppose we have total ignorance about the possible value of λ_A. In such a situation, as will be discussed in Section 6.5.1, Eq. (6.30), a convenient choice is given by the so-called improper prior $f^{\mathrm{prior}}(\theta) = 1/\theta$. An improper prior has to be used with care since this is not a pdf and, for example, $\mathsf{P}_t^{\mathrm{pred}}(A) \approx t \int \theta f^{\mathrm{prior}}(\theta)\, d\theta = +\infty$.

Updating the prior density. Now suppose we found that the time periods between the last 3 serious earthquakes were 92, 82, and 200 days. These observations are used to update $f(\theta)$, by means of (6.9). Since $f(x|\theta) = \theta \exp(-\theta x)$ the likelihood function $L(\theta)$ is given by

$$L(\theta) = f(220|\theta)\, f(82|\theta)\, f(92|\theta) = \theta^3 e^{-\theta(220+82+92)} = \theta^3 e^{-394\theta}.$$

Bayes' formula, Eq. (6.9), now gives

$$f^{\mathrm{post}}(\theta) = c\, L(\theta) f^{\mathrm{prior}}(\theta) = c\, \theta^2 e^{-394\,\theta}, \quad \theta > 0.$$

The density is recognized to be the pdf for the gamma distribution, *i.e.* $\Theta \in$ Gamma(3, 394) with $c = 394^3/2$.

Predictive probability from posterior density. The predictive probability of at least one serious earthquake next day is wanted. Using Eq. (6.5), for the posterior density $f^{\text{post}}(\theta)$ the predictive probability is given by

$$ \mathsf{P}_t^{\text{pred}}(A) \approx \mathsf{E}[P] = t\,\mathsf{E}[\Theta] = t\frac{3}{394} $$

(t in days). The quantiles of P are given by tq_α where q_α in their turn are quantiles of the intensity of earthquakes, *i.e.* of a Gamma(3,394) distribution.   □

In Example 6.9 we encountered a problem of how to choose the prior pdf $f^{\text{prior}}(\theta)$. This is needed to summarize the *a priori* knowledge about the parameter and also in order to be able to use Bayes' formula to compute the posterior density, *i.e.* to include data into the evaluation of uncertainty of the value of the parameter θ. We discuss different situations and give examples on how one can proceed to select the prior density. Another aspect is that in all Bayes' formulae there is a generic constant c that, at some stage, has to be given a value. The computation of the constant c can be problematic, especially when one has several parameters. In such a case the integral (sum) is multi-dimensional. One way to avoid such problems is to use the so-called *conjugated priors* described in Section 6.4.

Remark 6.1 (Recursive updating). When data arrive at different time instances it can be convenient to recursively update the density $f(\theta)$ as data arrive. This is possible by assumed conditional independence of data. It simplifies the Bayesian analysis, since we can always add some new data into the estimation procedure. Consider a model $f(x|\theta)$ and data $x_1, \ldots, x_n$. For any prior pdf $f^{\text{prior}}(\theta)$, the resulting posterior density, obtained using Eq. (6.9) or (6.6) n times each time a new observation x_i is received, will be the same as the one obtained by a single use of Eq. (6.9) with $L(\theta)$ computed for the whole data set $x_1, \ldots, x_n$.   □

6.4 Conjugated Priors

For a fixed family of distributions $F(x; \theta)$, *e.g.* normal, Weibull, and Poisson distributions, one looks for a corresponding family of densities that will be a convenient prior for θ. Here convenient means that the posterior density is of the same type as the prior density. This has several mathematical advantages, for example the normalization constant c in Eq. (6.6) can be easily found without cumbersome numerical integration.

There are tables with conjugated priors given in the literature, here we use only three of them (see the following subsections). Obviously, even if it is mathematically convenient to use conjugated priors to describe the uncertainty in parameters, these should be used only when they are close to our belief.

In this chapter three families of conjugated priors are presented:

Beta probability-density function (pdf):

$\Theta \in \text{Beta}(a, b)$, $a, b > 0$, if

$$f(\theta) = c\,\theta^{a-1}(1-\theta)^{b-1}, \quad 0 \le \theta \le 1, \quad c = \frac{\Gamma(a+b)}{\Gamma(a)\Gamma(b)}. \tag{6.10}$$

The expectation and variance of $\Theta \in \text{Beta}(a, b)$ are given by

$$\mathsf{E}[\Theta] = p, \quad \mathsf{V}[\Theta] = \frac{p(1-p)}{a+b+1}, \tag{6.11}$$

where $p = a/(a+b)$. Furthermore, the coefficient of variation

$$\mathsf{R}(\Theta) = \frac{1}{\sqrt{a+b+1}}\sqrt{\frac{1-p}{p}}. \tag{6.12}$$

A generalization of the beta pdf is the following two-dimensional Dirichlet pdf. If $\Theta = (\Theta_1, \Theta_2)$ has a Dirichlet distribution then Θ_1 and Θ_2, considered separately, have beta distributions, possibly with different parameters.

Dirichlet's pdf:

$\Theta = (\Theta_1, \Theta_2) \in \text{Dirichlet}(\mathbf{a})$, $\mathbf{a} = (a_1, a_2, a_3)$, $a_i > 0$, if

$$f(\theta_1, \theta_2) = c\,\theta_1^{a_1-1}\theta_2^{a_2-1}(1-\theta_1-\theta_2)^{a_3-1}, \quad \theta_i > 0, \theta_1 + \theta_2 < 1, \tag{6.13}$$

where $c = \frac{\Gamma(a_1+a_2+a_3)}{\Gamma(a_1)\Gamma(a_2)\Gamma(a_3)}$. Let $a_0 = a_1 + a_2 + a_3$; then

$$\mathsf{E}[\Theta_i] = \frac{a_i}{a_0}, \quad \mathsf{V}[\Theta_i] = \frac{a_i(a_0-a_i)}{a_0^2(a_0+1)}, \quad i = 1, 2. \tag{6.14}$$

Furthermore the marginal probabilities are Beta distributed, *viz.*

$$\Theta_i \in \text{Beta}(a_i, a_0 - a_i), \quad i = 1, 2. \tag{6.15}$$

Gamma pdf:

$\Theta \in \text{Gamma}(a, b), \quad a, b > 0, \quad$ if

$$f(\theta) = c\,\theta^{a-1}e^{-b\theta}, \quad \theta \geq 0, \quad c = \frac{b^a}{\Gamma(a)}. \tag{6.16}$$

The expectation, variance, and coefficient of variation for $\Theta \in$ Gamma(a, b) are given by

$$\mathsf{E}[\Theta] = \frac{a}{b}, \qquad \mathsf{V}[\Theta] = \frac{a}{b^2}, \qquad \mathsf{R}[\Theta] = \frac{1}{\sqrt{a}}. \tag{6.17}$$

The beta and gamma densities almost coincide for large values of b and with a much smaller than b, *i.e.* when a/b is close to zero. Further, the constant c, in the formulae for beta and gamma density, is computed using the following integrals:

$$\int_0^1 \theta^{a-1}(1-\theta)^{b-1}\,\mathrm{d}\theta = \frac{\Gamma(a)\Gamma(b)}{\Gamma(a+b)} \tag{6.18}$$

$$\int_0^\infty \theta^{a-1}e^{-b\theta}\,\mathrm{d}\theta = \frac{\Gamma(a)}{b^a} \tag{6.19}$$

$$\int_0^1 \int_0^1 \theta_1^{a_1-1}\theta_2^{a_2-1}(1-\theta_1-\theta_2)^{a_3-1}\,\mathrm{d}\theta_1\,\mathrm{d}\theta_2 = \frac{\Gamma(a_1)\Gamma(a_2)\Gamma(a_3)}{\Gamma(a_1+a_2+a_3)} \tag{6.20}$$

For $k = 1, 2, 3, \ldots$, we have $\Gamma(k) = (k-1)!$ and for any $a \geq 0$, $\Gamma(a+1) = a\Gamma(a)$.

In the following subsections we present three types of problems where beta, Dirichlet, and gamma priors, respectively, can be applied.

6.4.1 Unknown probability

Let us return to the problems discussed in Chapter 2 and Section 6.1. Consider a stream of events A and let B be an event (statement) describing a "scenario" of interest, for example:

A = "Fire ignition in a building", B = "Not all exit doors function properly"

In a Bayesian approach, the uncertainty of the probability of B is modelled by means of a random variable Θ with density $f(\theta)$. Here the outcomes of Θ are $\theta = p_B$.

Suppose that, as in a frequentist's approach, we observe n outcomes of the experiment A and find that the statement B was true k times. Clearly the value k is unknown in advance and hence it is modelled as a random variable K, which can take any of the values $0, 1, \ldots, n$. If θ was known (which means conditionally that $\Theta = \theta$), then K has a binomial probability-mass function (1.9), *i.e.*

$$P(K = k|\Theta = \theta) = \binom{n}{k} \theta^k (1 - \theta)^{n-k}, \quad k = 0, 1, \ldots, n.$$

The information that we have observed k of n times that B was true should be included in the prior density describing the likelihood of the possible values of $\theta = P(B)$. We see directly that Eq. (6.6) with $C = "K = k"$ can be used to compute the posterior density

$$f^{\text{post}}(\theta) = c\, P(K = k|\Theta = \theta) f^{\text{prior}}(\theta).$$

Note that here the information about the parameter θ consists of a pair (n, k), where n is the number of trials while k is the number of times B was true in the n trials. Now if the prior density is of beta type, *i.e.* $\Theta \in \text{Beta}(a, b)$, then

$$f^{\text{post}}(\theta) = \tilde{c}\theta^k(1-\theta)^{n-k}\theta^{a-1}(1-\theta)^{b-1} = \tilde{c}\theta^{a+k-1}(1-\theta)^{b+n-k-1}, \quad (6.21)$$

where $\tilde{c}$ is computed using Eq. (6.18), $1/\tilde{c} = \Gamma(a+k)\Gamma(b+n-k)/\Gamma(a+b+n)$. By this we proved:

The beta priors are conjugated priors for the problem of estimating the probability $p_B = P(B)$.

Let $\theta = p_B$. If one has observed that in n trials (results of experiments), the statement B was true k times and if the prior density $f^{\text{prior}}(\theta) \in \text{Beta}(a, b)$ then

$$f^{\text{post}}(\theta) \in \text{Beta}(\tilde{a}, \tilde{b}), \quad \tilde{a} = a + k, \quad \tilde{b} = b + n - k. \tag{6.22}$$

$$P^{\text{pred}}(B) = \int_0^1 \theta f^{\text{post}}(\theta)\, d\theta = \frac{\tilde{a}}{\tilde{a} + \tilde{b}}. \tag{6.23}$$

From Eq. (6.22), the Beta (a, b) prior means that our experience is equivalent to observing a times the event B in $a + b$ experiments.

Example 6.10 (Waste-water treatment: conjugated priors).   Recall Example 2.4. We studied there the probability $p_B = P(B)$ where

$B = $ "Cleaned waste-water passes the test".

This is now studied within the framework of the present chapter, with the probability p_B now described as a parameter θ, regarded as an r.v. with pdf $f(\theta)$.

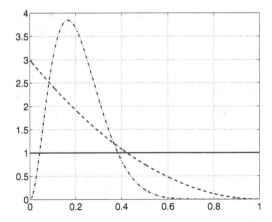

Fig. 6.2. Solid: Beta$(1,1)$ (prior distribution); Dashed: Beta$(1,3)$; Dashed-dotted: Beta$(3,11)$.

Total lack of knowledge is represented by a uniform prior density,

$$f^{\mathrm{prior}}(\theta) = 1, \quad 0 \le \theta \le 1.$$

We recognize this as $f^{\mathrm{prior}}(\theta) \in \mathrm{Beta}(1,1)$.

Assume that we observe that in the two last tests, we had two failures. Then by Eq. (6.22) we directly find the posterior density, $f^{\mathrm{post}}(\theta) \in \mathrm{Beta}(1,3)$ (Figure 6.2, dashed curve).

If the next 10 tests showed that water was successfully cleaned only 2 times, then this information gives the new (updated) posterior density $f^{\mathrm{post}}(\theta) \in \mathrm{Beta}(3,11)$ (Figure 6.2, dashed-dotted curve). The probability that the bacterial culture is efficient ($p_B \ge 0.5$), is computed from the density functions as $\mathsf{P}(\Theta \ge 0.5)$, and for the densities Beta$(1,1)$, Beta$(1,3)$ and Beta$(3,11)$ we find (by integration) $\mathsf{P}(\Theta \ge 0.5)$ to be 0.5, 0.125, 0.0112, respectively. $\qquad\qquad\square$

6.4.2 Probabilities for multiple scenarios

We here consider a generalization of the previous subsection and study the case of two excluding scenarios B_1, B_2, which cannot happen simultaneously. We assume that B_i are independent of the stream A. Let $p_1 = \mathsf{P}(B_1)$ and $p_2 = \mathsf{P}(B_2)$, which, by LLN, are frequencies of occurrences of B_1, B_2, respectively.

Assume that $p_1 + p_2 < 1$, then a third scenario is possible $B_3 =$ "Neither B_1 nor B_2 occurs", having probability $\mathsf{P}(B_3) = 1 - p_1 - p_2$. (Obviously, if $p_1 + p_2 = 1$, only one scenario B_1 needs to be considered, since $B_2 = B_1^c$, and the problem reduces to case discussed in the previous subsection.)

Now the parameter $\theta = (\theta_1, \theta_2)$, where $\theta_i = p_i$, is unknown. Suppose that n experiments were performed and let k_i be the number of time B_i

was true. In the Bayesian approach the uncertainty of the parameter θ will be modelled by means of a prior density $f^{\text{prior}}(\theta_1, \theta_2)$. We demonstrate here that it is convenient to choose Dirichlet priors, which are the conjugated priors for the problem of estimation of the unknown probabilities.

Suppose we observe n outcomes of the experiment and find that k_i times the statement B_i was true, $k_1 + k_2 + k_3 = n$. Clearly the values k_i are unknown in advance and hence are modelled as random variables K_i. If θ was known (which means conditionally that $\Theta = \theta$), then K has a multinomial probability-mass function Eq. (5.1):

$$P\big(K_1 = k_1, K_2 = k_2 \,|\, \Theta = (\theta_1, \theta_2)\big) = \frac{n!}{k_1!\, k_2!\, k_3!}\, \theta_1^{k_1} \theta_2^{k_2} (1 - \theta_1 - \theta_2)^{k_3}.$$

The information $C = ``K_1 = k_1, K_2 = k_2"$ should be included in the prior density and by means of Eq. (6.6), the posterior density is

$$f^{\text{post}}(\theta) = c\, P\big(K_1 = k_1, K_2 = k_2 \,|\, \Theta = (\theta_1, \theta_2)\big) f^{\text{prior}}(\theta).$$

Now it is easy to see that if the prior density is Dirichlet then also the posterior density belongs to the same class:

The Dirichlet priors are conjugated priors for the problem of estimating the probabilities $p_i = P(B_i)$, $i = 1, 2, 3$, such that B_i are disjoint and $p_1 + p_2 + p_3 = 1$.

Let $\theta_i = p_i$. Then, under assumptions of this section, if one has observed that the statement B_i was true k_i times in n trials and the prior density $f^{\text{prior}}(\theta_1, \theta_2) \in \text{Dirichlet}(\mathbf{a})$,

$$f^{\text{post}}(\theta_1, \theta_2) \in \text{Dirichlet}(\widetilde{\mathbf{a}}), \quad \widetilde{\mathbf{a}} = (a_1 + k_1, a_2 + k_2, a_3 + k_3), \quad (6.24)$$

where $k_3 = n - k_1 - k_2$. Further

$$P^{\text{pred}}(B_i) = E[\Theta_i] = \frac{\widetilde{a}_i}{\widetilde{a}_1 + \widetilde{a}_2 + \widetilde{a}_3}. \quad (6.25)$$

Example 6.11 (Chess players). (Continuation of Example 5.3.) Suppose that A and B are famous chess players who start to play a series of 8 matches against each other. We get results in the newspaper and wish to predict the result of the next game. We consider p_A, p_B (the probabilities that A and B win, respectively) to be unknown parameters. This lack of knowledge can be modelled using a prior distribution representing our knowledge of results from matches played earlier, ranking etc. If we have no idea about the capacity of the players, we may use uniform priors allowing all values of p_A, p_B, such that $p_A + p_B \le 1$, with equal likelihood. This can be established by choosing Dirichlet priors with parameters $a_i = 1$.

Suppose that after 4 matches, A won one match while B won two. Using Eq. (6.24) the posterior density is Dirichlet too with parameters $a_1 = 2$, $a_2 = 3$, $a_3 = 2$, and $a_0 = a_1 + a_2 + a_3 = 7$. Suppose we wish to find the probability that A wins in the next match. The predictive probability, which by means of the law of total probability, is given by

$$P^{\text{pred}}(\text{"A wins next match"}) = \int_0^1 P(\text{"A wins next match"}|\Theta_1 = \theta_1)$$

$$\cdot f^{\text{post}}(\theta_1)\, d\theta_1$$

$$= \int_0^1 \theta_1\, f^{\text{post}}(\theta_1)\, d\theta_1 = E[\Theta_1],$$

since $P(\text{"A wins next match"}|\Theta_1 = \theta_1) = \theta_1$. Now by Eq. (6.14), $E[\Theta_1] = a_1/a_0 = 2/7$. Similarly, $P^{\text{pred}}(\text{"B wins next match"}) = E[\Theta_2] = 3/7$ and draw has probability $1 - 2/7 - 3/7 = 2/7$.

Finally, if one wishes to know the predictive probability of $C =$ "The winner of the next match is A, the match thereafter B will win", the following calculations are needed: First, by conditional independence

$$P(A \cap B | \Theta_1 = \theta_1, \Theta_2 = \theta_2) = P(A | \Theta_1 = \theta_1, \Theta_2 = \theta_2)$$
$$\cdot P(B | \Theta_1 = \theta_1, \Theta_2 = \theta_2) = \theta_1 \theta_2$$

then

$$P^{\text{pred}}(C) = \int_0^1 \int_0^1 P(C | \Theta_1 = \theta_1, \Theta_2 = \theta_2) f^{\text{post}}(\theta_1, \theta_2)\, d\theta_1\, d\theta_2$$

$$= \int_0^1 \int_0^1 \theta_1 \cdot \theta_2 \cdot f^{\text{post}}(\theta_1, \theta_2)\, d\theta_1\, d\theta_2 = E[\theta_1 \theta_2]$$

$$= \int_0^1 \int_0^1 \theta_1 \theta_2 \frac{\Gamma(a_0)}{\Gamma(a_1)\Gamma(a_2)\Gamma(a_3)} \theta_1^{a_1-1} \theta_2^{a_2-1}$$

$$\cdot (1 - \theta_1 - \theta_2)^{a_3-1}\, d\theta_1\, d\theta_2$$

$$= \frac{\Gamma(a_0)}{\Gamma(a_1)\Gamma(a_2)\Gamma(a_3)} \frac{\Gamma(a_1+1)\Gamma(a_2+1)\Gamma(a_3)}{\Gamma(a_0+2)} = \frac{a_1 a_2}{a_0(a_0+1)},$$

where the last integral was computed by means of Eq. (6.20). Finally, for the specific values of parameters a_i, $P(C) = \frac{2 \cdot 3}{8 \cdot 7} = 0.107$.

Note that $P^{\text{pred}}(C) \neq \frac{2}{7} \cdot \frac{3}{7}$, which would be the case if Θ_1 and Θ_2 were independent. $\qquad\square$

6.4.3 Priors for intensity of a stream A

In Chapter 2 we gave conditions when a stationary stream A is Poisson. The intensity of A was denoted by λ_A, say, and if the stream is Poisson the number of events A that occur in an interval of length t is $N_A(t) \in \text{Po}(m)$, where $m = \lambda_A \cdot t$. This means that the expected number of times A occurs in a period of length t is equal to $\lambda_A \cdot t$.

In order to keep the same notation as in the previous section, we denote the unknown intensity λ_A by θ and write $N(t)$ instead of $N_A(t)$. Then we model our prior knowledge about θ by means of a random variable Θ with prior density $f^{\mathrm{prior}}(\theta)$. Now if θ were known (which means conditionally that $\Theta = \theta$), then $N(t)$ would have a Poisson probability-mass function

$$P(N(t) = k | \Theta = \theta) = \frac{(\theta t)^k}{k!} e^{-\theta t}, \quad k = 0, 1, 2, \ldots .$$

Our observations consist now of a pair: the exposure time t and the number of times $N(t)$ the initiation event A (accident) occurred under a time period t. This should be included in the prior density describing the likelihood of the possible values of θ, $i.e.$ the intensity of accidents λ. Again one can introduce a statement $C = $ "$N(t) = k$" and use Eq. (6.6) to compute the posterior density

$$f^{\mathrm{post}}(\theta) = c\, P(N(t) = k | \Theta = \theta) f^{\mathrm{prior}}(\theta).$$

Now if the prior density is of gamma type, $i.e.$ $\Theta \in \mathrm{Gamma}(a, b)$, then

$$f^{\mathrm{post}}(\theta) = c\,(\theta t)^k e^{-\theta t} \theta^{a-1} e^{-b\theta} = \widetilde{c}\, \theta^{a+k-1} e^{-\theta(b+t)}, \tag{6.26}$$

The constant $\widetilde{c}$ can be computed using Eq. (6.19), $1/\widetilde{c} = \Gamma(a+k)/(b+t)^{(a+k)}$. We summarize our findings:

The gamma priors are conjugated priors for the problem of estimating the intensity in a Poisson stream of events A. If one has observed that in time $\widetilde{t}$ there were k events reported and if the prior density $f^{\mathrm{prior}}(\theta) \in \mathrm{Gamma}(a, b)$, then

$$f^{\mathrm{post}}(\theta) \in \mathrm{Gamma}(\widetilde{a}, \widetilde{b}), \qquad \widetilde{a} = a + k, \quad \widetilde{b} = b + \widetilde{t}. \tag{6.27}$$

Further, the predictive probability of at least one event A during a period of length t is given by

$$P_t^{\mathrm{pred}}(A) \approx t E[\Theta] = t\, \frac{\widetilde{a}}{\widetilde{b}} \tag{6.28}$$

Remark 6.2 (Predictive probability, Poisson stream). In this remark we give an exact formula for the predictive probability of at least one event A in the period t. For a Poisson stream of A the number of times A occurs during a period t, given $\Theta = \theta$, is Poisson distributed with mean θt and

$$P_t(A | \Theta = \theta) = 1 - P(N_t(A) = 0) = 1 - e^{-\theta t}.$$

Hence, with $P = 1 - e^{-\Theta t}$, for any posterior density, which is $\text{Gamma}(\widetilde{a}, \widetilde{b})$ we have that

$$P_t^{\text{pred}}(A) = \mathsf{E}[P] = \int_0^\infty \left(1 - e^{-\Theta t}\right) f^{\text{post}}(\theta)\, d\theta$$

$$= 1 - \int_0^\infty e^{-\theta t} \frac{\widetilde{b}^{\widetilde{a}}}{\Gamma(\widetilde{a})} \theta^{\widetilde{a}-1} e^{-\widetilde{b}\theta}\, d\theta = 1 - \frac{\widetilde{b}^{\widetilde{a}}}{\Gamma(\widetilde{a})} \int_0^\infty \theta^{\widetilde{a}-1} e^{-(\widetilde{b}+t)\theta}\, d\theta$$

$$= 1 - \frac{\widetilde{b}^{\widetilde{a}}}{\Gamma(\widetilde{a})} \frac{\Gamma(\widetilde{a})}{(\widetilde{b}+t)^{\widetilde{a}}} = 1 - \left(\frac{\widetilde{b}}{\widetilde{b}+t}\right)^{\widetilde{a}}. \tag{6.29}$$

For t much smaller than $\widetilde{b}$, Eq. (6.29) gives similar value as Eq. (6.28). $\square$

6.5 Remarks on Choice of Priors

A critical issue in a Bayesian analysis is the choice of priors. These are subjective and everybody can have his own priors. The only restriction is that motivation should be given for the choice. Here we indicate some possible motivations for choosing specific values of the parameters in the beta and gamma priors in situations when not much is known about the values of parameters.

6.5.1 Nothing is known about the parameter θ

Beta priors

Suppose that $\theta = \mathsf{P}(B)$ for some statement B. If nothing is known about θ, uniform priors seem to be a reasonable choice. Fortunately the conjugated priors for this problem contain the uniform, *i.e.* $\text{Beta}(1,1)$.

Gamma priors

The choice of non-informative prior in the case when θ is the intensity of a stream of events A, say, (for example a stream of accidents) is more complicated. Since θ can take any non-negative value it is not obvious how to define the uniform priors.

If the priors are not probability densities these are called *improper* priors. Such priors can be used as long as the posterior odds form a true pdf. An often used improper prior is

$$f^{\text{prior}}(\theta) = 1/\theta, \qquad \theta > 0 \tag{6.30}$$

which could be denoted as $\text{Gamma}(0,0)$. In the following remark, we analyes how this is obtained from properties of the gamma distribution.

Remark 6.3 (Motivation for improper prior Gamma(0,0)). Suppose that, subjectively, values of the mean $E[\Theta]$ and a large coefficient of variation $R[\Theta]$ are assigned. Since large values of the coefficient of variation mean high uncertainty a possible choice of non-informative priors is to let $R[\Theta]$ increase to infinity. By means of Eq. (6.17) this implies that the parameter a tends to zero. Now, since $a \to 0$ and $E[\Theta]$ is constant, b converges to zero too. We note that the prior density $f^{\text{prior}}(\theta)$, suitably scaled ($c = 1$), converges to $1/\theta$, and the function in Eq. (6.30) is found.

Consequently, by increasing the coefficient of variation, information that $E[\Theta]$ was known becomes irrelevant. Such priors could be seen as non-informative. However, the integral of the function in Eq. (6.30) is equal to infinity and hence it is not a probability-density function.   □

Suppose now that the information is that during a time period t, no event A has been observed. Hence, by using Eq. (6.27), the improper prior in Eq. (6.30) would give the posterior odds Gamma$(0, t)$. Note that this is not a pdf and hence use of Eq. (6.30) is not recommendable, since the predictive probability cannot be computed. In such a situation, the use of *uniform* improper priors[2],

$$f^{\text{prior}}(\theta) = 1, \qquad \theta > 0, \tag{6.31}$$

which can be denoted as Gamma$(1, 0)$, could be applied. This prior results in the posterior pdf Gamma$(1, t)$, which is the exponential distribution with mean $1/t$.

6.5.2 Moments of Θ are known

In engineering, quite often unknown parameters, *e.g.* strength, are specified by assigning values for the expectation $E[\Theta]$; further, uncertainty is given by the coefficient of variation $R[\Theta] = D[\Theta]/E[\Theta]$. We interpret this approach as that one has a subjective opinion what θ is, *e.g.* $\theta = \theta_0$. For example, P("A flip results in Heads") $= \theta_0 = 1/2$ if the coin is fair. If we wish to allow for some uncertainty in our opinion then we can choose to have a random Θ such that $E[\Theta] = \theta_0$ with a specified coefficient of variation $R[\Theta]$.

If beta priors are chosen, the parameters a and b can be solved from Eqs. (6.11-6.12), while in the case of gamma priors, the formula (6.17) leads to the lowing values for the parameters a and b:

$$a = \frac{1}{R[\Theta]^2}, \quad b = \frac{1}{E[\Theta]R[\Theta]^2}. \tag{6.32}$$

Example 6.12 (Flight safety). Suppose we are interested in flight safety and follow reports about crashes. From our experience we believe that the average rate θ of "fatal accidents" is constant and close to 25 per year

---

[2]In principle, the uniform improper prior $f(\theta) = 1$, $\theta > 0$ is not a natural choice since this gives equal odds for low and high intensities.

(thus $E[\Theta] = 25$) and we add a vague statement about possible error in our prediction "± 20". If Θ were normally distributed, which is often the case if many data are available, then this vague statement could be interpreted as $D[\Theta] = 10$ or $R[\Theta] = 10/25 = 0.4$, which we assume in the following.

We choose to use gamma priors, $\Theta \in \text{Gamma}(a, b)$. Our "belief" is that $E[\Theta] = 25$ [years^{-1}] while $R[\Theta] = 0.4$ and hence using Eq. (6.32) we can compute the parameters $a = 6.25$ and $b = 0.25$. This corresponds to assumptions that nothing was known about the possible value of the rate of accidents and 6 accidents were observed in 3 months. The prior density $f^{\text{prior}}(\theta) \in \text{Gamma}(6.25, 0.25)$ is shown in Figure 6.3, left panel (dotted line).

Updating priors. In "Statistical Abstract of the United States" one can find the data for the number of crashes in the world during the years 1976-1985, which we denote as $k_1, \ldots, k_{10}$ with values

$$24, \quad 25, \quad 31, \quad 31, \quad 22, \quad 21, \quad 26, \quad 20, \quad 16, \quad 22,$$

respectively. These observations are now used to update our prior density. By means of Eq. (6.27) we know that $f^{\text{post}}(\theta) \in \text{Gamma}(6.25 + \sum k_i, 10.25)$. Since $\sum k_i = 238$, $f^{\text{post}}(\theta) \in \text{Gamma}(244.25, 10.25)$. In Figure 6.3, left panel, (solid line), we note that the density becomes narrower, reflecting better knowledge about the value of the parameter θ.

Consequently the probability of at least one accident tomorrow $P_t(A) \approx \Theta t$, where $t = 1/365$ [year] is very concentrated around the predictive probability

$$P_t^{\text{pred}}(A) \approx E[\Theta t] = \frac{244.25}{10.25} \frac{1}{365} = 0.065.$$

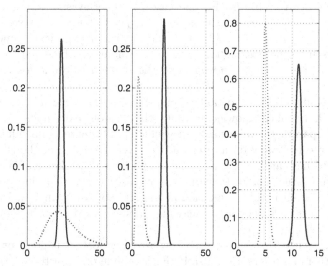

Fig. 6.3. The prior (dotted line) and posterior (solid line) densities $f^{\text{prior}}(\theta)$, $f^{\text{post}}(\theta)$ discussed in Example 6.12.

The uncertainty in the unknown probability $P_t(A)$ can also be described by the 95% credibility interval for $\Theta\, t$, $viz.$ $[t\theta_{0.975}, t\theta_{0.025}]$, where $\theta_{0.975}$ and $\theta_{0.025}$ are 0.975 and 0.025 quantiles of Θ (see Section 6.2.1 for the discussion on credibility intervals). For the Gamma(244.25, 10.25) the quantiles $\theta_{0.975} = 20.96$, while $\theta_{0.025} = 26.94$, which give the credibility interval for the probability $[20.96/365, 26.94/365] = [0.057, 0.074]$.

Influence of choice of prior. Clearly the prior information of 6 accidents in 0.25 years had been totally dominated by the data and has little influence on the posterior density. For example, suppose we had wrong ideas about the frequency of crashes and postulated that the mean was $\mathsf{E}[\Theta] = 5$ with the same coefficient of variation $\mathsf{R}[\Theta] = 0.4$. This would lead to the prior density $f^{\mathrm{prior}}(\theta) \in$ Gamma(6.25, 1.25) and the posterior density $f^{\mathrm{post}}(\theta) \in$ Gamma(244.25, 11.25), which is quite close to the previously computed posterior density. Data corrected our poor prior knowledge. The densities are shown in Figure 6.3, centre.

Finally, let us be very sure that the true frequency of accidents is close to 5 per year and choose a very low coefficient of variation $\mathsf{R}(\Theta) = 0.1$. Then the prior density $f^{\mathrm{prior}}(\theta) \in$ Gamma(100, 20) (*i.e.* one postulates that 100 accidents were observed in 20 years). Clearly the data of 238 accidents in 10 years cannot compensate for such wrong priors. The densities are shown in Figure 6.3, right panel.   □

Is it dangerous to choose the wrong prior density? Generally the answer is: theoretically "no", practically "it can be", as seen is the following paragraphs.

Suppose the random variable X has a distribution $F(x; \theta_0)$, where θ_0 is an unknown fixed parameter. Using the frequentist approach, we finally find the value of the parameter θ_0 as we get an infinite number of independent observations of X. This is also the case for the Bayesian approach if one has started with a prior density such that $f^{\mathrm{prior}}(\theta_0) > 0$, as seen in Section 6.6, Eq. (6.33). This formula tells us that if our experience (knowledge) does not exclude the possibility that the parameter can be equal to θ_0, then the Bayesian approach is *equivalent* to the frequentistic one for large data sets.

Consequently, it is important to choose wide (non-informative) priors if we do not have much knowledge about the random experiment — so that the possible parameter values are not excluded. This is recommendable if one expects to receive many data later on. However, Bayesian methods are most useful when there are few data available and hence the choice of the prior is an important issue. For instance, in many cases when a specific problem is studied, *e.g.*, intensity of fire ignition in a specific building (in a nuclear power plant), we do not expect many incidents. In such a situation it is important to carefully choose the priors.

Finally, even if the data will correct wrong priors, it is good practice to check whether the priors are sensible. For example one can try to "translate" the prior densities to the approximate amount of data the priors represent. For example, in the last example $E[\Theta] = 2$ and $R[\Theta] = 0.1$ and the gamma prior density corresponds to 100 accidents in 50 years. All available data about the crashes cannot compensate for so wrong priors; 10 years is a much shorter time than 50 years.

6.6 Large number of observations: Likelihood dominates prior density

In this section, we return to the discussion from the previous section about importance of a good choice of the prior density $f^{\text{prior}}(\theta)$. Earlier we claimed that data can correct the wrong prior density as long as the true parameter is not excluded. This is a consequence of the fact that the posterior density $f^{\text{post}}(\theta)$, given in Eq. (6.9), becomes proportional to the likelihood function $L(\theta)$. Hence, when a large number of data are available, the choice of the prior density is less important in the analysis[3].

Distributions dependent on a single parameter

Assume first that the chosen class of distributions to model the random variable X depends only on one parameter, *i.e.* θ in $F(x; \theta)$ is one-dimensional (for example X is binomial, Poisson, exponential, or Rayleigh distributed). Suppose we have observed a large number n of values of X: $X_1 = x_1$, $X_2 = x_2, \ldots, X_n = x_n$ and that the conditional density (probability-mass function) $f(x|\theta)$ satisfies the regularity assumptions required in Theorem 4.1. In such a case, we know that the ML estimator Θ^* is consistent, *i.e.* converges to the unknown parameter θ_0, say, and the error $\mathcal{E} = \theta_0 - \Theta^*$ is approximately normally distributed, *i.e.* for large values of n

$$P(\mathcal{E} \leq x) = P(\theta_0 - \Theta^* \leq x) \approx \Phi\left(\frac{x}{\sigma_{\mathcal{E}}^*}\right),$$

where $\sigma_{\mathcal{E}}^* = 1/\sqrt{-\ddot{l}(\theta^*)}$, $l(\theta)$ is the log-likelihood function and θ^* is the ML estimate of θ. In order to shorten the notation we write this property as $\mathcal{E} \in \text{AsN}(0, (\sigma_{\mathcal{E}}^2)^*)$, (see Eq. (4.18)). Next, we demonstrate how the asymptotic normality of the estimation error $\mathcal{E}$ relates to the properties of the posterior density $f^{\text{post}}(\theta)$.

Again, let θ_0 be the unknown value of the parameter. If the prior density $f^{\text{prior}}(\theta)$ is a smooth function and does not exclude the possibility of θ_0, *i.e.* $f^{\text{prior}}(\theta_0) > 0$, then for large n the posterior pdf $f^{\text{post}}(\theta) \approx \text{N}(\theta^*, (\sigma_{\mathcal{E}}^2)^*)$.

---

[3]Note that there are situations, not met in this book, when parameters are vectors and data have little information about some of the parameters. In such situations the choice of priors can remain essential for the final result.

Posterior pdf for a large number of observations.

If $f^{\mathrm{prior}}(\theta_0) > 0$ then

$$\Theta \in \mathrm{AsN}(\theta^*, (\sigma_{\mathcal{E}}^2)^*) \tag{6.33}$$

as $n \to \infty$, where θ^* is the ML estimate of θ and $(\sigma_{\mathcal{E}}^2)^* = -1/\ddot{l}(\theta^*)$.

Remark 6.4. To prove Eq. (6.33), Taylor's formula is employed for the log-likelihood function $l(\theta)$ and this is expanded around the ML estimate θ^*, *viz.*

$$l(\theta) \approx l(\theta^*) + \dot{l}(\theta^*)(\theta - \theta^*) + \frac{1}{2}\ddot{l}(\theta^*)(\theta - \theta^2)^*.$$

Further, since the likelihood function $L(\theta) = e^{l(\theta)}$ and $\dot{l}(\theta^*) = 0$, the following approximation is obtained

$$L(\theta) \approx \exp\left(l(\theta^*) + \dot{l}(\theta^*)(\theta - \theta^*) + \frac{1}{2}\ddot{l}(\theta^*)(\theta - \theta^2)^*\right) \tag{6.34}$$

$$= c \exp\left(\frac{1}{2}\ddot{l}(\theta^*)(\theta - \theta^*)^2\right).$$

As n increases, $\ddot{l}(\theta^*)$ decreases to minus infinity. The decay is so fast that the prior density can be replaced by a constant $f^{\mathrm{prior}}(\theta) \approx f^{\mathrm{prior}}(\theta^*)$ and hence

$$f^{\mathrm{post}}(\theta) \approx c \exp\left(\frac{1}{2}\ddot{l}(\theta^*)(\theta - \theta^*)^2\right) = c \exp\left(-\frac{1}{2}((\theta - \theta^*)^2/(\sigma_{\mathcal{E}}^2)^*)\right), \tag{6.35}$$

where c is just the normalizing constant and we have that $f^{\mathrm{post}}(\theta) \approx N(\theta^*, (\sigma_{\mathcal{E}}^2)^*)$.   □

We now apply the result to the earthquake data.

Example 6.13 (Prediction of earthquake tomorrow). Continuation of Examples 6.8 and 6.9. In Example 4.6 62 recorded periods were given between serious earthquakes. The mean period was 437.2 days. In that example, it was demonstrated that the variability of periods between earthquakes can be adequately modelled by means of an exponentially distributed random variable X. The variable has pdf $f_X(x) = \theta e^{-\theta x}$, where $\theta = \lambda_A$ is the intensity of earthquakes.

Posterior distribution. In Chapter 4, $a = 1/\theta$ was used as a parameter of the exponential distribution. The ML estimate of a is $a^* = \bar{x}$. A similar derivation will lead to the ML estimate $\theta^* = 1/\bar{x}$ and $\ddot{l}(\theta) = -n/\theta^2$. Since $\bar{x} = 437.2$ we have that $\theta^* = 1/437.2 = 0.0023$, while

$$(\sigma_{\mathcal{E}}^2)^* = \frac{(\theta^*)^2}{n} = 8.4 \cdot 10^{-8}.$$

Consequently $f^{\mathrm{post}}(\theta) \approx N(0.0023, 8.4 \cdot 10^{-8})$.

Predictive probability. The predictive probability for a serious earthquake tomorrow is $P_t^{\text{pred}}(A) \approx t\,E[\Theta] = 0.0023$. This value should be compared with the predictive probability $P_t^{\text{pred}}(A) = 0.0076$, computed in Example 6.9, where only three observations of X were available to derive the posterior pdf. The new value is about three times smaller.

The predictive probability is an average value of $P_t(A) \approx t\,\Theta$ and the coefficient of variation is a simple measure of its variability. For the gamma posterior pdf used in Example 6.9 is found $R[\Theta] = 1/\sqrt{3} = 0.577$, while including 59 further observations gives

$$R[\Theta] = \frac{D[\Theta]}{E[\Theta]} = \frac{\sqrt{8.4 \cdot 10^{-8}}}{0.0023} = 0.126.$$

Comparison of posterior distributions. Finally we compare the "asymptotic" normal posterior pdf used in this example with the previously used gamma posterior pdf. Since the pdfs are very concentrated around their means let us first change units from days to years. Then $f^{\text{post}}(\theta) \approx N(0.0023 \cdot 365, 365^2 \cdot 8.4 \cdot 10^{-8})$, while, for $f^{\text{prior}}(\theta) = 1/\theta$ used in Example 6.9, the gamma posterior density has parameters $a = 62$ and $b = (437.2/365) \cdot 62 = 74.26$; $f^{\text{post}}(\theta) \in \text{Gamma}(62, 74.26)$. In Figure 6.4, left panel, the two posterior densities are compared. The solid line is the gamma prior while the dotted shows the asymptotically normal one. We can see that for this data set the posterior densities are very close and can be used equivalently.   □

Example 6.14 (Flight safety). Continuation of Example 6.12 in which a Bayesian method was used to measure uncertainty of the value of the intensity of flight crashes. The posterior density used was $\text{Gamma}(244.25, 10.25)$. The risk for crash during a time period of length t is measured by $P = t\,\Theta$ and

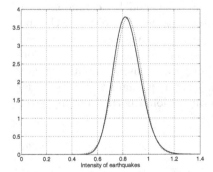
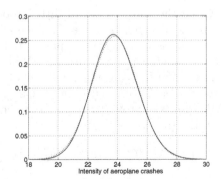

Fig. 6.4. *Left:* Comparison of posterior pdf for intensity of earthquakes Θ in Example 6.13. Solid line: Gamma($62, 74.26$) distribution. Dotted line: Asymptotic normal posterior pdf $N(0.8395, 0.0112)$. *Right:* Comparison of posterior pdf for intensity of airplane crashes Θ in Example 6.14 (right plot). Solid line: Gamma($244.25, 10.25$) distribution. Dotted line: Asymptotic normal posterior pdf $N(23.8, 2.38)$.

hence the predictive probability of at least one crash during a period of time t, measured in years, is approximately equal to $t \cdot 244.25/10.25$.

The posterior density reflects our experience of the number of crashes during the decade 1976–1985. Since crashes are not rare events the ten years of observations seem to constitute a large data set for which the asymptotic normality of the posterior density should be applicable. We investigate this claim next.

First, in order to use Eq. (6.33) we need to recall that if the intensity of crashes θ were known then the number of crashes during different years is independent, $Po(\theta)$ distributed variables. Let $x_1, \ldots, x_{10}$ be the observed number of crashes during the years 1976–1985. Then the ML estimate of θ is $\theta^* = \sum x_i/n$ and $\ddot{l}(\theta^*) = -n^2/\sum x_i$, $n = 10$ (see Example 4.8 for computational details). Hence the posterior density is approximately $N(\theta^*, -1/\ddot{l}(\theta^*))$ and since $\sum x_i = 238$ we have that the posterior density is $N(23.8, 2.38)$.

The posterior densities Gamma(244.25,10.25) and $N(23.8, 2.38)$ are compared in Figure 6.4, right panel, where one can see that the densities almost coincide.

Next, using the normal posterior density, the 95%-credibility interval for the probability of at least one accident in the period $t = 1$ day is computed:

$$[t(23.8 - 1.96 \cdot \sqrt{2.38}), \ t(23.8 + 1.96 \cdot \sqrt{2.38})] = [0.057, 0.074].$$

As expected the interval is almost identical to the one derived in Example 6.12 using the gamma posterior pdf. However, the normal posterior pdf is more convenient to use than the gamma. For example the quantiles of the normal variable are given in any textbook while, in order to get quantiles of the gamma distributed variable, a dedicated software is needed.   □

Distributions dependent on more than one parameter

Often the chosen class of distributions $F(x; \theta)$ to model the random variable X depends on more than one parameter; *e.g.* normal, Weibull, Gumbel distributions all have two parameters $\theta = (\theta_1, \theta_2)$. Then also $\Theta = (\Theta_1, \Theta_2)$ is a two-dimensional variable[4]. What can be said about the posterior density $f^{\text{post}}(\theta_1, \theta_2)$ as the number of observations n increases? Actually, similar results as for the one-dimensional situation are true, namely the posterior density is approximately equal to the two-dimensional normal density given in Eq. (5.5):

$$\Theta = \text{AsN}(\theta_1^*, \theta_2^*, (\sigma_{\mathcal{E}_1}^2)^*, (\sigma_{\mathcal{E}_2}^2)^*, \rho_{\mathcal{E}_1 \mathcal{E}_2}^*), \quad \Theta \approx \theta^* + \mathcal{E}, \tag{6.36}$$

as $n \to \infty$. Here the estimates of variances $(\sigma_{\mathcal{E}_1}^2)^*$, $(\sigma_{\mathcal{E}_2}^2)^*$ and the correlation $\rho_{\mathcal{E}_1 \mathcal{E}_2}^*$ can be computed using Eqs. (5.13-5.15). The asymptotic normality of

---

[4]The probability-density function and some other properties of two-dimensional variables were introduced in Chapter 5.

$f^{\mathrm{post}}(\theta)$ was already known by Laplace in 1810 [45] and was rigorously proved by Le Cam in 1953 [46].

In Chapter 4, the variable $\mathcal{E}$ was used to model the estimation error due to finite sample size. We have emphasized that the estimate and the error distribution completely describe the uncertainty of θ^*. In a Bayesian analysis the uncertainty is modelled by assuming that the parameter is an outcome of a random variable Θ. Now when a large number of data are available one has that approximately $\mathsf{E}[\Theta] = \theta^*$, while the deviation from the mean $\Theta - \mathsf{E}[\Theta]$ has the same distribution as $\mathcal{E}$. Consequently, in this situation, the classical inference and the Bayesian one give similar answers.

6.7 Predicting Frequency of Rare Accidents

In the previous section we discussed the situation when many data are available and the data dominate the priors, *i.e.* the posterior density becomes proportional to the likelihood function. In this section we discuss the diametrically different situation when the observations are very few. What is meant by "few" may depend on the safety level required, as illustrated by an example concerning safety of transports of nuclear fuel waste. The problem is discussed by Kaplan and Garrick [42].

Example 6.15 (Transport of nuclear fuel waste). Spent nuclear fuel is transported by railroad. From historical data, one knows that there were 4 000 transports without a single release of radioactive material. Since fuel waste is highly dangerous, one has discussed the possibility of constructing a special (very safe and expensive) train to transport the spent fuel.

One problem was the definition of an acceptable risk p^{acc} for an accident, *i.e.* one wishes the probability of an accident θ, say, to be smaller than p^{acc}. Since θ is unknown and uncertainty of its value is modelled by a random variable Θ the issue is to check, on basis of available data and experience, whether the probability $\mathsf{P}(\Theta < p^{\mathrm{acc}})$ is high.

A number between 10^{-8} and 10^{-10} was first proposed for p^{acc}, *i.e.* the average waiting time for an accident is 10^8 to 10^{10} transports (mean of geometric distribution). In such a scale the experienced 4000 safe transports look clearly negligible and hence the conclusion was: if one wishes to transport the waste with the required reliability, one needs to develop transport systems with maximum reliability.

We turn now to the problem of how the information about 4 000 transports affects our belief about the risk for accidents. Suppose accidents happen independently. Then[5]

$$\mathsf{P}(\text{"No accidents for 4\,000 transports"} \mid \Theta = \theta) = (1 - \theta)^{4000} \approx \mathrm{e}^{-4000\,\theta},$$

---

[5]Here we use that for small θ, $\theta \approx 1 - \mathrm{e}^{-\theta}$. See also Remark 6.5 on computation of the probability of no accidents in n transports, $(1-\theta)^n$, for small θ and large n.

and the posterior density $f^{\text{post}}(\theta) = cf^{\text{prior}}(\theta)e^{-4000\,\theta}$ will be close to zero for any reasonable choice of the prior density and $\theta > 10^{-3}$. This agrees with the conclusion of Kaplan and Garrick that the information of 4000 release-free transport is quite informative:

> "The experience of 4000 release-free shipments is not sufficient to distinguish between release frequencies of 10^{-5} or less. However, it is sufficient to substantially reduce our belief that the frequency is on the order of 10^{-4} and virtually demolish any belief that the frequency could be 10^{-3} or greater".

$\square$

Remark 6.5. We investigate here a technique for calculations of probabilities, which is useful in applications, where events are studied that are unlikely to happen, but the *exposure* to the risk is long.

Assume that the risk for an accident is $1/1000$ and that we expose ourselves for the risk 1000 times. Then the probability that no accident will happen is

$$\left(1 - \frac{1}{1000}\right)^{1000} \approx e^{-1}, \quad \text{since} \quad \lim_{n\to\infty}\left(1 - \frac{a}{n}\right)^{n} = e^{-a}. \tag{6.37}$$

Finally, if we require a safety level of 10^{-8}, then the chance for accidents in the first 4000 transports is simply

$$1 - \left(1 - \frac{1}{10^{8}}\right)^{4000} \approx 1 - \left[e^{-1}\right]^{4000/10^{8}} \approx 4000/10^{8} = 4 \cdot 10^{-5}, \tag{6.38}$$

i.e. negligible.
$\square$

Streams of initiation events and scenarios

In Example 6.15 we studied the problem of estimation of the frequency λ, say, of very rare accidents. In such cases a direct estimation of frequencies is difficult, if it is even possible, because the period when data are gathered usually very short compared to the return period $T = 1/\lambda$. Similar problems occur in the evaluation of risk for failure of existing structures; *e.g.* collision of a ship with a particular bridge, an aeroplane crashing into a nuclear power plant, etc. Here accidents are not repeatable, since when these happen new safety measures are often introduced, changing λ. In both situations, in order to overcome shortness of data, system analysis is often performed in the form of events and/or failure trees.

We do not go further into this matter and consider only the simplest case, introduced in Section 2.5: We refer to accidents as occurrences of an *initiation event* A, with intensity λ_A, followed by an event B describing the *scenario* leading to a hazard. Consequently, an accident happens when

A and B occur simultaneously. If B is independent of the stream of A, which is often assumed, then the intensity of accidents $\lambda = \lambda_A P(B)$ (cf. Eq. (2.10)).

The risk of an accident is often measured by means of $P_t(A \cap B)$, the probability of at least one accident in $t = 1$ year. Often the acceptable risk p^{acc}, say, is 10^{-2} or smaller, dependent on the consequences an accident might have. (In Chapter 8 we will further discuss the choice of the values of p^{acc}.) Since

$$P_t(A \cap B) \leq t \lambda_A P(B) = p, \tag{6.39}$$

say, p is a conservative estimate of the risk. In the case where p is small and two accidents cannot happen simultaneously, we also have that $P_t(A \cap B) \approx t \lambda_A P(B)$. Consequently p is often used as a measure of risk.

Predictive probability for a single stream of initiation events

Since the intensity λ_A is unknown, we can model it as a random variable Θ_1, having a gamma pdf if conjugated priors are used. Moreover, $P(B)$, denoted by Θ_2, has a beta pdf as conjugated prior. As B is independent of the stream A, we also assume that Θ_1 and Θ_2 are independent and the unknown intensity of accidents $\Theta = \Theta_1 \Theta_2$. The reason for such a decomposition is that more data may be rendered available to update the prior densities $f(\theta_1)$ and $f(\theta_2)$. The probability of at least one accident in a period t is thus given by

$$P_t(A \cap B) \approx P = \Theta_1 \Theta_2 t. \tag{6.40}$$

The predictive probability is then approximated by

$$P_t^{\mathrm{pred}}(A \cap B) \approx E[P] = E[t\Theta_1\Theta_2] = tE[\Theta_1]E[\Theta_2]. \tag{6.41}$$

A measure of the precision of the estimate is given by the coefficient of variation of $\Theta = \Theta_1\Theta_2$. This is evaluated easily by assumed independence of parameters Θ_1, Θ_2. Note first that

$$R[P] = \frac{\sqrt{V[t\Theta_1\Theta_2]}}{E[t\Theta_1\Theta_2]} = \frac{\sqrt{V[\Theta_1\Theta_2]}}{E[\Theta_1\Theta_2]}$$

and hence, since $V[X] = E[(X - E[X])^2] = E[X^2] - E[X]^2$,

$$R[P] = \frac{\sqrt{E[\Theta_1^2]E[\Theta_2^2] - E[\Theta_1]^2 E[\Theta_2]^2}}{E[\Theta_1]E[\Theta_2]} = \sqrt{\frac{E[\Theta_1^2]}{E[\Theta_1]^2}\frac{E[\Theta_2^2]}{E[\Theta_2]^2} - 1} \tag{6.42}$$

$$= \sqrt{(R[\Theta_1]^2 + 1)(R[\Theta_2]^2 + 1) - 1} \tag{6.43}$$

Example 6.16 (Fire ignition). In a cinema the exit doors are checked once a month to insure that they work properly. Suppose that in the last 5 years a fire has started twice in the cinema. Additionally, no malfunctions of exit doors were filed during this period. On the basis of this information, we give a measure of risk for at least one incident, that is "fire ignition in the cinema and not all exit doors can be opened" during a period of one year ($t = 1$).

Suppose that no information about the fire intensity for the particular cinema is available and hence improper priors $1/\theta_1$ will be used for Θ_1. The information of 2 fires in 5 years will be included in the priors leading to the posterior density $f^{post}(\theta_1) \in \text{Gamma}(2, 5)$. The unknown probability $\Theta_2 = P(B)$ will have a uniform prior and hence the posterior density $f^{post}(\theta_2) \in \text{Beta}(1, 12 \cdot 5 + 1)$.

Let us now approximate with P the probability of at least one serious accident in period of length t, $P_t(A \cap B)$. Using Eq. (6.11) and Eq. (6.17) we have that $E[\Theta_1] = 2/5$ while $E[\Theta_2] = 1/(12 \cdot 5 + 2) = 1/62$ and hence

$$E[P] = tE[\Theta_1]E[\Theta_2] = 1 \cdot \frac{2}{5} \cdot \frac{1}{62} = 0.0065.$$

$\square$

Problems

6.1. A beta-distributed r.v. Θ has the density function

$$f(\theta) = c\theta^{a-1}(1-\theta)^{b-1}, \quad 0 \le \theta \le 1,$$

where c is a normalization constant. Show by direct calculation that in the special case of parameters $a = b = 1$, we obtain a uniform distribution.

6.2. A gamma distributed r.v. Θ has the density function

$$f(\theta) = c\theta^{a-1}e^{-b\theta}, \quad \theta \ge 0,$$

where c is a normalization constant. Show by direct calculation that in the special case of $a = 1$, we obtain an exponential distribution.

6.3. Detection of possible leakages at sections in a pipeline system is performed by some specialized equipment. One wants to study the intensity of faults per km. A suggested prior distribution for this intensity is $\text{Gamma}(1, 100)$.

(a) What is the expected value of the prior distribution?
(b) The examination starts, and 12 imperfections are found along 500 km of pipeline. Find the posterior distribution.
(c) Find the average number of imperfections per km, as given by the posterior distribution. Compare with your answer in (a).

6.4. Time intervals between successive failures of the air-conditioning system for a fleet of Boeing 720 planes have been recorded, see Proschan [63]. The data below consider plane 7914 (times in hours):

$$50 \quad 44 \quad 102 \quad 72 \quad 22 \quad 39 \quad 3 \quad 15 \quad 197 \quad 188 \quad 79 \quad 88$$
$$46 \quad 5 \quad 5 \quad 36 \quad 22 \quad 139 \quad 210 \quad 97 \quad 30 \quad 23 \quad 13 \quad 14$$

Assume that the times T between failures of components are exponentially distributed, i.e.

$$P(T > t) = e^{-\lambda t}, \quad t > 0.$$

The intensity λ is unknown and will be modelled by means of an r.v. Λ.

(a) Use the data set to derive the posterior density for Λ. *Hint.* Use Example 6.13.
(b) Suppose we are interested in the probability that the air-conditioning system will work for longer than 24 hours, $p = P(T > 24) = \exp(-24\lambda)$. Compute the predictive probability $\mathsf{P}^{\mathrm{pred}}(T > 24) = \mathsf{E}[P]$, where $P = \exp(-24\Lambda)$. *Hint.* Check that P is lognormally distributed.

6.5. Suppose the waiting time T (in minutes) for you to get in contact at a calling centre for traffic information is exponentially distributed as follows:

$$F_T(t) = 1 - e^{-\lambda t}, \quad t > 0.$$

Based on previous experience, one suggests a Gamma(1,15) distribution for the intensity Λ.

(a) Suppose we started with uniform improper priors. What does the Gamma(1,15) distribution mean, in terms of experience of waiting?
(b) One has observed the following waiting times: 10 minutes, 5 minutes, and 2 minutes. Based on these observations, update the prior distribution — in other words, calculate the posterior distribution for Λ.
(c) Find the expected value of the posterior distribution.
(d) Suppose we are interested in the probability of waiting for a time period longer than t ($t = 1, 5, 10$ min), that is, $p = P(T > t) = \exp(-\lambda t)$. Compute the predictive probability $\mathsf{P}^{\mathrm{pred}}(T > t) = \mathsf{E}[P]$, where $P = \exp(-\Lambda t)$. *Hint.* Make use of Eq. (6.19).

6.6. A man plays five times on an automatic gaming machine, and, surprisingly, he wins every time. Let p denote the probability that the player will win in a single game.

(a) What is the classical estimate of p?
(b) Adopt now a Bayesian attitude and model the parameter p as a random variable P. Assume that the prior distribution of P is uniform (continuous distribution). What is the posterior distribution of P?
(c) What is the predictive probability to win next time?

6.7. The famous boat race between teams from the universities in Cambridge and Oxford was premiered in 1829. In 2004, the 150th race took place on the Thames.

(a) Suppose you have no idea about the capacities of the teams. Suggest a suitable Dirichlet prior.
(b) From the start and up to 2004, Cambridge won 78 times and Oxford 71. Over the years, there has been one dead heat (in 1877). Update your prior density using this information.
(c) Calculate the probability that Oxford will win the next race.

6.8. In a mine, drainage water and subsoil water is stored in a dam, to be treated before it is dumped in a nearby river. Unfortunately, now and then the dam will flood and then untreated, pollutant water is released in the river.

Suppose the instants for such releases are described by a Poisson process with unknown (but constant) intensity λ. Releases occur on average once in four years and the uncertainty of Λ is given by the coefficient of variation $R[\Lambda] = 2$.

(a) Choose an appropriate Gamma prior and estimate the risk for at least one release in 6 months.
(b) During a two-year period, one flooding of the dam, leading to release of dangerous water, occurred. Use Bayes' formula to update the probability distribution of λ and compute the predictive probability of flooding in 6 months.

6.9. Recall Example 6.12 on flight safety. By including the information of the observed crashes in 1976–1985, the posterior density $f^{\text{post}}(\theta) \in \text{Gamma}(244.25, 10.25)$ was found, where θ is the intensity of crashes in unit year^{-1}.

Use the result derived in Remark 6.2 to compute the predictive probability of no crashes during a one-week period. Compare this probability with the one obtained by the approximation in Eq. (6.28).

6.10. Assume that T is exponentially distributed, i.e.

$$P(T \leq t) = 1 - e^{-\lambda t}, \quad t > 0$$

where $\lambda = 1/E[T]$ is an unknown constant. Suppose there are n independent observations $t_1, t_2, \ldots, t_n$. Demonstrate that the gamma distribution is conjugated prior for λ. *Hint.* See Example 6.9.

6.11. Suppose the number of perished in motorcycle accidents (see Problem 4.9) is Poisson distributed with mean m.

(a) Using results from asymptotic theory, give the posterior density for m. *Hint.* See Example 6.14.
(b) Give the 0.95-credibility interval for m.

6.12. In this problem we discuss again accidents with tank trucks (cf. Problem 2.13). Suppose we want to evaluate the risk of a traffic accident involving tank trucks in the Swedish region of Dalecarlia for one day, say, tomorrow. Denote this event by C.

(a) Suppose that your experience is quite vague and can be summarized that no accidents have been observed the last month. Compute the predictive probability of C and give a measure of the uncertainty by means of the coefficient of variation. *Hint.* Use uniform improper priors, Gamma(1,0).
(b) Suppose in years 2002-2004 2, 0, and 2 accidents were observed. Update the prior density and recompute $E[P]$, $R[P]$.
(c) In order to increase the precision of the derived probability, one plans to use data from Problem 2.13. Perform the analysis and compute $P^{\text{pred}}(C)$, $R[P]$. *Hint.* Make use of Eq. (6.42).

7

Intensities and Poisson Models

In this chapter we return to a great extent to the notions of intensities. In the first section, the failure intensity is introduced; this gives the distribution of the waiting time to the first event. This intensity is of particular interest when lifetimes (of components or humans) are considered. Estimation procedures and statistical problems are discussed. Relative risks and risk exposure are the main topics of Section 7.2. In Section 7.3, models for Poisson counts are considered, leading to an introduction to the often-used Poisson regression models. This makes modelling possible of situations where the relation between an intensity and some explanatory, non-random variables, is given by a regression equation.

In Section 7.4, we introduce the notion of Poisson point process (PPP), an extension of Poisson streams discussed in Chapter 2. This enables modelling of events that can occur in spatial locations or at space and time locations, discussed in Section 7.5. Finally, we study superposition and decomposition of Poisson processes.

7.1 Time to the First Accident — Failure Intensity

7.1.1 Failure intensity

Before presenting new notions, let us revisit Example 4.1 (lifetimes of ball bearings) to analyse refined probabilistic modelling of lifetimes.

Example 7.1 (Lifetimes of ball bearings). In safety analysis, studies are often made of data of a type describing time to the first occurrence of an event. Time can sometimes be measured in rather strange units like the number of revolutions to failure, if lifetimes of ball bearings are studied (cf. Example 4.1, where an experiment with 22 observed lifetimes was presented).

An important issue is obviously to find a suitable distribution to describe the variability of lifetimes. In Example 4.1, the data were described using the empirical distribution F_n, while in Example 9.1 a Weibull distribution will be used to model variability of lifetimes. In this chapter we introduce another (equivalent) means to describe data: the so-called *failure intensity* $\lambda(s)$. The intensity measures risk for failure of a component of age s. For example, consider the risk that a ball bearing that has been used for 30 millions of revolutions will break in the next one. If the risk for failure increases with age, which is the case with ball bearings, then we say that lifetime of ball bearings has increasing failure rate (IFR).

□

Let T denote a waiting time for the first failure (accident, death, etc.) Suppose the value of T cannot be predicted and hence is modelled as a random variable. Let $F(t) = \mathsf{P}(T \leq t)$ be the probability that the failure happens in the interval $[0, t]$. One sometimes speaks about the *survival function*

$$R(t) = \mathsf{P}(T > t) = 1 - F(t)$$

which can equivalently be used to describe statistical properties of the lifetime.

The properties of the distribution $F(t)$ (or survival function $R(t)$) are often discussed in safety analysis, where failure times (life times) of components or structures are of interest. In such analysis, one is not only limited to failures that can be traced to accidents caused by environmental actions but also can be related to wear and other ageing processes. The distribution $F(t)$ may also reflect variability of quality (or strength) in some population of components: an element is chosen randomly from a population and then the lifetime of the chosen element is observed. Generally, any r.v. taking only positive values can be a model for the lifetime of members in some population.

Next we introduce a very important characterization of T called the *failure-intensity function*, (for short, the failure intensity), alternatively, the *hazard function*.

Definition 7.1 (Failure-intensity function). *For an r.v. $T \geq 0$ there is a function $\Lambda(t)$, called the **cumulative failure-intensity function**, such that*

$$R(t) = \mathrm{e}^{-\Lambda(t)}, \quad t \geq 0.$$

If T has a density, then

$$R(t) = \exp\left(-\int_0^t \lambda(s)\,\mathrm{d}s\right)$$

*where the function $\lambda(s) = \frac{\mathrm{d}\,\Lambda(s)}{\mathrm{d}\,s}$ is called the **failure-intensity function**.*

The failure-intensity function defines the distribution of T uniquely. If the distribution is of the continuous type, the failure intensity can also be calculated by

$$\lambda(s) = \frac{f(s)}{1 - F(s)}. \tag{7.1}$$

It can also be demonstrated that

$$\lambda(s) = \lim_{t \to 0} \frac{P(T \le s + t \mid T > s)}{t},$$

which means that for small values of t, $\lambda(s) \cdot t$ is approximately the probability that an item of age s will break within the period of time t.

Generally, failure-intensity functions are classified as *IFR* (increasing failure rate), where components wear with time, or *DFR* (decreasing failure rate) where the weak components fail first so the ones that rest are the strongest: consequently failures occur less frequently. Often both mechanisms are present simultaneously and we observe an increasing failure rate for the old components due to damaging processes. This is often experienced by owners of old cars. In the following somewhat artificial example the IFR and DFR failure intensities are given.

Example 7.2 (Strength of a wire). Suppose the strength R of a particular wire is modelled as a Weibull distribution, that is, with a distribution function

$$F_R(r) = 1 - e^{-(r/a)^c}, \quad r \ge 0.$$

The wire is used under water and is exposed to a load increasing with time, due to growth of the organic material attached to its surface. The rate of growth is considered constant, γ; hence, during a period of length t, the load has increased by γt (the initial weight is neglected).

At the lifetime T, when the weight exceeds the strength, obviously $R = \gamma T$ or equivalently, $T = R/\gamma$. Hence the lifetime distribution is given by

$$F_T(t) = P(T \le t) = P(R/\gamma \le t) = F_R(\gamma t) = 1 - e^{-(\gamma t/a)^c}.$$

Since $R(t) = e^{-(\gamma t/a)^c}$, the cumulative failure-intensity function $\Lambda(t) = (\gamma t/a)^c$ and hence $\lambda(t) = c\frac{\gamma}{a}(\gamma t/a)^{c-1}$. Suppose that in some units, $a = 1$ and $\gamma = 0.1$. For different choices of the shape parameter c, the failure-intensity function is presented in Figure 7.1; from top to bottom $c = 0.8$, $c = 1.0$, and $c = 1.2$. Note that, depending on the choice of c, the function might be classified as IFR, DFR, or have a constant failure intensity.   □

In the previous examples failure intensity described the properties of populations of some components. Principally, it was used to model the uncertainty of properties like quality or strength of a component. A different situation is met in the following example.

Example 7.3 (Constant failure intensity). Consider periods in days between serious earthquakes worldwide (presented in Example 1.1). This data set was investigated in many aspects in Chapter 4. Now assume that we at some fixed date s (say, today) start counting the time until the next earthquake. As in Chapter 2, we thus consider a stream of events with intensity λ_A

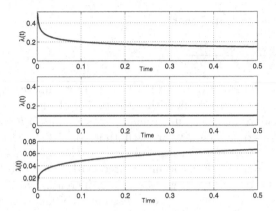

Fig. 7.1. Failure-intensity functions, Weibull distribution. From top to bottom, $c = 0.8$, $c = 1.0$, $c = 1.2$.

and the event $A=$"Earthquake occurs". Recall the properties I-III in Chapter 2 (page 40), which we assume to be satisfied in our situation; then Theorem 2.5 gives

$$P(T > t) = P(N(s, \, s+t) = 0) = e^{-\lambda_A t}$$

and hence $\Lambda(t) = \lambda_A t$, giving by differentiation $\lambda(t) = \lambda_A$. Thus, the failure-intensity function is *constant* and equal to the intensity of the stream.

If the intensity of events A is *non-stationary*, $\lambda_A(s)$ say, then similar calculations lead to the failure-intensity function $\lambda(t) = \lambda_A(s+t)$. $\quad\square$

Often one is interested in whether the time to failure is longer than t, if we know the age of the component to be t_0, say, *i.e.* we wish to compute the probability $P(T > t_0 + t \, | \, T > t_0)$. This conditional probability can be easily computed if the failure intensity is known:

$$P(T > t_0 + t \, | \, T > t_0) = \frac{P(T > t_0 + t \text{ and } T > t_0)}{P(T > t_0)}$$

$$= \frac{P(T > t_0 + t)}{P(T > t_0)} = \frac{e^{-\int_0^{t_0+t} \lambda(s)\,ds}}{e^{-\int_0^{t_0} \lambda(s)\,ds}}$$

$$= e^{-\int_{t_0}^{t_0+t} \lambda(s)\,ds}. \tag{7.2}$$

Note that for $\lambda(s) = \lambda$ (being constant),

$$P(T > t_0 + t \, | \, T > t_0) = e^{-\lambda t},$$

that is, old components have the same distribution for their remaining life as new ones. This is sometimes stated as "memorylessness" of the exponential distribution.

We exemplify the use of Eq. (7.2) with an example from life insurance where the "components" are humans. The variability of lifetimes, and hence the failure intensity, depends on the choice of population. For example when considering the lifetimes of inhabitants in two countries, there can be different diseases common for each country, habits of smoking, traffic situation, frequency of catastrophes like earthquakes, etc. which leads to different functions $\lambda(s)$.

Example 7.4 (Life insurance). Let T be a lifetime for a human. In life insurance, $P(T > t)$ is specified by standard tables, based on observed lifetimes of a huge number of people; one example is the Norwegian N-1963 standard. A popular choice of $\lambda(s)$ is the Gompertz-Makeham distribution (with roots to Makeham [53]), given by the failure-rate function

$$\lambda(s) = \alpha + \beta c^s,$$

s measured in years. For example, for N-1963, the estimates are

$$\alpha^* = 9 \cdot 10^{-4}, \quad \beta^* = 4.4 \cdot 10^{-5}, \quad c^* = 10^{0.042}.$$

We want to solve the following problems:

(i) Calculate the probability that a person will reach the age of at least seventy.
(ii) A person is alive on the day he is thirty. Calculate the conditional probability that he will live to be seventy.

For problem (i), we obtain the solution as

$$P(T > 70) = \exp\left\{-\int_0^{70} \lambda(s)\,ds\right\} = 0.63.$$

The solution to problem (ii) is given by Eq. (7.2) as

$$P(T > 70 | T > 30) = \exp\left\{-\int_{30}^{70} \lambda(s)\,ds\right\} = 0.65.$$

$\square$

Combining different risks for failure

In real life, there are often several different types of risks that may cause failures; one speaks of different failure modes. Each of these has an intensity $\lambda_i(s)$ and a lifetime T_i. We are interested in the distribution of T: the time instant when the first of the modes happen. If T_i are independent then the event $T > t$ is equivalent to the statement that all lifetimes T_i exceed t, i.e. $T_1 > t, T_2 > t, \ldots, T_n > t$ and hence

$$P(T > t) = P(T_1 > t) \cdot \ldots \cdot P(T_n > t) = e^{-\int_0^t \lambda_1(s)\,ds} \cdot \ldots \cdot e^{-\int_0^t \lambda_n(s)\,ds}$$
$$= e^{-\int_0^t \lambda_1(s)\,ds - \ldots - \int_0^t \lambda_n(s)\,ds} = e^{-\int_0^t \lambda_1(s) + \ldots + \lambda_n(s)\,ds} \tag{7.3}$$

which means that the failure intensity, including the n independent failure modes, is $\lambda(s) = \sum \lambda_i(s)$.

Remark 7.1. In the special case when the failures can be related to external actions (accidents causing failures) constituting independent streams A_i, each with constant intensity λ_i, Eq. (7.3) was already derived in Chapter 2. There, a stream $A = A_1 \cup \ldots \cup A_n$ was considered, with the interpretation that at least one of A_i happens. The intensity λ_A is equal to the sum of intensities λ_i, (see Eq. (2.9)). If the streams are Poisson then the stream A is also Poisson (see Theorem 7.1, p. 188), and hence

$$\mathsf{P}(T \le t) = 1 - e^{-\lambda_A t},$$

i.e. T is exponentially distributed with intensity $\lambda_A = \lambda_1 + \cdots + \lambda_n$.   □

7.1.2 Estimation procedures

Earlier in this chapter, we have introduced the notions failure intensity and survival function when studying the distribution of the time T to failure for some items. In this section we discuss how these functions can be estimated from data. Obviously, a standard (parametric) method is to assume that $F(t)$ belongs to a class of distributions $F(t; \theta)$, estimate parameters, and finally calculate $\lambda(s)$. We here instead present a *non-parametric method*, commonly used in applications with lifetime data.

In reliability studies as well as in clinical trials in the medical sciences, it is not always possible to wait for all units to reach their final "lifetime" (lifetime could mean time for failure, or death, or the appearance of a certain condition). An intricate issue is that *censored* data may occur; for example, an item may not have reached its lifetime until the study is finished or is lost during the time (*e.g.* people move). Efficient estimation procedures need to take censoring aspects into account.

In this section, we review some commonly used tools within statistical analysis of survival or reliability data: the Nelson–Aalen estimator for estimation of the cumulative failure-intensity function $\Lambda(t)$, and the log-rank test for testing hypotheses about the failure-intensity functions of two samples. For further reading, we refer to Klein and Moeschberger [43] where a thorough presentation of methods in survival analysis is given.

Nelson–Aalen estimator

This estimator was first presented by Nelson [57] and later refined by Aalen [1]. It estimates the cumulative failure-intensity function

$$\Lambda(t) = \int_0^t \lambda(s) \, ds.$$

The Nelson–Aalen estimator is considered to have good small-sample performance, *i.e.* when n is small, when estimating the survival function.

Introduce the following notation:

t_i: Time points for failures
d_i: Number of failures at time t_i
n_i: Number of items at risk at time t_i, *i.e.* number of items not yet failed
   prior to failure time t_i.

The estimator is given by

$$\Lambda^*(t) = \sum_{t_i \leq t} \frac{d_i}{n_i} \qquad (7.4)$$

and thus $R^*(t) = \exp(-\Lambda^*(t))$. (Note that $R^*(t) \neq 1 - F_n(t)$.)

If censoring is present, the values of n_i will be affected, leading to a change
in the value of the estimated survival function.

Example 7.5 (Cycles to failure). In an experiment, the number of cycles
to failure for reinforced concrete beams was measured in seawater and air [37].
The observations (in thousands) were as follows:

Seawater: 774   633   477   268   407   576   659   963   193
Air: 734   571   520   792   773   276   411   500   672

Parametric model. A Weibull distribution is often used to model the strength
of a material, and plots of the observations in Weibull probability paper indicate that Weibull might be a good choice. With

$$F_T(t) = 1 - e^{-(t/a)^c}, \quad t \geq 0$$

one finds by statistical software the ML estimates $a^* = 620$, $c^* = 2.63$ for
seawater conditions. Based on these, an estimate of the cumulative failure-
intensity function can easily be computed and is shown as the solid curve in
Figure 7.2.

Non-parametric model. The following table gives the Nelson–Aalen estimate
of the cumulative failure-intensity for seawater (creation of the corresponding
scheme for air is left as an exercise):

i	t_i	n_i	d_i	$\Lambda^*(t_i)$
1	193	9	1	0.1111
2	268	8	1	0.2361
3	407	7	1	0.3790
4	477	6	1	0.5456
5	576	5	1	0.7456
6	633	4	1	0.9956
7	659	3	1	1.3290
8	774	2	1	1.8290
9	963	1	1	2.8290

In Figure 7.2, the Nelson–Aalen estimate is shown (the stair-wise function).
From the plot it can be judged that we have a case, which can be considered
IFR.   □

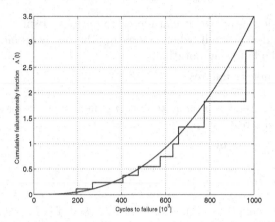

Fig. 7.2. Cumulative failure-intensity function, concrete beams in sea water. *Curve:* Weibull estimate. *Stair-wise function:* Non-parametric Nelson–Aalen estimate.

Log-rank test

Finally, we present a statistical test, called the *log-rank test*, for comparison of the intensities $\lambda_1(t)$ and $\lambda_2(t)$ in two groups (1 and 2). The aim is to test the hypothesis

$$H_0: \quad \lambda_1(t) = \lambda_2(t).$$

The test can be generalized to more than two groups, but we content ourselves in this exposition to the simplest case and refer to the literature for more specialized studies. (Note that two groups can have different number of elements.)

Consider the time points for failures $t_1, t_2, \ldots, t_D$, both groups considered. Introduce the following notation:

d_{i1}: Number of failures in group 1 at times t_i

d_{i2}: Number of failures in group 2 at times t_i

d_i: $d_i = d_{i1} + d_{i2}$

n_{i1}: Number of items in group 1 at risk at time t_i, *i.e.* number of items not yet failed prior to failure time t_i.

n_{i2}: Number of items in group 2 at risk at time t_i, *i.e.* number of items not yet failed prior to failure time t_i.

n_i: $n_i = n_{i1} + n_{i2}$

The test quantity is

$$Q = \frac{1}{s^2} \left(\sum_{i=1}^{D} d_{i1} - \sum_{i=1}^{D} d_i \frac{n_{i1}}{n_i} \right)^2$$

where

$$s^2 = \sum_{i=1}^{D} \frac{d_i}{n_i} \cdot \frac{n_i - d_i}{n_i} \cdot \frac{n_{i1} n_{i2}}{n_i - 1}.$$

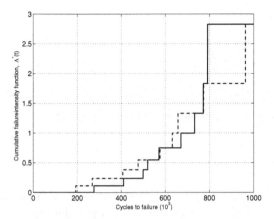

Fig. 7.3. Cumulative failure-intensity functions (Nelson–Aalen estimates). *Solid:* air; *Dashed:* seawater.

The test is similar to the χ^2 test and is as follows: If $Q \geq \chi^2_\alpha(1)$, reject H_0. (Note that since for $X \in N(0,1)$, $X^2 \in \chi^2(1)$; hence, $\chi^2_\alpha(1) = \lambda^2_{\alpha/2}$.)

Example 7.6 (Cycles to failure). Consider again the experiment mentioned in Example 7.5. In Figure 7.3, the cumulative failure-intensity functions are shown for air (solid) and seawater (dashed). Does seawater seem to lessen the number of cycles to failure, or in other words, can we reject the hypothesis that $\lambda_1(s) = \lambda_2(s)$, where group 1 corresponds to seawater conditions, group 2 to air?

From the 18 observations of lifetimes, the quantities needed are computed and presented in the following table:

t_i	n_{i1}	d_{i1}	n_{i2}	d_{i2}	n_i	d_i
193	9	1	9	0	18	1
268	8	1	9	0	17	1
276	7	0	9	1	16	1
407	7	1	8	0	15	1
411	6	0	8	1	14	1
477	6	1	7	0	13	1
500	5	0	7	1	12	1
520	5	0	6	1	11	1
571	5	0	5	1	10	1
576	5	1	4	0	9	1
633	4	1	4	0	8	1
659	3	1	4	0	7	1
672	2	0	4	1	6	1
734	2	0	3	1	5	1
773	2	0	2	1	4	1
774	2	1	1	0	3	1
792	1	0	1	1	2	1
963	1	1	0	0	1	1

From this table, we find

$$\sum_{i=1}^{18} d_{i1} = 9, \quad \sum_{i=1}^{18} d_i \frac{n_{i1}}{n_i} = 8.02,$$

and $s^2 = 4.16$. It follows that $Q = 0.23$. Hence $\chi^2_{0.05}(1) = 3.84$ and we do not reject the hypothesis about equal failure intensity.   □

7.2 Absolute Risks

In the previous section we introduced the concept of failure intensity $\lambda(s)$, which describes variability of lifelengths in a population of components, objects or human beings. Extensive statistical studies are needed to estimate $\lambda(s)$. More often, observed information is not sufficient to determine the failure intensity. In this section we consider situations when information about failures is less detailed: instead of knowing the times for failures t_i, access is available only to the total number; for example, failures during a specified period of time (or in a certain geographical region). Let us call failures "accidents", and suppose that these cause serious hazards for humans. *Absolute risk* is meant as the chance for a person to be involved in a serious accident (fatal), or of developing a disease, over a time period. Chances for accidents due to different activities are often compared. A full treatment of such issues is outside of the scope of this book and hence we only mention some aspects of the problem.

Poisson assumption

Let N be the number of deaths due to an activity, in a specified population (a country), and period of time (often one year). The distribution of N may not be easy to choose. For example, if N is the number of accidents that occur independently with small probability then N may have a Poisson distribution, $N \in \text{Po}(\mu)$, where $\mu = \text{E}[N]$. This is a consequence of the approximation of the binomial distribution by the Poisson distribution (the law of small numbers). For instance, it seems reasonable to model the number of commercial air-carrier crashes during one year by a Poisson variable. However, the number of people killed in those accidents is not Poisson distributed, since usually a large number of people are killed in a single accident. Since N, for different activities, can have different types of distributions, risks are often compared by means of averages. However, as demonstrated next, such comparisons have to be made with care.

Example 7.7 (Number of deaths in traffic). In year 1998 it was reported about 41 500 died in traffic accidents in the United States while in Sweden the number was about 500 [7]. In order to compare these numbers, one needs to compare the size of populations in both countries. A fraction of the numbers of deaths by the size of population, giving the frequencies of death, is often used to measure risk for death. In US the frequency was about 1 in 6000, circa $1.7 \cdot 10^{-4}$; while in Sweden, 1 in 17000, circa $0.6 \cdot 10^{-4}$, which is nearly three times lower. (Comparisons of chances to die in traffic accidents between countries can be difficult since statistics may use different definitions and have different accuracy.) □

The last example turns our attention to a problem often discussed in the literature of reliability and risk analysis, namely when risks are acceptable (or tolerable).

Tolerable risks

Often a distinction is made between the so-called "voluntary risks" and the "background risks". Clearly accidents due to an activity like mountaineering are obviously a voluntary risk, while the risk for death because of a collapse of a structure is an example of a background risk and is much smaller. (In United Kingdom one estimates that one hour of climbing has twice as high probability for a fatal accident than for a fatal accident in 100 years caused by structural failures, see table in [77].)

In the literature indicators of tolerable risks can be found, see *e.g.* Otway *et al.* [59]. The magnitudes of the risks specified in Table 7.1 are meant approximatively: the number of fatal accidents during a year divided by the size of the population exposed for the hazard. (Fatal accidents in traffic belongs to the second category of hazards.)

Example 7.8 (Perished in traffic). Continuation of Example 7.7. The estimated chances of dying in traffic in the U.S. was nearly three times as high as in Sweden. When looking for explanation for the difference, the first thing to be explored is the total exposure of the populations for the hazard, in other words if an average inhabitant of the U.S. spends more time in a car

Table 7.1. Indicators of tolerable risks.

Risk of death per person per year	Characteristic response
10^{-3}	Uncommon accidents; immediate action is taken to reduce the hazard
10^{-4}	People spend money, especially public money to control the hazard (*e.g.* traffic signs, police, laws)
10^{-5}	Parents warn their children of the hazard (*e.g.* fire, (drowning, fire arms, poison)
10^{-6}	Not of great concern to average person; aware of hazard, but not of personal nature; act of God.

than a person in Sweden does. For traffic-related accidents, exposure is often measured by total vehicle kilometres.

Neglecting that the exposures are estimates and hence uncertain numbers, we found that in 1998 the intensity $\lambda = \mathsf{E}[N]/t$ in the U.S. was about 1 person per 100 million km driven, while in Sweden, λ is 1 per 125 million km. The conclusion is that a person who drives 0.01-million km during one year has a chance of the order 10^{-4}, $0.8 \cdot 10^{-4}$, respectively, of dying as a result of traffic accidents in both countries. In other words, the chances are quite similar.   □

In our setup the absolute risk was derived for an average member of the population (a person chosen at random). However, the natural question is whether the same risk is valid for some subpopulations: geographical, stratified by age, income, etc. We return to this kind of question when Poisson regression is presented. Here we end with an example where we compare risk for fire in an average school in a country as a whole compared with a school in a smaller urban region.

Example: Intensity of fire ignitions in schools in Sweden

In published statistical tables ([74], [76]) one can find that in 2002 there were $k = 13\,053$ educational buildings in Sweden and $n = 422$ fires were recorded. (We ignore the fact that these two numbers are uncertain, taken from different statistical tables.) As is common practice in fire safety, the assumption is made that the stream of fire ignitions in a school is Poisson, and that ignitions in different schools happen independently (see examples in Chapter 2).

Constant intensity

The simplest approach is to assume that intensities in all schools are constant and equal to λ (per school). As derived in Example 4.8, the ML estimate is

$$\lambda^* = \frac{n}{k} = \frac{422}{13\,053} = 0.032 \quad [\text{year}^{-1}].$$

Now the probability of at least one fire in a school in three years, $P_t(A)$, $t = 3$ years, can be estimated as

$$P_t(A) = 1 - e^{-\lambda^* t} = 1 - e^{-0.097} = 0.092.$$

The expected number of fires in an average school during a three-year period is found to be $\mu = 3 \cdot 0.032 = 0.096$.

Validation of the model: Schools in Stockholm

Here we use a small data set presented in [69] and further analysed in [68]. Data contain the number of fires n_i for 20 schools in Stockholm, Sweden.

These have been chosen, at random, from Stockholm Fire Department files containing reports from actions in 2000-2002. The number of fires in each of the schools was n_i, $i = 1, \ldots, 20$

1 1 3 1 1 3 1 2 1 1 1 1 1 1 1 2 1 1 1 1

We now investigate if the risk for fire in schools in Stockholm differs from the average risk for the country, *i.e.* circa 0.1 fires on average during three years. (We neglect some uncertainties in the estimate $\mu^* = 0.032$ found above and assume stationarity of fire ignitions in years 2000-2002.) Our model is that the number of fires (during three years) in Stockholm schools is independent Poisson distributed variables with mean μ_S, which has to be estimated. We suspect that $\mu_S > \mu$. However, there is a small difficulty here: namely the fire department files contain only addresses of schools where the fire started. Thus, schools with zero fires are not present in the data (see Remark 4.3 where the inspection paradox was discussed); hence the average of the data is an obviously biased estimate of μ_S. In order to resolve the problem we need to work with conditional probabilities, conditionally that one knows that there was already a fire in a school.

We proceed as follows. Using data, we derive the ML estimate θ^* of the three-year average $\theta = \mu_S$. The asymptotic normality of the estimation error is used to construct a 0.95-confidence interval for μ_S. If the country average (here consider as known constant) lies outside the interval we can reject the hypothesis that the intensity of Stockholm school fires is the same as the average in the country.

ML estimate of μ_S

Let N be the number of fires observed in schools that had at least one fire during the period. Clearly N may take values $1, 2, \ldots$ with probabilities

$$P(N = n) = \frac{\theta^n}{n!} \frac{e^{-\theta}}{1 - e^{-\theta}},$$

where θ is the unknown average number of fires in a school in 3 years. Suppose $n_1, \ldots, n_k$ are independent observations from k schools, then the likelihood-, log likelihood-, and the derivatives functions are given by

$$L(\theta) = \prod_{i=1}^{k} P(N = n_i) = \prod_{i=1}^{k} \frac{\theta^{n_i}}{n_i!} \frac{e^{-\theta}}{1 - e^{-\theta}},$$

$$l(\theta) = -\sum_{i=1}^{k} \ln(n_i!) + \ln(\theta) \sum_{i=1}^{k} n_i - k\theta - k\ln(1 - e^{-\theta}),$$

$$\dot{l}(\theta) = k\frac{\bar{n}}{\theta} - \frac{k}{1 - e^{-\theta}}, \quad \ddot{l}(\theta) = -k\left(\frac{\bar{n}}{\theta^2} - \frac{e^{-\theta}}{(1 - e^{-\theta})^2}\right), \tag{7.5}$$

where $\bar{n} = \sum_{i=1}^{k} n_i/k$. The ML estimate θ^* is the solution to the equation $\theta^* = \bar{n}(1 - \mathrm{e}^{-\theta^*})$ while $\sigma_{\mathcal{E}}^* = 1/\sqrt{-\ddot{l}(\theta^*)}$. The estimate θ^* can be found by means of numerical procedures or a graphical method to solve the equation. For the data the solution is $\theta^* = 0.5481$ while $\sigma_{\mathcal{E}}^* = 0.2151$. Since with approximately 0.95 confidence

$$\mu_S \in \left[0.5481 - 1.96 \cdot 0.2151, \; 0.5481 + 1.96 \cdot 0.2151\right] = \left[0.13, 0.97\right]$$

we reject the hypothesis that $\mu_S = 0.096$, *i.e.* that the schools in Stockholm has the same yearly average number of fire as the country as whole.

7.3 Poisson Models for Counts

As we have seen the number of accidents N_i in different populations may vary; it also can change from year to year. Sometimes the differences can be explained as the result of random variability, *i.e.* when N_i are independent outcomes of the same random experiment. However, often the independence can be questionable or properties of the "experiment" changes with time, hence N_i are not iid.

In this section we study this type of problems closer. We do not treat it in full generality but assume that N_i are independent Poisson distributed variables, counting a number of failures (accidents) in different populations or time periods. Since the Poisson distribution has only one parameter, it means that our model is fully specified if $\mu_i = \mathsf{E}[N_i]$ are estimated.

Let N be the number of people killed in traffic, for instance, the next year. We assume that $N \in \mathrm{Po}(\mu)$ where $\mu = \mathsf{E}[N]$. In order to be able to make any statement of type $\mathsf{P}(N > 400)$, μ needs to be estimated. This is usually done using historical data. Denote by N_i the number of people killed in year i. We assume that N_i are independent, Poisson distributed with mean $\mu_i = \mathsf{E}[N_i]$. We have access to historical data and we know that $N_i = n_i$.

Using the historical data, we wish to find a pattern of how μ_i varies in order to extrapolate the variability to the future, *i.e.* the unknown value μ. Obviously, if there is no clear pattern in μ_i, the ML estimate μ^* and the historical data can hardly be used to predict future. However, if the mechanism generating accidents can be assumed to be *stationary*, then $\mu_i = \mu$ for all i. The average value $\bar{n} = \sum n_i/k$ is the ML estimate μ^* of μ.

In this section, we first briefly consider two data sets with regard to possible constant μ over time. For a more thorough analysis, tests are then introduced: to test for Poisson distribution and $\mu_i = \mu$ (constant mean). Finally, for the situation with μ_i not constant, the expected value is modelled as a function of other, explanatory variables.

Example 7.9 (Flight safety). In Example 6.12 flight safety was studied. From "Statistical Abstract of the United States", data for the number of crashes in the world during the years 1976-1985 are found:

24   25   31   31   22   21   26   20   16   22

Here a model for constant mean number of accidents for the period seems sensible.   □

Sometimes *trends* are observed in n_i: these seem to increase (or decrease) over time. A possible model can be that the mean changes linearly *viz.*

$$\mu_i = \mu + \beta \cdot i$$

where β is a constant and i the year. Historical data are used to find estimates μ^* and β^*. However, often more complicated models for the variable mean has to be used.

Example 7.10 (Traffic accidents in Sweden). Suppose we are interested in the number of deaths related to traffic accidents in Sweden. From official statistics [7], we find that during the years 1990-2004 the following number of people died due to accidents in Sweden:

772   745   759   632   545   531   508   507   492   536   564
551   532   529   480

We can see that number of deaths is decreasing and obviously one cannot assume that the data are observations of independent Poisson variables with constant mean.   □

7.3.1 Test for Poisson distribution – constant mean

In the following subsection we test whether data contradict the iid Poisson model for N_i, *i.e.* $\mu_i = \mu$. However, first we present a useful approximation of the Poisson distribution valid for large populations, commonly assumed to be valid for $\mu > 15$.

For $N \in \text{Po}(\mu)$ when μ is large, a very effective tool is to approximate the Poisson distribution by a normal distribution (the so-called normal approximation).

Normal approximation of Poisson distribution.
Let N be a Poisson distributed random variable with expectation μ,

$$N \in \text{Po}(\mu).$$

If μ is large (in practice, $\mu > 15$), we have approximately that

$$N \in \text{N}(\mu, \mu). \tag{7.6}$$

Example 7.11 (Accidents in mines). Consider Example 2.11, page 39. We there estimated the intensity of accidents in mines $\lambda = 3$ year^{-1} and argued that the stream of accidents is Poisson. Suppose we want to calculate the probability of at least 80 accidents during 25 years, that is, $P(N(25) \geq 80)$. Since the stream is Poisson, $N(25) \in \mathrm{Po}(3 \cdot 25) = \mathrm{Po}(75)$. For simplicity of notation, let $N = N(25)$; we compute

$$P(N \geq 80) = 1 - P(N \leq 79) = 1 - P(N = 0) - P(N = 1) - \ldots - P(N = 79),$$

which might be cumbersome[1]. An alternative solution is to employ the normal approximation instead to evaluate the probability:

$$P(K \geq 80) \approx 1 - \Phi((79.5 - 75)/\sqrt{75}) = 1 - \Phi(0.52) = 0.30.$$

$\square$

Suppose we have k observations $n_1, \ldots, n_k$ of Poisson distributed quantities $N_1, \ldots, N_k$. Our assumption is that all $N_i \in \mathrm{Po}(\mu)$, *i.e.* stationarity (homogeneity) is present.

For small values of μ but large k we can use the χ^2 test presented in Section 4.2.2 to validate the model.

In the case when μ is large, to test whether data do not contradict the assumption of stationarity or constant mean, often the following property of a Poisson distribution is used: $V[N] = E[N] = \mu$. In the case of a Poisson distribution, the ratio $V[N]/E[N]$ is obviously equal to 1. The test to be presented below is based on this fact. If μ is large, by Eq. (7.6) $N \in N(\mu, \mu)$ and we can estimate $E[N]$ by $\bar{n}$ and $V[N]$ by s_{k-1}^2. As confidence interval for $\theta = V[N]/E[N]$ can be constructed, *viz.*

$$\frac{\bar{n}}{s_{k-1}^2} \frac{\chi_{1-\alpha/2}^2(k-1)}{k-1} \leq \frac{V[N]}{E[N]} \leq \frac{\bar{n}}{s_{k-1}^2} \frac{\chi_{\alpha/2}^2(k-1)}{k-1} \tag{7.7}$$

with approximate confidence $1 - \alpha$. If $\theta = 1$ is not in that interval, the hypothesis that N is Poisson distributed is rejected. For further reading about tests of this type, see Brown and Zhao [6].

Remark 7.2. The assumption of equal variance and mean is not always satisfied working with real data. If $V[X] > E[X]$, *overdispersion* is present. Additional statistical tests for this are found in the literature (cf. [18]).

In the following example we study how the hypothesis that μ_i are constant over time can be validated.

Example 7.12 (Flight safety). This is a continuation of Example 7.12 where number of crashes of commercial air carriers in the world during the years 1976-1985 were presented. Let us assume that the flight accidents form

---

[1]Some of the probabilities can be hard to compute and Stirling's formula $n! \approx \sqrt{2\pi}\, n^{n+0.5} e^{-n}$ needs to be used.

a Poisson stream and hence n_i are independent observations of $\text{Po}(\mu)$ distributed variables.

A point estimate is $\mu^* = \bar{n} = 23.8$. Since $\mu^* > 15$, the model implies that n_i can be considered independent observations of an $\text{N}(\mu, \mu)$ distributed variable (cf. Eq. (7.6)). Consequently, we expect for the ML estimate of the variance $s_{k-1}^2 \approx \bar{n}$. For this data set, $s_{k-1}^2 = 22.2$, which is close to 23.8.

Next, the confidence interval given in Eq. (7.7) is computed:

$$\left[\frac{23.8}{22.2} \cdot \frac{2.7}{9}, \; \frac{23.8}{22.2} \cdot \frac{19.02}{9} \right] = [0.32, \; 2.26]$$

and hence the hypothesis of constant μ is not rejected. □

7.3.2 Test for constant mean – Poisson variables

Suppose it can be assumed that data are observations of independent Poisson distributed variables but we suspect that the mean is not constant. More precisely, we check if the data do not contradict the assumption that $\text{E}[N_i] = \mu$. The test we wish to use is based on a quantity called *deviance* and is based on log-likelihood values. The specific of the test is that we do not need to assume that the mean μ is high.

Statistical test using deviance

Let N_i be independent Poisson distributed variables and consider two models: a more general model, where no restriction are put on the means $\mu_i = \text{E}[N_i]$, and a simpler where all means are equal, *i.e.* $\mu_i = \mu$. Let n_i be the observed values of N_i. Using the ML method the optimal estimates $\mu_i^* = n_i$ if the general model is assumed while the ML estimate is $\mu^* = \sum n_i/k$ for the simpler, more restrictive model.

Since the more general model contains the simpler, the log-likelihood function $l(\mu_1^*, \ldots, \mu_k^*)$ must be higher than $l(\mu^*)$. Higher values of the log-likelihood function means that the observed data are more likely to occur under the model, hence the increase of the function is a measure of how much better the more complex model explains the data. It can be shown that the following test quantity, called *deviance*,

$$\text{DEV} = 2 \cdot \left(l(\mu_1^*, \ldots, \mu_k^*) - l(\mu^*) \right), \tag{7.8}$$

for large k is $\chi^2(k-1)$ distributed if the simpler model is true[2]. Thus if $\text{DEV} > \chi_\alpha^2(k-1)$, the difference between log-likelihoods cannot be explained by the statistical variability and hence the simpler model should be rejected. Straightforward calculations lead to the following formula

$$\text{DEV} = 2 \sum_{i=1}^{k} n_i \left(\ln(\mu_i^*) - \ln(\mu^*) \right) = 2 \sum_{i=1}^{k} n_i \left(\ln(n_i) - \ln(\bar{n}) \right), \tag{7.9}$$

where for $n_i = 0$ we let $n_i \ln(n_i) = 0$.

---

[2]The test can also be used for small k if μ is large.

Example 7.13 (Daily rains). This is continuation of Example 2.14 where the data n_i, $i = 1, \ldots, 12$, are numbers of daily rains exceeding 50 mm observed in month i, during years 1961-1999. We suspect that the simplest model of constant mean $\mu_i = \mu$, estimated to be $\mu^* = \bar{n} = 3.67$, is not correct. Let us compute the deviance by Eq. (7.9)

$$\text{DEV} = 2\{4(\ln(4) - \ln(3.67)) + \cdots + 10(\ln(10) - \ln(3.67))\} = 19.64.$$

The value 19.64 should be compared with the 0.05 quantile found as $\chi^2_{0.05}(11) = 19.68$. Obviously this is a boarder case. Although DEV is slightly below the quantile we decide that with approximative confidence 0.95 the hypothesis of the means $\mu_i = \mu$ can be rejected.   □

Example 7.14 (Motorcycle data). Consider the data set from Problem 4.9 where the numbers of killed motorcyle riders in Sweden 1990-1999, are reported. We suspect that the simplest model that $\mathsf{E}[N_i] = \mu_i = \mu$ explains well the data and wish to test it against the more complex model that $\mathsf{E}[N_i] = \mu_i$.

$$\text{DEV} = 2\sum_{i=1}^{10} n_i\big(\ln(n_i) - \ln(\bar{n})\big) = 5.5,$$

since $\bar{n} = 33.1$. The value 5.5 should be compared with the 0.05 quantile found as $\chi^2_{0.05}(9) = 16.92$. We conclude that the more complex model does not explain data better than the simpler one does.   □

7.3.3 Formulation of Poisson regression model

As seen in the previous subsection, often the assumption of constant mean μ for the number of accidents N_i has to be rejected. In such a situation it is desirable to find a model for the variability of the mean μ_i. A standard approach is to find (or select from available data) a collection of *explanatory* variables (quantities) that influence means. A method to find a functional relation between the explanatory variables and the means is the so-called Poisson regression.

Regression techniques are widely used in statistical applications found in most sciences, a standard reference is the book by Draper and Smith [22]. The random outcomes of an experiment Y_i (called responses or dependent variables) of the ith experiment have means related to a vector of p, say, explanatory[3] variables $x_1, x_2, \ldots, x_p$.

A regression model

Consider a sequence of Poisson distributed counting variables N_i, $i = 1, \ldots, k$, for example the number of accidents (failures) occurring in year i. Let n_i be the observed values of N_i. Suppose that for each i one observes

---

[3]Several names exist in the literature: independent variables, regressor variables, predictor variables.

p different variables characterizing the population, or mechanisms generating accidents. Consequently, data consist of n_i and a vector $x_{i1}, x_{i2}, \ldots, x_{ip}$, $i = 1, \ldots, k$. In addition in some models an extra quantity t_i, say, measuring the exposure for risk is selected and the model for $\mu_i = \mathsf{E}[N_i]$ is written down as follows[4]

$$\mu_i = t_i \exp(\beta_0 + \beta_1 x_{i1} + \ldots + \beta_p x_{ip}). \tag{7.10}$$

As before, one assumes that $N_i \in \mathrm{Po}(\mu_i)$ are independent and hence the ML estimates of the parameters β_i are readily available. The algorithm is given in Section 7.3.4.

Example 7.15. The simplest regression model is derived when $p = 0$, *i.e.* there are no explanatory variables x_{ij} at all. Then with $\lambda = \exp(\beta_0)$ the model is $\mu_i = t_i \lambda$. The ML estimate of the unknown intensity λ and standard deviation of the estimation error are given by

$$\lambda^* = \frac{\sum_{i=1}^{k} n_i}{\sum_{i=1}^{k} t_i}, \qquad \sigma_{\mathcal{E}}^* = \sqrt{\lambda^* / \sum t_i}. \tag{7.11}$$

Obviously if all exposures t_i are equal, $t_i = 1$, then $\mu = \lambda$ giving the estimate $\mu^* = \bar{n}$. $\qquad\square$

The model in Eq. (7.10) is convenient for studying the influence of a variable x_{ij} on the mean μ_i. The *rate ratio* defined as

$$RR_j = \exp(\beta_j), \quad j = 1, \ldots, p \tag{7.12}$$

measures multiplicative increase of intensity of events when x_{ij} increases by one unit. The rate ratio is estimated by $RR_j^* = \exp(\beta_j^*)$, where β_j^* is the ML estimate of β_j. Using asymptotic normality of ML estimators, confidence intervals for RR_j can easily be given.

Example 7.16 (Traffic accidents in Sweden). This is continuation of Example 7.10 where we presented the number of people killed in traffic in Sweden in years 1990-2004. Constant work on improving safety in traffic, new legislations, technical improvements in cars (ABS, airbags, etc.) as well as better standards of roads should result in a decrease of the death rate. However, the increase in traffic volume has contrary effects.

In the report [7] the following model was proposed, $\mu_i = a \cdot b^i \cdot x_i^c$, where $i = 1, 2, \ldots$ are the years and a, b, and c unknown parameters. Further x_i is the traffic index in year i. Since we do not have access to the traffic index we consider first a simplified model when $c = 0$, $\mu_i = a \cdot b^i$. However, we use the equivalent formulation from Eq. (7.10)

$$\mu_i = \exp(\beta_0 + x_{i1}\beta_1),$$

---

[4]The functional form in Eq. (7.10) follows the set-up of so-called generalized linear models [56].

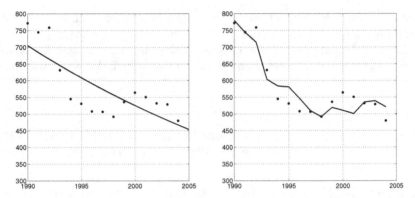

Fig. 7.4. Number of deaths because of traffic in Sweden, 1990–2004. *Left:* Simple Poisson regression, yearly trend. *Right:* Poisson regression taking into account yearly trend and traffic volume.

where[5] $x_{i1} = i - 8.0$. The parameters β_0, β_1 are estimated using the ML algorithm giving

$$\beta_0^* = 6.35, \quad \beta_1^* = -0.0294.$$

The estimated values $\mu_i^* = \exp(6.37 - 0.0294 \cdot (i - 8.0))$ are given in Figure 7.4 (left), solid line. These constitute a regression curve and are compared with observed values n_i shown as dots.

We can see that data n_i oscillate quite regularly around the regression curve μ_i^*, which contradicts the assumed independence of N_i. However, the model can still be a useful, crude description of the data. The most important property of this model is that it indicates that the average number of deaths decreases with $RR_1^* = \exp(\beta_1^*) = 0.97$ by 3%. (This was one of the conclusions of the VTI report [7].) □

By taking further explanatory variables in Eq. (7.10) more sophisticated models can be proposed. In Example 7.16 we had two parameters ($p = 1$ while $k = 15$) and we concluded that a more complex model would be needed to adequately describe the traffic data. However, a higher number of parameters β_j will lead to higher uncertainty of the estimate μ_i^*. In the limiting case when $p \geq k - 1$ there are at least as many parameters to estimate as there are observations n_i. Consequently, the estimates $\mu_i^* = n_i$ can be used as well instead of $\mu_i = \exp(\beta_0^* + \sum \beta_j^* x_{ij})$.

Clearly, more complex models better explain the observed variability in data; however, as the number of parameters increases the estimated values

---

[5]The values of the explanatory variables are centred in order to obtain more well-conditioned covariance matrices, hence $x_{i1} = i - 8.0$ since $(1/15) \sum_{i=1}^{15} i = 8.0$.

often become more uncertain. When combining both types of uncertainty—
(1) the uncertainty of the future outcome of the experiment; (2) the uncertainty of the parameter— the computed measures of risks can be more uncertain for the complex model than for the simpler one[6]. This leads us to
the next important issue, the model selection. We do not go deep into this
matter, but just indicate how the different models can be compared using the
already-introduced quantity, deviance (for the simplest case see Eq. (7.8-7.9)).

Model selection and use of deviance

The above-discussed Poisson regression is a very versatile approach to model
variability of counts. Applications are found in most sciences: technology,
medicine, etc. In this subsection we further discuss these models, more precisely, the number of explanatory variables to be used. Illustrating examples
will be given.

One way of comparing different models is to analyse the value of the logarithm of the ML function $l(.)$ for different choices of explanatory variables.
Let us consider two models: a more general model, with p explanatory variables, and a simpler where only $q < p$ of the variables x_i are used. (Here
$q = 0$ if no explanatory variables x are used.) Denote by $\beta_\mathbf{p}$, $\beta_\mathbf{q}$, the β parameters in the two models. Using the ML method optimal estimates $\beta_\mathbf{p}^*$ and
$\beta_\mathbf{q}^*$ are chosen. Since the more general model contains all the parameters of
the simpler one (and some additional) the log-likelihood function $l(\beta_\mathbf{p}^*)$ must
be higher than $l(\beta_\mathbf{q}^*)$. Since higher values of the log-likelihood function means
that the observed data are more likely to occur (if the model is true), the
increase of the function is a measure of how much better the more complex
model explains the data. It can be shown that the following test quantity,
called deviance,

$$\text{DEV} = 2 \cdot \big(l(\beta_\mathbf{p}^*) - l(\beta_\mathbf{q}^*)\big), \tag{7.13}$$

for large k is $\chi^2(p-q)$ distributed if the less complex model is true. Thus if
$\text{DEV} > \chi_\alpha^2(p-q)$, the difference between log-likelihoods cannot be explained
by the statistical variability and hence the simpler model should be rejected.
In other words, the more complex model fits data significantly better. Further
discussion of this type of χ^2 test can be found in [82], page 345 or [10],
Section 8.2.

Now the computation of the deviance DEV is relatively simple if the ML
estimates $\beta_\mathbf{p}^*$, $\beta_\mathbf{q}^*$ are given. Using $\beta_\mathbf{p}^*$, $\beta_\mathbf{q}^*$, the estimates of $\mu_i = \mathsf{E}[N_i]$ can
be readily computed

$$\mu_i^* = t_i \cdot \exp(\beta_0^* + \beta_1^* x_{i1} + \cdots + \beta_l^* x_{il}),$$

---

[6]We return to this problem in Chapter 10 where 100-year values will be
estimated.

where $l = p$ and $l = q$, respectively. Denote by μ_{iS}^* the estimates derived using β_q^* while μ_{iC}^* the ones derived using β_p^*. Then

$$\text{DEV} = 2 \sum_{i=1}^{k} n_i \left(\ln(\mu_{iC}^*) - \ln(\mu_{iS}^*) \right). \tag{7.14}$$

Example 7.17 (Traffic accidents in Sweden). This is a continuation of Example 7.16 where we concluded that the proposed model for the expected number of perished in traffic in one year is too simple. We believe that the systematical variability (see Figure 7.4, left panel) of n_i around the estimated regression could be explained by changes in the amount of traffic. It is obvious that the years where the observations are below the average correspond to the years when traffic growth was slower.

In the report [7], estimates of the total vehicle kilometres during 1990-2004 in 10^9 kilometres, where $i = 1$ corresponds to year 1990, were also reported. The estimates y_i, say, are as follows:

64.3   64.9   65.5   64.1   64.9   66.1   66.5   66.7   67.4   69.6

70.6   71.6   74.0   75.4   76.1

Now the new, more complex, model for μ_i (with $p = 2$) is

$$\mu_i = \exp(\beta_0 + \beta_1 x_{i1} + \beta_2 x_{i2}),$$

where $x_{i1} = i - 8.0$ while $x_{i2} = y_i - 68.5$. The parameters β are estimated using the ML algorithm giving

$$\beta_0^* = 6.35, \quad \beta_1^* = -0.082, \quad \beta_2^* = 0.063.$$

In Figure 7.4 (right panel) we can see the estimated values of μ_i as a solid line together with observations marked as dots. The two rate ratios $RR_1 = \exp(\beta_1)$ and $RR_2 = \exp(\beta_2)$ are estimated to be $RR_1^* = 0.92$ and $RR_2^* = 1.065$. The rough interpretation of the ratios is that the safety improvements led to a yearly decrease of about 8% of the expected number of perished in the traffic but the increase in the traffic volume by 10^9 km increases the expectation by ca 6.5%. Since on average the traffic volume increases by $0.84 \cdot 10^9$ km, this leads to a yearly decrease of the expected number of perished by about 3% (the same as given for the simpler model in Example 7.16). The more complete model seems to give more insight into the problem; however, we should also check whether the more complicated model explains the data significantly better than the simpler one does.

Consequently, let us compute the deviance. Again, let μ_{iS}^* denote the estimated averages μ_i^* presented in Figure 7.4 (left panel) for the simpler regression, $p = 1$, while μ_{iC}^* be the corresponding estimates μ_i^* presented in Figure 7.4 (right panel) for the more complex regression, $p = 2$. Then the deviance given by Eq. (7.14) is equal to

$$\text{DEV} = 2 \sum_{i=1}^{15} n_i(\ln(\mu_{iC}^*) - \ln(\mu_{iS}^*)) = 59.75$$

which could be compared with the 0.001 quantile found as $\chi^2_{0.001}(1) = 10.83$. Since DEV > 10.83, we reject with high confidence the hypothesis that the more complex model explains the data equally well as the simpler one. □

Example 7.18 (Derailments in Sweden). In [73], statistics for derailments in Sweden are given. Authorities are interested in the impact of usage of different track types. Data consist of derailments of passenger trains during 1 January 1985 – 1 May 1995, where n_i is the number of derailments on track type i and t_i is the corresponding exposure in 10^6 train kilometres. The following numbers are extracted from [73]. The observations n_i, t_i are given in columns two and three, respectively;

$i = 1$ 15   421   [Welded track with concrete sleepers]
$i = 2$ 28   80   [Welded track with wooden sleepers]

A statistical test is needed to test for possible differences in safety; below, we use the deviance. The numbers of derailments that occur at tracks of type i, denoted by N_i, is assumed to be independent and Poisson distributed. Further, let $\mu_i = \mathsf{E}[N_i] = \lambda_i t_i$, where t_i are exposures measured in 10^6 train km (tkm). The simpler model is that $\lambda_1 = \lambda_2 = \lambda$ while the more complex is that λ_1 and λ_2 are different. We are interested in the rate ratio $RR = \lambda_2/\lambda_1$.

Eq. (7.11) gives the estimate $\lambda^* = (n_1+n_2)/(t_1+t_2) = 0.0858\ [10^{-6}\text{tkm}^{-1}]$; consequently, $\mu_{1S}^* = \lambda^* t_1 = 36.1$ and $\mu_{2S}^* = \lambda^* t_2 = 6.9$. Next, for the complex model $\mu_{iC}^* = n_i$ and hence using Eq. (7.13)

$$\text{DEV} = 2\big(15(\ln(15) - \ln(36.1)) + 28(\ln(28) - \ln(6.9))\big) = 52.1.$$

Since the more complex model has two parameters while the simpler has only one, one should compare the computed deviation with the quantile $\chi^2_{0.001}(1) = 10.83$. Consequently, with very high confidence, we reject the simplest model. Hence in the following we consider only the more complex model.

The rate ratio RR. The rate ratio measures how the increase of intensity of events changes between the two populations, here $RR = \lambda_2/\lambda_1$ and is estimated by

$$RR^* = \frac{\lambda_2^*}{\lambda_1^*} = \frac{28 \cdot 421}{15 \cdot 80} = 9.8,$$

i.e. the risk for derailment is nearly ten times higher for the second type of track.

The Poisson-regression model. The estimations μ_i can also be described as a Poisson-regression problem since we can write

$$\mu_i = t_i \exp(\beta_0 + \beta_1 x_i). \tag{7.15}$$

Here x_i is a dummy variable taking only two values: defined to be zero when $i = 1$ and one when $i = 2$. The parameter estimate β_1^* could be computed using the ML algorithm, however, here we take a shortcut and use that RR^* has already been estimated. Since $RR^* = 9.8$, we find $\beta_1^* = \ln(9.8) = 2.28$.

Any statistical software would compute the estimates β_i^* and give the matrix with $-[\ddot{l}(\beta^*)]^{-1}$ needed for computations of standard deviations of the estimation error, $\sigma_{\mathcal{E}_i}^*$. However, since in this simple example these can be easily derived analytically we present the complete solution for illustration of the methodology.

The main purpose of these computations is to derive an asymptotic confidence interval for RR. (Asymptotic normality of ML estimators is utilized.) When the estimate $\sigma_{\mathcal{E}}^*$ associated with β_1^* is computed then, with approximately 0.95 confidence, $\beta_1^* - 1.96\sigma_{\mathcal{E}}^* < \beta_1 < \beta_1^* + 1.96\sigma_{\mathcal{E}}^*$ and hence

$$\exp(\beta_1^* - 1.96\sigma_{\mathcal{E}}^*) < RR < \exp(\beta_1^* + 1.96\sigma_{\mathcal{E}}^*).$$

What remains is computation of the estimated variance $(\sigma_{\mathcal{E}}^2)^*$. The variance is the second element of the diagonal of $\boldsymbol{\Sigma} = [-\ddot{l}(\beta_0^*, \beta_1^*)]^{-1}$. Now, the matrix of second-order derivatives can be computed using Eq. (7.17) when the estimates μ_i^* are known. From the definition of x_{ij}, Eq. (7.17) gives

$$[\ddot{l}(\beta_0^*, \beta_1^*)] = -\begin{pmatrix} \sum \mu_i^* & \mu_2^* \\ \mu_2^* & \mu_2^* \end{pmatrix} = -\begin{pmatrix} 43 & 28 \\ 28 & 28 \end{pmatrix}.$$

Consequently

$$\boldsymbol{\Sigma} = \begin{pmatrix} 0.0667 & -0.0667 \\ -0.0667 & 0.1024 \end{pmatrix},$$

and hence with approximately 0.95 confidence

$$\exp(2.28 - 1.96\sqrt{0.1024}) < RR < \exp(2.28 + 1.96\sqrt{0.1024}),$$

$5.2 < RR < 18.3$. Thus, rail type 1 is, with high confidence, at least five times safer to use than rail type 2 is.

$\square$

7.3.4 ML estimates of $\beta_0, \ldots, \beta_p$

For simplicity of derivations, let us introduce $x_{i0} = 1$ and let

$$\mathsf{E}[N_i] = \mu_i = t_i \exp\left(\sum_{j=0}^{p} \beta_j x_{ij}\right),$$

where $N_i \in \mathrm{Po}(\mu_i)$, $i = 1, \ldots, k$. Clearly N_i may take values $0, 1, 2, \ldots$ with probabilities

$$P(N_i = n) = \frac{\mu_i^n}{n!} e^{-\mu_i}.$$

Denote by n_i the observed N_i, *i.e.* the number of events that occurred in a period of time t_i. The likelihood-, log likelihood-, and the derivative functions are given by

$$L(\beta) = \prod_{i=1}^{k} P(N_i = n_i) = \prod_{i=1}^{k} \frac{\mu_i^{n_i}}{n_i!} e^{-\mu_i},$$

$$l(\beta) = -\sum_{i=1}^{k} \ln(n_i!) + \sum_{i=1}^{k} n_i \ln(\mu_i) - \sum_{i=1}^{k} \mu_i,$$

$$\dot{l}(\beta) = \sum_{i=1}^{k} \frac{d\mu_i}{d\beta} \left(\frac{n_i}{\mu_i} - 1 \right). \tag{7.16}$$

Now Eq. (7.16), with β replaced by β_j can be used to compute the derivatives of the log-likelihood functions. Since $\partial \mu_i / \partial \beta_j = x_{ij} \mu_i$ the derivatives and second-order derivatives of the log-likelihood are given by

$$\frac{\partial l(\beta)}{\partial \beta_j} = \sum_{i=1}^{k} (n_i - \mu_i) x_{ij}, \quad \frac{\partial^2 l(\beta)}{\partial \beta_j \partial \beta_m} = -\sum_{i=1}^{k} \mu_i \, x_{ij} x_{im}. \tag{7.17}$$

As before the ML estimate of $\beta_{\mathbf{p}}^* = (\beta_0^*, \ldots, \beta_p^*)$ are solutions to the system of $(p + 1)$ non-linear equations in β_j, *viz.* $\sum_{i=1}^{k} (n_i - \mu_i) x_{ij} = 0$. Often these cannot be solved analytically, but a numerical method, *e.g.* the recursive Newton–Raphson algorithm, can be used:

- The algorithm starts with a guess β^0, say, of the values of the vector β, for example

$$\beta_0^0 = \ln(\sum n_i) - \ln(\sum t_i), \qquad \beta_i^0 = 0, \qquad i > 0.$$

- If the values of the parameters after the mth iteration are denoted by β^m then the N–R algorithm renders the new estimates by the following formula

$$\beta^{m+1} = \beta^m - [\ddot{l}(\beta^m)]^{-1} \dot{l}(\beta^m),$$

where $[\ddot{l}(\beta)]$ is a matrix with derivatives $\frac{\partial^2 l(\beta)}{\partial \beta_j \partial \beta_m}$ while $\dot{l}(\beta)$ is a column vector of $\frac{\partial l(\beta)}{\partial \beta_j}$.

- The algorithm stops when all components in the vector $\dot{l}(\beta^{m+1})$ are small enough.

7.4 The Poisson Point process

The Poisson point process is an important tool, widely used not only in applications to risk and safety analysis, but also in telecommunication engineering, financial, and insurance mathematics. Applications to risk analysis and accidents were present already in the 1920s, cf. [30]. In Section 2.6.1, we introduced a Poisson stream of events, which is here renamed *Poisson point process (PPP)* in order to generalize the notion from a line (time) to higher-dimensional spaces.

We start with an alternative definition of a PPP on the line, *i.e.* in the case when the PPP is a Poisson stream of events A, say, and review some basic properties of a PPP. Of particular interest is the distribution of the time intervals T_i between the occurrences of A.

Definition 7.2 (Poisson Point process (PPP)). *If the time intervals $T_1, T_2, \ldots$ between occurrences of an event are independent, exponentially distributed variables with common failure intensity λ, then the times $0 < S_1 < S_2 < \ldots$ when the event A occurs form a **Poisson point process** with intensity λ.*

Let us recall the notation $N_A(s,t)$, $N_A(t)$ from Definition 2.2. (In the following the subscript A is omitted.) For fixed values s, t the random variable $N(s,t)$ is the number of times an event A occurred in the time interval $[s, s+t]$ while $N(t)$ is understood as $N(0,t)$. The variable $N(t)$ can also be seen as a function of time which (see Figure 7.5), is called a Poisson process.

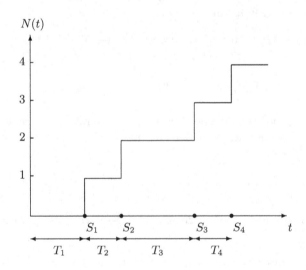

Fig. 7.5. Illustration of a Poisson process.

We summarize the important properties of a PPP:

Let λ be the intensity of a PPP. Then

- The time to the first event, T, is exponentially distributed:

$$P(T > t) = e^{-\lambda t}.$$

- Times between events, T_i, are independent and exponentially distributed:

$$P(T_i > t) = e^{-\lambda t}.$$

- The number of events $N(s, t) \in Po(m)$, i.e. is Poisson distributed with $m = \lambda t$.
- The number of events in disjoint time intervals are independent and (obviously) Poisson distributed.

Remark 7.3. If we assume that real-world phenomenon can be modelled by means of a Poisson point process then the intensity λ is the only parameter that is needed to compute probabilities of interest, since $N(t) \in Po(\lambda t)$. If the mean $E[N(t)] = \lambda t$ is small, then

$$P(N(t) = 0) = e^{-\lambda t} \approx 1 - \lambda t, \qquad P(N(t) = 1) \approx \lambda t = E[N(t)],$$

and the probability of more than one accident is of smaller order.   □

Typical applications of a PPP often are to model variability of counting and book-keeping of times, for example between cars passing a checkpoint. The Poisson model implies that in any time period t, say, the number of cars that have been registered in the period $N(t)$, say, is Poisson distributed with mean equal to λt.

In safety analysis of complex systems, e.g. an electrical power network in a country, transients that occur in the system need to be analysed as consequences of different types of failures (accidents). The failures are modelled using Poisson streams and the safety of the system is investigated by means of suitable (numerical) simulations of transients. The possibility of analytical computations is limited by the complexity of a system. One of the inputs is times of failures and hence Poisson streams with a given intensity λ need to be simulated.

Simulation of a Poisson point process

Since the intensity of events λ is constant, we expect that there are no specific patterns regarding the positions of the points in a PPP. This somewhat unprecise statement can be illuminated by the following method to simulate a PPP.

Step 1 First choose an interval of length t, for example $[0, t]$.

Step 2 Then, by some Monte Carlo method, generate the number of points in the interval $N(t)$, *i.e.* random numbers with distribution $\text{Po}(\lambda t)$ (see Chapter 3 for details). Denote the generated number by n (for instance, if $n = 10$, then there are 10 points in $[0, t]$).

Step 3 What remains to find are the exact locations of the n points. These should be totally random. In fact, the locations are independent and uniformly distributed variables. By this we mean that we need to simulate n values u_i of uniformly distributed (between zero and one) random numbers. Then the positions of the events are given by $t \cdot u_i$ (not ordered).

It is important to be able to motivate the correctness of the assumption that the sequence of events forms a Poisson stream. According to Section 2.6.1, conditions I-III, one needs to motivate that the mechanism generating accidents is stationary. One often limits oneself to check if the intensity of accidents is constant (see Examples 7.13-7.14). Next, one needs to argue that the number of accidents in disjoint intervals is independent and, finally, that two or more events cannot happen exactly at the same moment. Here the reason for the use of a PPP consists mainly of general arguments. This type of "validation" is often used when events occur rarely and hence use of statistical tests is limited.

Remark 7.4 (Barlow–Proschan test). Actually the property used in Step 3 in the simulation algorithm, that times when accidents occur are uniformly distributed can be used to construct a test whether the ordered observed times $0 < S_1 < S_2 < \ldots < S_n$ do not contradict the assumption that those are the first n times of the PPP. It can be shown that the statistic

$$Z = \frac{1}{S_n} \sum_{i=1}^{n-1} S_i$$

is approximately normally distributed. From Step 3, it can be seen that Z has the distribution of the sum of $n - 1$ uniformly distributed random variables U_i. Consequently, a table of means and variances gives that $\mathsf{E}[Z] = (n-1)/2$ and $\mathsf{V}[Z] = (n-1)/12$ and hence, with approximately probability $1 - \alpha$

$$\frac{1}{2}(n-1) - \lambda_{\alpha/2}\sqrt{\frac{n-1}{12}} < Z < \frac{1}{2}(n-1) + \lambda_{\alpha/2}\sqrt{\frac{n-1}{12}}. \tag{7.18}$$

Now having observed the times s_i, $i = 1, \ldots, n$, the value of $z = \sum_{i=1}^{n-1} s_i/s_n$ is computed. If z is outside the interval given in Eq. (7.18) then the hypothesis that the times s_i are outcomes of Poisson point process is rejected. This procedure is called Barlow–Proschan's test. □

Example 7.19 (Periods between earthquakes). Let us reconsider times between earthquakes t_i, first encountered in Example 1.1, later discussed *e.g.*

in Example 4.6, where a χ^2 test was used to test for exponentially distributed time intervals. Here we make use of Barlow–Proschan's test outlined earlier. Obviously $s_k = \sum_{i=1}^{k} t_i$, $k = 1, \ldots, n$ and hence

$$z = \frac{\sum_{k=1}^{n-1} \sum_{i=1}^{k} t_i}{\sum_{i=1}^{n} t_i}. \tag{7.19}$$

For the data, $n = 62$ and we find $z = 31.06$. The interval $[30.5 - 1.96\sqrt{61/12}, 30.5 + 1.96\sqrt{61/12}] = [26.1, 34.9]$ contains z and hence the hypothesis that times for major earthquakes forms a PPP cannot be rejected.

□

7.5 More General Poisson Processes

Earlier in this chapter, we have used the Poisson point process to describe when events occur in time, *i.e.* a Poisson stream. However, applications do not have to be restricted to events occurring in time. Consider for example cracks along an oil pipeline and think about how a PPP can be applied. The concept can be generalized even more.

A general Poisson process

Let $N(B)$ denote the number of events (or accidents) occurring in a region B. Consider the following list of assumptions (cf. Section 7.4):

(A) More than one event cannot happen simultaneously.
(B) $N(B_1)$ is independent of $N(B_2)$ if B_1 and B_2 are disjoint.
(C) Events happen in a stationary (in time) and homogeneous (in space) way, more precisely, the distribution of $N(B)$ depends only on the size $|B|$ of the region: for example $N(B) \in \text{Po}(\lambda|B|)$.

The process for which we can motivate that (A–B) are true is called a Poisson process. It is a stationary process with constant intensity λ if (A–C) holds.

An illustration of a Poisson process in the plane is given in Figure 7.6.

Example 7.20 (Japanese black pines). In Figure 7.7 are shown the locations of Japanese black pine samplings in a square sampling region in a natural region. The observations were originally collected by Numata [58] and the data are used as a standard example in the textbook by Diggle [20]. Having adequate biological information about the species region and other relevant information one could may be assume the validity of assumptions (A-C) leading to the Poisson model for the locations of the trees.

As statisticians we can also validate the model, *i.e.* check if some statistics do not contradict the assumed PPP. First, let us estimate the intensity λ of pines.

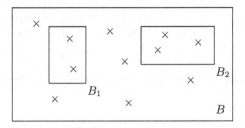

Fig. 7.6. Illustration of a Poisson process in the plane. Here $N(B) = 11$ while $N(B_1) = 2$, $N(B_2) = 3$.

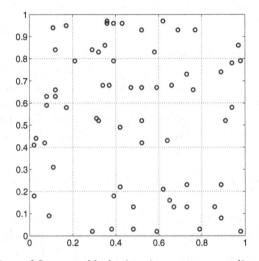

Fig. 7.7. Locations of Japanese black pines in a square sampling region.

The region studied was 5.7×5.7 m^2, which we refer to as one area unit (au) in the following. There are 65 pines in a region of 1 au, and hence the estimate of the estimate of intensity $\lambda^* = 65$ au^{-1}. We divide the region in 25 smaller squares, each of size $0.2 \cdot 0.2 = 0.04$ au. Since we assumed homogeneity of trees, we expect on average $0.04 \cdot 65 = 2.6$ trees in each of such smaller regions. Obviously the true number differs from the average and their variability is modelled as 25 independent Po(2.6) distributed variables.

From Figure 7.7 are found 1, 5, 4, 11, 2, 1, 1 regions containing 0, 1, 2, 3, 4, 5, 6 pines. The probability-mass function for Po(2.6) is $p_k = 2.6^k \exp(-2.6)/k!$ and hence one expects to have $25 \cdot p_i$ smaller regions to contain k plants. The expectations are 1.9, 4.8, 6.3, 5.4, 3.5, 1.8, 0.8, respectively; how close is this to what the model predicts? We use a χ^2 test:

$$Q = \frac{(1 - 25p_0)^2}{25p_0} + \frac{(5 - 25p_1)^2}{25p_1} + \frac{(4 - 25p_2)^2}{25p_2} + \frac{(11 - 25p_3)^2}{26p_3}$$
$$+ \frac{(2 - 25p_4)^2}{25p_4} + \frac{(1 - 25p_5)^2}{25p_5} + \frac{(1 - 25p_6)^2}{25p_6} = 8.1.$$

Since $\chi^2_{0.05}(7 - 1 - 1) = 11.07$, the hypothesis about Poisson distribution cannot be rejected (see Eq. (4.3)).   □

Example 7.21 (Bombing raids on London). During the bombing raids on London in World War II, one discussed whether the impacts tended to cluster or if the spatial distribution could be considered random. This was not merely a question of academic interest; one was interested in whether the bombs really targeted (as claimed by Germans) or fell at random[7]. An area in the south of London was divided into 576 small areas of $1/4$ km^2 each; the Poisson distribution was found to be a good model. For further discussion, consult Chapter VI.7 in the classical book by Feller [25].   □

7.6 Decomposition and Superposition of Poisson Processes

The Poisson process is a mathematical tool in risk analysis to describe the occurrence of events of particular interest in some application. We now go one step further: to a given event, additional properties can be related.

Example 7.22. In Example 1.11, the event $A =$"Fire starts" was considered. The stream of A is often modelled as a PPP. Now the fire was furthermore classified at the arrival by means of two scenarios: $B =$"Fire with flames" or (if B was false) "Smoke without flames". The date of the fire is written down and is marked with a star in case scenario B followed fire ignition, *i.e.* fire with flames was recorded, was true. Otherwise, when "merely smoke" was recorded, a dot is marked.

As it was shown in Eq. (2.14), if the scenario B were independent of stream of ignition then the point process of "stars" (dates of fires with flames) is a PPP too. Consider one type of fire, *e.g.* the one marked with stars[8]. In this section we discuss generalizations of the presented splitting of a PPP into point processes of stars and dots.   □

Consider an event A that is true at point S_i and suppose that S_i form a PPP with intensity λ. Consider for instance a Poisson process in the plane, as used

---

[7]This problem has even influenced literary texts, as the following excerpt from Pynchon's *Gravity's Rainbow*, [64], Part 1, Chapter 9:

> Roger has tried to explain to her the V-bomb statistics; the difference between distribution, in angel's-eye view, over the map of England, and their own chances, as seen from down there. She's almost got it, nearly understands his Poisson equation..."...Couldn't there be an equation for us too,..."..."...There is no way, love, not as long as the mean density of the strikes is constant ..."

[8]The same is valid for the point process of dots, since if B is independent of the stream A then the complement B^c is independent too.

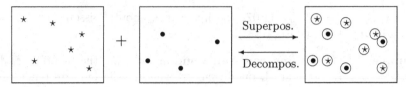

Fig. 7.8. Superposition (decomposition) of Poisson processes.

in Example 7.20. Let B be a scenario (a statement that can be true or false when A occurs, *i.e.* at points S_i). Now at each point S_i (when A occurs) we put a mark "star" if B is true. All remaining S_i (when B is false) are marked by dots (see Figure 7.8). If B is independent of the PPP A, then the point processes of stars and dots are independent Poisson and have intensities $P(B)\lambda$, $(1 - P(B))\lambda$, respectively.

It is not surprising that the reverse operation of superposition of two (or more) independent Poisson processes gives a Poisson process.

Theorem 7.1. Superposition Theorem: *Assume that we have two independent Poisson point processes S_i^I and S_i^{II} with intensities λ^I, λ^{II}, respectively. Consider a point process S_i, which is a union of the point processes S_i^I and S_i^{II}. (If S_i^I, S_i^{II} are marked by stars and dots, respectively, replace all symbols with a ring ($\circ$) and let S_i be positions of rings.) The point process of S_i is a superposition of the two processes and is a PPP itself, with intensity $\lambda = \lambda^I + \lambda^{II}$.*

For further reading about decomposition and superposition, including proofs, see the books by Gut [33] or Çinlar [11].

Problems

7.1. Assume that the lifetime process for humans has the death-rate function

$$\lambda(t) = a + b \cdot e^{t/c}, \qquad t > 0,$$

where $a = 3 \cdot 10^{-3}$, $b = 6 \cdot 10^{-5}$, and $c = 10$. The unit of time is 1 year.

(a) Calculate the probability that a person will reach the age of at least fifty.
(b) A person is alive on the day he is thirty. Calculate the conditional probability that he will live to be fifty.

7.2. Consider the experiment presented in Example 7.6. Use the Nelson–Aalen estimator to estimate the cumulative failure-intensity function of the observed lifetimes for concrete beams in air.

7.3. At time $t = 0$, a satellite is put into orbit. Two transmitters have been installed. At $t = 0$, both of them are working, but they break down independently with constant failure rate λ each. When both transmitters have failed to work, the satellite is out of order. Find the failure rate for the whole transmitter system.

7.4. The random variable Z is Poisson distributed and has a coefficient of variation of 0.50. Calculate $P(Z = 0)$.

7.5. The number of cars passing a street corner is modelled by a Poisson process with intensity $\lambda = 20$ h^{-1}. Calculate (approximately) the probability that more than 50 cars will pass during two hours (2 h).

7.6. Consider an oil pipeline. Suppose the number of imperfections $N(x)$ along a distance x can be modelled by a Poisson process, that is, $N(x) \in \text{Po}(\lambda x)$, where λ is the intensity (km^{-1}). Let $\lambda = 1.7$ km^{-1}.

(a) Calculate the probability that there are more than 2 imperfections along a distance of 1 km.
(b) Calculate the probability that two consecutive imperfections are separated by a distance longer than 1200 m.

7.7. Consider again the data set with time intervals between failures given in Problem 6.4.

(a) Test if data do not contradict the assumption of a PPP.
(b) Modelling the occurrences of failures of the air-conditioning system as a PPP, use the observations to estimate the intensity λ for plane 7914.

7.8. The number of defects ("specks") in plates is described by a Poisson distribution. One has the following observations: 30 plates are of colour 1 and 45 plates of colour 2.

Colour 1:	1	3	1	0	0	0	2	1	1	0	2	0	0	2	0
	1	0	2	0	0	2	0	0	1	1	0	0	1	0	0

(Observations from $\text{Po}(m_1)$)

Colour 2:	0	0	0	0	0	2	0	0	1	1	1	0	0	0	0
	0	1	0	0	1	0	0	1	0	1	0	0	0	0	0
	0	0	0	1	0	0	0	0	2	1	0	1	1	1	0

(Observations from $\text{Po}(m_2)$)
    Give an estimate of $m_1 - m_2$ and estimate the standard deviation of the proposed estimator of $m_1 - m_2$. Compute a 0.95-confidence interval for $m_1 - m_2$.

7.9. A group of parachutists is launched randomly over a region. Suppose the mean intensity of parachutists is λ per unit area and assume a Poisson model; that is, the number of people in a region of area A is Poisson distributed with mean λA.
    For a randomly selected person in this region, let R denote the distance to the nearest neighbour.

(a) Find the distribution for R. *Hint:* Note that $P(R > r)$ is the same as the probability of seeing no people within a circle of radius r.
(b) Give the expected value, $E[R]$ (cf. Problem 3.7).
(c) Suppose that a group of 20 people are launched over a region of size 1 km^2. An estimate of λ is then $2 \cdot 10^{-5}$ m^{-2}. Use the previous results to compute the average distance between the parachutists.

7.10. Consider flying-bomb hits on London, discussed in Example 7.21. The total number of small areas was 576 and the total number of hits was 537. In [12], the following numbers are found (reprinted in [25]):

k	0	1	2	3	4	≥ 5
n_k	229	211	93	35	7	1

where n_k is the number of areas with exactly k hits.

Test for a Poisson distribution using a χ^2 test.

7.11. Consider the data set of hurricanes, given in Problem 4.12.

(a) Based on the given 55 yearly observations, estimate the intensity of hurricanes. Compute the probability of more than 10 hurricanes in a given year using normal approximation. Use this probability to compute the expected number of years with more than 10 hurricanes during a 55-year period.

(b) The question of a possible increase over time of the average number of hurricanes has been much discussed in media as well as in the specialized research literature on climatology. We here investigate this complex issue by a simple Poisson-regression model:

$$\mathsf{E}[N_i] = \exp(\beta_0 + \beta_1 x_i), \qquad i = 1, \ldots, 55$$

where the explanatory variable x is time in years $x = 0, \ldots, 54$. A constant intensity over time means $\beta_1 = 0$. We want to test for a possible trend, *i.e.* the null hypothesis is $\beta_1 = 0$.

A software package returns the values of log-likelihood functions $l(\beta_0^*, \beta_1^*) = -123.8366$ (with $\beta_0^* = 1.8000$, $\beta_1^* = 1.4 \cdot 10^{-4}$) and $l(\beta_0^*, 0) = -123.8374$ (with $\beta_0^* = 1.8038$). Calculate the deviance and draw conclusions.

7.12. In Figure 7.9 are shown the locations of 71 pines in a square sampling region in Sweden. Use the division into 25 small squares given in the figure and perform

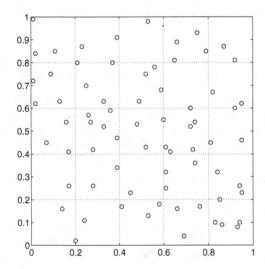

Fig. 7.9. Locations of pines in a square sampling region at a location in Sweden.

calculations as in Example 7.20 to investigate whether the pines are distributed in the plane according to a Poisson process.

Hint. Some observations fall at the border between squares. In that situation, let the observation belong to the higher limit.

7.13. Consider lorries travelling over a bridge. Assume that the times S_i of arrivals form a Poisson process with intensity 2 000 per day. Consider a scenario B = "A lorry transports hazardous material". Assume that the scenario is independent of the stream of lorries. (This would not be the case if a chemical company is usually sending a convoy of lorries with hazardous material to the same destination). From statistics, one has found that with probability $p = 0.08$, a lorry transports hazardous materials and with probability $q = 0.92$, other material is transported.

(a) In one week (Monday–Friday), on average 10 000 lorries will travel over the bridge. What is the probability that during a week a number of 300 more than the average will pass?

(b) What is the probability that during a week (Monday–Friday) there are more than 820 transports of hazardous materials?

8

Failure Probabilities and Safety Indexes

In Section 6.7 we discussed the problem of estimating risks for very rare accidents, which are seldom observed but can have serious consequences. In that situation the applicability of direct estimation of the probabilities using the empirical frequencies of such accidents is limited due to lack of data or large uncertainty in values of the computed measures of risks. An alternative method to compute risk, here the probability of at least one accident in one year, is to identify streams of events A_i, which, if followed by a suitable scenario B_i, leads to the accident. Then the risk for the accident is approximately measured by $\sum \lambda_{A_i} \mathsf{P}(B_i)$ where the intensities of the streams of A_i, λ_{A_i}, all have units [year^{-1}]. An important assumption is that the streams of initiation events are independent and much more frequent than the occurrences of studied accidents. Hence these can be estimated from historical records. (Estimations of intensities λ_i were discussed in the previous chapter.) What remains is computation of probabilities $\mathsf{P}(B_i)$.

We consider cases when the scenario B describes the ways systems can fail, or generally, some risk-reduction measures fail to work as planned. Hence $\mathsf{P}(B)$ describes the chances of a "failure", which we write explicitly in the notation $P_{\mathrm{f}} = \mathsf{P}(B)$. We are particularly interested in situations when, as often seen in safety of engineering structures, B can be written in a form that as a function of uncertain values (random variables) exceeds some critical level u^{crt}

$$B = \text{``} h(X_1, X_2, \ldots, X_n) > u^{\mathrm{crt}} \text{''}$$

Hence the main subject of this chapter is to study distributions of *functions of random variables* X_i with known distributions. Some of the variables X_i may describe uncertainty in model, parameters, etc. while others may describe genuine random variability of the environment. One thus mixes the variables X with distributions interpreted in the frequentist's way with variables having subjective probability distributions. Hence the interpretation of what the *failure probability*

$$P_{\mathrm{f}} = \mathsf{P}(B) = \mathsf{P}(h(X_1, X_2, \ldots, X_n) > u^{\mathrm{crt}}) \tag{8.1}$$

means is difficult and will depend on properties of the analysed risk scenario.

As mentioned earlier, in this chapter we focus on computations of P_f as defined in Eq. (8.1), hence with

$$Z = h(X_1, X_2, \ldots, X_n), \tag{8.2}$$

formally, the failure probability is given by

$$P_f = \mathsf{P}(Z > u^{\mathrm{crt}}) = 1 - F_Z(u^{\mathrm{crt}}).$$

At first, one might think it is a simple matter to find the failure probability P_f, since only the distribution of a single variable Z needs to be found. However, that is not the case. Here Z is a *function* of other variables and computation of its distribution is usually not a simple task. We give some examples in Section 8.1 when the distribution of Z can be computed. However, often that will not be possible or, if the information of the distribution of X_i is too uncertain, not really recommendable. In such situations we may use safety indices, introduced in Section 8.2, instead of poorly computed probabilities. For complicated problems, even the safety indices cannot be computed exactly. Thus we discuss, in Section 8.3, how Gauss' formulae can be employed to compute approximations for the value of an index. Gauss' formulae can also be used to approximately compute confidence intervals, the so-called delta method. This is presented in the final section.

8.1 Functions Often Met in Applications

The reliability of an engineering system may be defined as the probability of performing its intended function or mission. The level of performance of a system will obviously depend on the properties of the system. Often the problem can be formulated on the form *supply* versus *demand*, *i.e.* the (supply) capacity of a system must meet certain (demand) requirements.

A typical example is an imposed load on a structure. Here, the strength of the material, including material constants and geometry of the structure, is an example of variables of supply type. The load is regarded as a demand. In civil engineering a situation is often considered where variables can be classified as describing *strengths* of the system (higher strength means lower probability of failure). Other variables can be called *loads*, since higher loads will lead to higher probability of failure.

In this section, we discuss the distribution of Z in Eq. (8.2) for some standard types of functions and common families of distributions. In some of the examples, we study applications involving variables having interpretations as strengths or loads.

8.1.1 Linear function

Example 8.1 (Load and strength). Consider for simplicity a system with a single random strength R and a load S. The system will fail when the strength is lower than the load, hence we study

$$Z = R - S$$

and the statement "System fails" is true when $Z < 0$, *i.e.* $R < S$. We wish to find the distribution of a linear combination of random variables of supply-and-demand type. Generally the probability $P(R < S)$ has to be computed by means of numerical integration. If R and S are independent, Eq. (5.23) can be employed, that is

$$P(R < S) = \int P(R < s) f_S(s) \, \mathrm{d}s = \int F_R(s) f_S(s) \, \mathrm{d}s.$$

Alternatively, one can simulate independent random numbers r_i and s_i and estimate the frequency of cases when $r_i < s_i$. That frequency becomes an estimate of the probability $P(R < S)$. Often in reliability applications, the case is encountered that S is Gumbel distributed, R is Weibull.   □

We now give an example where Eq. (5.23) is used to obtain an expression for the distribution of the sum.

Example 8.2 (Crack propagation, time to failure). Consider crack growth in some specimen. The time to failure, T, due to cracking is the sum of two times, $T = T_1 + T_2$:

$T_1 = $ "Time to initiation of a microscopic crack",

$T_2 = $ "Time for the crack to grow a fixed distance and cause failure"

If the component is supposed to be used for a period of time t_0, failure occurs if $T < t_0$.

We may model T_1 as an exponential random variable with mean $1/\lambda$ (the initiation is caused by an accident[1]). Further, in well-controlled experiments, T_2 is often well modelled by a Gumbel distribution with parameters dependent on how extensive the load is and the place where the crack was initiated.

The probability of failure in the sense of above is thus given by Eq. (5.23), *viz.*

$$P(T \le t_0) = P(T_2 \le t_0 - T_1) = \int_0^{t_0} P(T_2 < t_0 - t_1) f_{T_1}(t_1) \, \mathrm{d}t_1$$

$$= \int_0^{t_0} \exp(-\mathrm{e}^{-(t_0 - t_1 - b)/a}) \lambda \mathrm{e}^{-\lambda t_1} \, \mathrm{d}t_1$$

which can be computed by numerical integration if the parameters a, b, and λ are known.   □

In the following example, we formulate a safety criterion where a sum of random variables appears. We focus on the distribution of sums of random variables in a moment.

Example 8.3 (Hooke's law). By Hooke's law, the elongation ϵ of a fibre is proportional to the force F, that is, $\epsilon = K^{-1} F$ or $F = K\epsilon$. Here K, called Young's modulus, is uncertain and modelled as a random variable with mean m and variance σ^2.

---

[1] For example, the load exceeded the fatigue limit, or a change in the geometries of the object due to the accident causes higher stress concentrations.

Consider a wire containing 1000 fibres with individual independent values of Young's modulus K_i. A safety criterion is given by $\epsilon \leq \epsilon_0$. With $F = \epsilon \sum K_i$ we can write

$$P(\text{"Failure"}) = P\left(\frac{F}{\sum K_i} > \epsilon_0\right) = P(\epsilon_0 \sum K_i - F < 0).$$

Hence, in this example, we have

$$h(K_1, \ldots, K_{1000}, F) = \epsilon_0 \sum K_i - F$$

which is a linear function of K_i and F. Here, F is an external force (load) while $\sum K_i$ is the strength of the material. $\qquad \square$

An important linear function to study is the sum of n random variables

$$Z = X_1 + \cdots + X_n.$$

We restrict ourselves to independent X_i. If all X_i have the same distribution $F(x)$, say, with mean $m = E[X]$ and variance $\sigma^2 = V[X]$ then for large n the distribution of the sum $P(Z \leq z)$ can be approximately computed using the Central Limit Theorem presented in Theorem 4.4. For a small number of summands n and (or) when variables have different distributions it is usually hard to compute the distribution of the sum. There are, however, some exceptions.

Normal variables

The most important case when the sum of random numbers is particularly easy to handle is when X_i are normally distributed. From Chapter 3 we know that any normally distributed random variable $Z \in N(m, \sigma^2)$ is defined by two parameters, location m and scale σ, and hence only these have to be specified.

Theorem 8.1. *If $X_1, \ldots, X_n$ are independent normally distributed random variables, i.e. $X_i \in N(m_i, \sigma_i^2)$, then their sum Z is normally distributed too, i.e. $Z \in N(m, \sigma^2)$, where*

$$m = m_1 + \cdots + m_n, \qquad \sigma^2 = \sigma_1^2 + \cdots + \sigma_n^2. \tag{8.3}$$

This property extends to dependent variables; here we present the case when $n = 2$. Suppose $X_1, X_2 \in N(m_1, m_2, \sigma_1^2, \sigma_2^2, \rho)$; then for any constants a, b, and c the variable

$$Z = c + aX_1 + bX_2 \in N(m, \sigma^2), \tag{8.4}$$

where

$$m = c + a\,m_1 + b\,m_2, \qquad \sigma^2 = a^2\sigma_1^2 + b^2\sigma_2^2 + 2\,a\,b\,\sigma_1\sigma_2\rho. \tag{8.5}$$

Example 8.4 (Hooke's law). Consider again the wire, composed of 1000 fibres. Assume that $F \in \mathrm{N}(m_F, \sigma_F^2)$ is independent of K_i. By the central limit theorem, we find that $\sum K_i$ is approximately $\mathrm{N}(1000m, 1000\sigma^2)$ where $\mathrm{E}[K_i] = m$, $\mathrm{V}[K_i] = \sigma^2$. Introducing $Z = \epsilon_0 \sum K_i - F$, we have that $Z \in \mathrm{N}(m_Z, \sigma_Z^2)$ where

$$m_Z = 1000\, m\, \epsilon_0 - m_F, \qquad \sigma_Z^2 = 1000\, \epsilon_0^2\, \sigma^2 + \sigma_F^2.$$

Hence

$$\mathrm{P}(\text{"Failure"}) = \mathrm{P}(Z < 0) = \Phi\left(-\frac{m_Z}{\sigma_Z}\right).$$

□

Gamma variables

For independent gamma distributed random variables $X_1, X_2, \ldots, X_n$, where $X_i \in \mathrm{Gamma}(a_i, b)$, $i = 1, \ldots, n$, one can show that

$$\sum_{i=1}^{n} X_i \in \mathrm{Gamma}(a_1 + a_2 + \cdots + a_n, b).$$

That is, the sum of gamma variables with common parameter b is again gamma distributed. Recall from Section 3.3.1 that $X \in \mathrm{Gamma}(1, b)$ is an exponentially distributed r.v. (with expectation $\mathrm{E}[X] = 1/b$). Hence, the sum of iid exponentially distributed random variables is Gamma distributed.

Example 8.5. Suppose we are exposed to some risk of accidents with intensity λ, which can be well approximated by means of a Poisson point process, *e.g.* the distances between accidents are independent exponentially distributed with mean equal to the return period $1/\lambda$. The mission that will take time t has capacity to survive $n-1$ accidents, *i.e.* it fails if $T = T_1 + T_2 + \cdots + T_n < t$. Now, the event $T < t$ is equivalent to the event $N(t) \geq n$, where $N(t)$ is the number of accidents in period t. It follows that

$$P_{\mathrm{f}} = \mathrm{P}(T < t) = 1 - e^{-\lambda t} \sum_{k=0}^{n-1} \frac{(\lambda t)^k}{k!}. \tag{8.6}$$

We demonstrated that for $T \in \mathrm{Gamma}(n, \lambda)$, Eq. (8.6) gives the cdf for T.

□

Poisson variables

Already in Chapter 2, the superposition of Poisson streams was discussed. In the simpler situation, just considering random variables, one can prove that a sum of independent Poisson variables, $K_i \in \text{Po}(m_i)$, $i = 1, \ldots, n$, is again Poisson distributed:

$$\sum_{i=1}^{n} K_i \in \text{Po}(m_1 + \cdots + m_n).$$

Recall the more general results of superposition and decomposition of Poisson processes in Section 7.6.

8.1.2 Often used non-linear function

As mentioned in the introduction, failures of some systems can be described as the value of a function of random variables exceeding a threshold. We here present an example of such a situation, which leads us to closer studies of the lognormal distribution.

Lognormal variables

Assume that in year 2000, one has invested K [€] in a stock portfolio and one wonders what its value will be in year 2020. Denote the value of the portfolio in year 2020 by Z and let X_i be factors by which this value changed during a year $2000 + i$, $i = 0, 1, \ldots, 20$. Obviously the value is given by

$$Z = K \cdot X_0 \cdot X_1 \cdot \ldots \cdot X_{20}.$$

Here "failure" is subjective and depends on our expectations, for example we may be interested in an increase of our savings and hence "failure" means that $Z < K$. In order to estimate the risk (probability) for failure, one needs to model the properties of X_i. Whether the factors X_i are independent and follow the same distribution is not easy to know. One can only study historical data and develop a model for X_i under assumption that the future will follow the same model as the past.

A variable Z, which is a multiplication of different factors, has found many applications in engineering and hence finding the distribution of Z is an important problem. This is often done by means of logarithmic transformation

$$\ln Z = \ln K + \ln X_1 + \cdots + \ln X_n.$$

In order to compute the distribution of $\ln Z$ we need to find the distribution of a sum of random variables. Obviously, if the distribution of $\ln Z$ is known, i.e. there is a function $F(r)$, say, such that $\mathsf{P}(\ln Z \leq r) = F(r)$, then the distribution of Z is given by

$$F_Z(z) = \mathsf{P}(Z \leq z) = \mathsf{P}(\ln Z \leq \ln z) = F(\ln z).$$

Often one can assume that X_i are iid and n is large. Then the Central Limit Theorem shows that $\ln Z$ is approximatively normally distributed. This is an important situation, and the distribution presented here is widely used in applications.

Definition 8.1 (Lognormal distribution). *A variable* Z *such that*

$$\ln Z \in N(m, \sigma^2)$$

is called a **lognormal variable.**

Using the distribution Φ of a $N(0, 1)$ variable (see Eq. (3.6)) we have that

$$F_Z(z) = P(Z \le z) = P(\ln Z \le \ln z) = \Phi\left(\frac{\ln z - m}{\sigma}\right). \tag{8.7}$$

Moreover, it can be proven that for a lognormally distributed variable Z,

$$E[Z] = e^{m + \sigma^2/2}, \tag{8.8}$$

$$V[Z] = e^{2m} \cdot (e^{2\sigma^2} - e^{\sigma^2}), \tag{8.9}$$

$$D[Z] = e^m \sqrt{e^{2\sigma^2} - e^{\sigma^2}} = e^{m + \sigma^2/2} \cdot \sqrt{e^{\sigma^2} - 1}. \tag{8.10}$$

Note that the coefficient of variation $R[Z] = \sqrt{\exp(\sigma^2) - 1}$ is only a function of σ^2; solving for σ^2, we obtain $\sigma^2 = \ln(1 + R[Z]^2)$. With σ^2 known, m can be computed if $E[Z]$ is given. However, m is much easier to find in the case when the median of Z is specified. For a normal variable $\ln Z \in N(m, \sigma^2)$ the parameter m is both mean and median, thus

$$0.5 = P(\ln Z \le m) = P(Z \le e^m),$$

and hence the median of Z is $\exp(m)$.

We now give an example where the product of independent lognormally distributed variables is presented. In this example we estimate the risk for "failure" of cleaning spill water in a chemical industry.

Example 8.6 (Concentration of pollutants). Suppose the spilled water in a chemical factory is treated before it is dumped into a nearby lake. Let X denote the concentration of a pollutant feeding into the treatment system, and Y the concentration of the same pollutant leaving the system. Suppose that for a day, X has a lognormal distribution with median 4 mg/l and coefficient of variation $R[X] = 0.2$.

Because of the erratic nature of biological and chemical reactions, the efficiency of the treatment system is unpredictable. Hence the fraction of pollutant remaining untreated, denoted by K, is also a random variable. Assume K

Fig. 8.1. Treatment system for spill water; X: concentration of pollutant before treatment; Y: concentration of pollutant after treatment; K: efficiency of treatment.

is lognormal with median of 0.15 and coefficient of variation $R[K] = 0.1$. We also assume that X and K are independent.

We answer several questions:

(i) What is the distribution of $Y = K \cdot X$?

The lognormal variable is defined by the property that its logarithm is normally distributed. Hence $\ln K \in N(m_1, \sigma_1^2)$ and $\ln X \in N(m_2, \sigma_2^2)$. Since K and X are independent, their logarithms are independent too. Consequently

$$\ln Y = \ln KX = \ln K + \ln X \in N(m, \sigma^2),$$

is a sum of two independent normal variables and hence it is also normal

$$\ln Y \in N(m_1 + m_2, \sigma_1^2 + \sigma_2^2).$$

What remains is to find the parameters $m_1, m_2, \sigma_1, \sigma_2$ from the specifications of the problem. Note that $m_2 = E[\ln X]$ and $\sigma_2 = D[\ln X]$ are not simply equal to $E[X]$, $D[X]$, respectively. Using the relations in Eqs. (8.8-8.10) we find

$$m_1 = \ln 0.15, \quad \sigma_1^2 = \ln(1 + 0.1^2), \quad m_2 = \ln 4, \quad \sigma_2^2 = \ln(1 + 0.2^2),$$

and finally

$$m = \ln 4 + \ln 0.15 = -0.51, \qquad \sigma = \sqrt{\ln(1 + 0.2^2) + \ln(1 + 0.1^2)} = 0.22.$$

(ii) Suppose the maximal concentration of the pollutant permitted to be dumped into the lake is specified to be 1 mg/l. What is the probability (failure probability) that on a normal day this specified standard will be exceeded?

What is needed to calculate is $P(Y > 1)$. This is simple since

$$P(Y > 1) = P(\ln Y > 0) = 1 - \Phi(-m/\sigma) = 1 - 0.99 = 0.01.$$

$\square$

Model uncertainty

Lognormal distributions are often used to describe model uncertainties. Consider a quantity Z_{mod}, which is modelled by $g(X_1, \ldots, X_n)$ where $X_1, \ldots, X_n$ are uncertain parameters or measured quantities. If the true value z can be obtained from an experiment, when the values $X_i = x_i$ are known, we have $z = k \cdot g(x_1, \ldots, x_n)$. The quantity $k = z/g(x_1, \ldots, x_n)$ is called a model uncertainty factor. Since the fraction k varies, we write the relation $Z = K \cdot g(X_1, \ldots, X_n)$.

A common model is a lognormal random variable K. Suppose this is specified by its median, $k_{0.5}$ say, and its coefficient of variation $a = \mathsf{R}[K]$. If $a < 0.2$ then

$$K \approx k_{0.5}\, e^{aX}$$

where $X \in \mathrm{N}(0,1)$. Values of the median $k_{0.5}$ are then interpreted as follows:

- $k_{0.5} = 1$: the model is unbiased
- $k_{0.5} < 1$: the model is conservative (gives often too large estimates)
- $k_{0.5} > 1$: the model is unconservative (gives often too small values)

8.1.3 Minimum of variables

The weakest-link principle, used for instance in mechanics, means that the strength of a structure is equal to the strength of its weakest part. In other words we may say that "failure" occurs if the minimum strength of some component is below a critical level u^{crt}:

$$\min(X_1,\ldots,X_n) \le u^{\mathrm{crt}}.$$

If X_i are independent with distributions F_i, then

$$
\begin{aligned}
\mathsf{P}(\min(X_1,\ldots,X_n) \le u^{\mathrm{crt}}) &= 1 - \mathsf{P}(\min(X_1,\ldots,X_n) > u^{\mathrm{crt}}) \\
&= 1 - \mathsf{P}(X_1 > u^{\mathrm{crt}},\ldots,X_n > u^{\mathrm{crt}}) \\
&= 1 - (1 - F_1(u^{\mathrm{crt}})) \cdot \ldots \cdot (1 - F_n(u^{\mathrm{crt}})).
\end{aligned}
$$

The computations are particularly simple if all X_i are Weibull distributed.

Example 8.7 (Strength of a wire). In laboratory, experiments have been performed with 5-centimeter-long wires with strengths X_i, $i = 1,\ldots,n$. The average strength is $m_X = 200$ kg and the coefficient of variation $\mathsf{R}[X] = 0.20$. From experience, one knows that such wires have Weibull-distributed strengths,

$$F_{X_i}(x) = 1 - e^{-(x/a)^c}, \quad x \ge 0,$$

and the relation

$$a = \frac{\mathsf{E}[X]}{\Gamma(1 + 1/c)}$$

is valid. Consider now the distribution of the strength X of a l-m long wire. This can be seen as a chain composed of $k = 20\,l$ 5-cm-long metre wires. Hence, the distribution of $X = \min(X_1, X_2, \ldots, X_k)$ is

$$\mathsf{P}(X \le x) = 1 - (1 - (1 - e^{-(x/a)^c}))^k = 1 - e^{-k(x/a)^c} = 1 - e^{-(x/a_k)^c},$$

that is, a Weibull distribution with a new scale parameter $a_k = a/k^{1/c}$. The change of scale parameter due to minimum formation is called *size effect* (larger objects are weaker).

If we want to calculate the probability that a wire of length 5 m will have a strength less than 50 kg, we need values of the parameters a and c. From the experiments, the coefficient of variation $R[X] = 0.20$ is known, as well as the expectation $E[X] = 200$ kg. From tables or by numerical computation values of a and c can be obtained (cf. Table 4 in appendix). In our case, we find $c = 5.79$ and $a = E[X]/\Gamma(1 + 1/c) = 200/0.9259 = 216.01$ and hence $a_k = 216.01/100^{1/5.79} = 97.51$. Thus

$$P(X \leq 50) = 1 - e^{-(50/97.51)^{5.79}} = 0.021.$$

$\square$

The distribution of the *maximum* of random variables is studied deeper in Chapter 10, where topics from statistical extreme-value theory are discussed.

8.2 Safety Index

A safety index is used in risk analysis as a safety measure, which is high when the probability of failure P_f is low. This measure is a more crude tool than the probability, and it is used when the uncertainty in P_f is too large or when there is not sufficient information to compute P_f.

8.2.1 Cornell's index

Let us return to the simplest case, $Z = R - S$, introduced in Example 8.1. As illustrated before, the distribution of a sum (and difference) of two random variables often cannot be given by an analytical formula but has to be computed using numerical methods. In the special, but very important, case where the variables R and S are independent and normally distributed, i.e. $R \in N(m_R, \sigma_R^2)$ and $S \in N(m_S, \sigma_S^2)$, then also $Z \in N(m_Z, \sigma_Z^2)$, where $m_Z = m_R - m_S$ and $\sigma_Z = \sqrt{\sigma_R^2 + \sigma_S^2}$, and thus

$$P_f = P(Z < 0) = \Phi\left(\frac{0 - m_Z}{\sigma_Z}\right) = \Phi(-\beta_C) = 1 - \Phi(\beta_C),$$

where

$$\beta_C = \frac{m_Z}{\sigma_Z}$$

is the so-called *Cornell's safety index*. The index measures the distance from the mean $m_Z = E[Z] > 0$ to the unsafe region (that is zero) in the number of standard deviations. For illustration, see Figure 8.2 where we have chosen

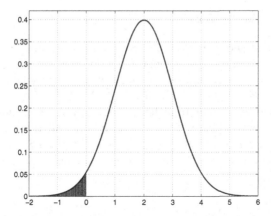

Fig. 8.2. Illustration of safety index. Here: $\beta_C = 2$. Failure probability $P_f = 1 - \Phi(2) = 0.023$ (area of shaded region).

$m_Z = 2$ and $\sigma_Z = 1$ in a normal distribution. For these values of parameters, we can immediately deduce the value of β_C, by just inspection of the figure. Recall the interpretation of measuring the distance from the mean expressed in standard deviations; we thus find $\beta_C = 2$. In this particular situation we have a failure probability $P_f = 1 - \Phi(2) \approx 0.023$.

For R and S that are not normally distributed it may be much more difficult to get the distribution of Z (principally, one has to compute an integral). More importantly, our knowledge about the distributions of R and S can be very uncertain, making the whole issue of computation of the distribution of Z questionable. However, in such a situation and also in the general case, *i.e.* when Z is defined by Eq. (8.2) and is a function of may be hundreds of strength and load variables, we may still compute Cornell's index.

Even for general Z, Cornell's safety index $\beta_C = 4$ still means that the distance from the mean of Z to the unsafe region is 4 standard deviations. Observe that usually $P_f \neq 1 - \Phi(\beta_C)$, and we have no exact relation between the index β_C and the failure probability P_f. There exists however a conservative estimate of P_f (we do not prove it here even if the proof is not especially difficult), namely

$$P(\text{"System fails"}) = P(Z < 0) \leq \frac{1}{1 + \beta_C^2}. \tag{8.11}$$

Clearly, the higher the safety index, the safer the system. The bounds are valid for any variable Z, $E[Z] \geq 0$, independent of its distribution and are hence quite conservative. For example, if the safety index is 3, then our bound tells us that failure probability is less than 1 per 10. If we knew that Z was a normally distributed variable, then $\beta_C = 3$ corresponds to the failure probability below 2 per 1000.

8.2.2 Hasofer-Lind index

What we have just shown is that Cornell's index is a quite crude measure of reliability. It has one more deficiency: it is not unique. Let us explain why. The statement "System fails" is equivalent to the failure set, which is characterized by the inequality $Z = h(R_1, \ldots, R_k, S_1, \ldots, S_n) < 0$. The failure set is defined uniquely but the function h (and hence the variable Z) is not unique. For example, in the simplest case of one strength and load variable we can write $P(\text{"System fails"}) = P(R/S - 1 < 0)$, and hence $\widetilde{Z} = R/S - 1$ could be used instead of Z and one could compute the safety index $\widetilde{\beta}_C$, say. Obviously, the failure probability

$$P_{\mathrm{f}} = P(Z < 0) = P(\widetilde{Z} < 0), \quad \text{but} \quad \beta_C \neq \widetilde{\beta}_C.$$

In this introductory section, we use Cornell's index only as an example of a notion to measure risk. In practice, β_C is seldom used.

We have demonstrated that the value of Cornell's safety index may depend on the choice of function h. This undesirable property can be remedied by the so-called Hasofer–Lind index, here denoted by β_{HL} and presented by Hasofer and Lind in [35]. This index measures distance from expectations of the strength and load variables to the unsafe region in a way that is independent of a particular choice of the h function. The Hasofer–Lind safety index is commonly used in reliability analysis, although quite advanced computer software is needed for its computation. However:

> In the special case when h is a linear function, the Hasofer–Lind index β_{HL} is equal to Cornell's index β_C.

For a more general discussion of safety indexes with applications to structural engineering, consult Ditlevsen and Madsen, [21].

8.2.3 Use of safety indexes in risk analysis

Here we sketch a common application of the safety indexes (Hasofer–Lind) in risk analysis related to design of structures. The material is based on the recommendations proposed by *Joint Committee on Structural Safety* (see [79]). For β_{HL}, one has approximately that $P_{\mathrm{f}} \approx \Phi(-\beta_{\mathrm{HL}})$. Clearly, a higher value of the safety index implies lower risk for failure and also a more expensive structure. In order to propose the so-called target safety index one needs to consider both costs and consequences. Possible *classes of consequences* are:

Minor Consequences This means that risk to life, given a failure, is small to negligible and economic consequences are small or negligible (*e.g.* agricultural structures, silos, masts).

Moderate Consequences This means that risk to life, given a failure, is medium or economic consequences are considerable (*e.g.* office buildings, industrial buildings, apartment buildings).

Table 8.1. Safety index and consequences.

Relative cost of safety measure	Minor consequences of failure	Moderate consequences of failure	Large consequences of failure
Large	$\beta_{\mathrm{HL}} = 3.1$	$\beta_{\mathrm{HL}} = 3.3$	$\beta_{\mathrm{HL}} = 3.7$
Normal	$\beta_{\mathrm{HL}} = 3.7$	$\beta_{\mathrm{HL}} = 4.2$	$\beta_{\mathrm{HL}} = 4.4$
Small	$\beta_{\mathrm{HL}} = 4.2$	$\beta_{\mathrm{HL}} = 4.4$	$\beta_{\mathrm{HL}} = 4.7$

Major Consequences This means that risk to life, given a failure, is high or that economic consequences are significant (*e.g.* main bridges, theatres, hospitals, high-rise buildings).

Obviously, the cost of risk prevention, etc. also has to be considered (see Table 8.1) where we present target reliability indexes ("target" means that one wishes to design the structures so that the safety index for a particular failure mode will have the target value). Here the so-called "ultimate limit states" are considered, which means failure modes of the structure — in everyday language: that one cannot use it anymore. This kind of failure concerns mainly the maximum load-carrying capacity as well as the maximum deformability.

In order to give some intuition what "target safety level", proposed in the table, means we now have a brief discussion of the problem.

8.2.4 Return periods and safety index

As mentioned before, if Z were a normally distributed variable, then the failure probability $P_{\mathrm{f}} = 1 - \Phi(\beta_{\mathrm{HL}})$. Consequently $\beta_{\mathrm{HL}} = 3.1$ gives P_{f} corresponding to one per thousand. We can think about the value $1/1000$ as a nominal value that can be used to compare different solutions (constructions) at the design stage of a construction. (Higher value of index means safer structure.)

It is important to remember that the values of β_{HL} contain time information. An important issue is that the safety index considers a measure of safety for *one year*, *i.e.* $P_{\mathrm{f}} = P_t(A)$ where $t = 1$ year. As discussed in Chapter 2, the severity of the event A can be measured using its return period, *i.e.* if $P_{\mathrm{f}} = 0.01$, A is called a 100-year event.

The safety index $\beta_{\mathrm{HL}} = 3.1$ implies that the intensity of accidents is $1/1000$ [year^{-1}], or equivalently, the return period is 1000 years. Corresponding return periods to the other values of β_{HL} in Table 8.1 can be found:

Safety index β_{HL}	3.1	3.3	3.7	4.2	4.4	4.7
Return period (years)	10^3	$2 \cdot 10^3$	10^4	10^5	$2 \cdot 10^5$	10^6

(Note that these are nominal values.) Since most buildings follow these design recommendations and we do not observe failures very frequently, it means that the method is not too unconservative.

Note that if there are 100 objects having return period of 1000 years between failures ($\beta_{HL} = 3.1$) then (under assumption of independence of failures between the objects) the return period of a failure of one object is only 10 years. The intensity of failure in the populations of 100 objects will be 100 times higher than the intensity of failure for an individual object, *i.e.* equal to $100/1000$ giving a return period of 10 years.

Finally, a structure may contain n different failure modes. If we assume that those are independent and have failure probabilities $P_f(i)$, $i = 1, \ldots, n$ then the return period of failure of the structure

$$T_f = \frac{1}{\lambda}, \quad \text{where} \quad \lambda = \sum_{i=1}^{n} P_f(i) = \sum_{i=1}^{n} \lambda_i,$$

and $1/\lambda_i$ are return periods for accident of type i (cf. Theorem 2.4, page 39).

Note that the number n is usually under-estimated, since there often exist failure modes that were not taken into account. That is why safety indices corresponding to return periods of millions of years are used. (If you have 1000 independent failure modes, each with return period 1 million years, the nominal return period for the whole structure will be only 1000 years.)

8.2.5 Computation of Cornell's index

Although Cornell's index β_C has some deficiencies it is still an important measure of safety and its inverse, the coefficient of variation, also called relative uncertainty, is frequently computed in practical situations. We turn now to computation of β_C, which in general has to be done approximately.

Let us return to the random variable from Eq. (8.2)

$$Z = h(R_1, \ldots, R_k, S_1, \ldots, S_n),$$

such that $E[Z] > 0$. Assume that only expected values and variances of the variables R_i and S_i are known. We also assume that all strength and load variables are independent. In order to compute β_C we need to find

$$E[h(R_1, \ldots, R_k, S_1, \ldots, S_n)], \qquad V[h(R_1, \ldots, R_k, S_1, \ldots, S_n)].$$

We have presented formulae for computation of the variance of a linear function of random variables (see Eq. (5.11)). However, the function h is usually much more complicated and computation of Cornell's index

$$\beta_C = \frac{E[h(R_1, \ldots, R_k, S_1, \ldots, S_n)]}{\left[V[h(R_1, \ldots, R_k, S_1, \ldots, S_n)]\right]^{1/2}}$$

can only be done by means of some approximations. The main tools are the so-called *Gauss' formulae*, which is presented and discussed next. For the one-dimensional case, see Eqs. (8.14-8.15); while for a more general case, cf. Eqs. (8.16-8.17). In the physics literature, one speaks about the law of propagation of error.

8.3 Gauss' Approximations

We first state Gauss' approximation for a function of one random variable.

> **Theorem 8.2 (Gauss' approximation, one variable).** *Let X be a random variable with $\mathsf{E}[X] = m$ and $\mathsf{V}[X] = \sigma^2$. Further, let h be a function with continuous derivative. Then*
>
> $$\mathsf{E}[h(X)] \approx h(m) \quad and \quad \mathsf{V}[h(X)] \approx (h'(m))^2 \sigma^2. \tag{8.12}$$

A motivation for the result in Eq. (8.12) follows. Choose a fixed point x_0, and write Taylor's formula to approximate h around x_0 by a polynomial function

$$h(x) \approx h(x_0) + h'(x_0)(x - x_0) + \frac{1}{2}h''(x_0)(x - x_0)^2.$$

Now, let us choose x_0 to be a "typical value" $x_0 = \mathsf{E}[X] = m$, say. Then using Eq. (3.18) we have that

$$\mathsf{E}[h(X)] \approx h(m) + h'(m)\mathsf{E}[(X - m)] + \frac{1}{2}h''(m)\mathsf{E}[(X - m)^2]$$

$$= h(m) + \frac{1}{2}h''(m)\mathsf{V}[X] \tag{8.13}$$

since $\mathsf{E}[(X - m)] = 0$ and $\mathsf{V}[X] = \mathsf{E}[(X - m)^2]$ (see Eq. (3.19)). Since even the function h can be uncertain, for example it can be derived empirically from some measurements using statistical methods like regression or smoothing, the second derivative $h''(m)$ can be corrupted by errors. Hence, one often disregards the term $\frac{1}{2}h''(m)\mathsf{V}[X]$ in Eq. (8.13) and uses a simplified form of Gauss' approximation

$$\mathsf{E}[h(X)] \approx h(\mathsf{E}[X]). \tag{8.14}$$

If the function h is approximately linear in a neighbourhood where its argument is calculated, Eq. (8.14) is a good approximation of the expectation.

We turn now to the variance and, by again using the Taylor expansion around $x_0 = m$, we have that

$$\mathsf{V}[h(X)] \approx \mathsf{V}\big[h(m) + h'(m)(X - m)\big] = \big(h'(m)\big)^2 \mathsf{V}[X] \tag{8.15}$$

where we have used Eq. (3.20) and the fact that m, $h(m)$, and $h'(m)$ are constants.

The more general case when h is a function of several variables, follows from a multi-dimensional version of Taylor's formula. For transparency of the formulae, we consider first a function h of two variables X and Y.

Theorem 8.3 (Gauss' approximation, two variables). *Let X and Y be independent random variables with expectations m_X, m_Y, respectively. For a smooth function h the following approximations*

$$E[h(X,Y)] \approx h(m_X, m_Y), \tag{8.16}$$

$$V[h(X,Y)] \approx \left[h_1(m_X, m_Y)\right]^2 V[X] + \left[h_2(m_X, m_Y)\right]^2 V[Y], \tag{8.17}$$

where

$$h_1(x,y) = \frac{\partial}{\partial x} h(x,y), \qquad h_2(x,y) = \frac{\partial}{\partial y} h(x,y),$$

*are called **Gauss' formulae.***

Remark 8.1. Gauss' approximation formulae have been derived for independent, or rather uncorrelated, variables X and Y. If X and Y are correlated the derivation from Taylor's formula to Gauss' approximation is not correct, simply one term is missing. The correct formula is as follows

$$E[h(X,Y)] \approx h(m_X, m_Y), \tag{8.18}$$

$$V[h(X,Y)] \approx \left[h_1(m_X, m_Y)\right]^2 V[X] + \left[h_2(m_X, m_Y)\right]^2 V[Y]$$
$$+ 2h_1(m_X, m_Y) h_2(m_X, m_Y) \text{Cov}[X,Y]. \tag{8.19}$$

$\square$

Using the general versions of the formulae (8.16-8.17), Cornell's index can be approximately computed by the following formula

$$\beta_C \approx \frac{h(m_{R_1}, \ldots, m_{R_k}, m_{S_1}, \ldots, m_{S_n})}{\left[\sum_{i=1}^{k+n} \left[h_i(m_{R_1}, \ldots, m_{R_k}, m_{S_1}, \ldots, m_{S_n})\right]^2 \sigma_i^2\right]^{1/2}}, \tag{8.20}$$

where σ_i^2 is the variance of the ith variable in the vector of loads and strengths $(R_1, \ldots, R_k, S_1, \ldots, S_n)$, while h_i denote the partial derivatives of the function h. (Here loads and strengths are mutually independent.)

As soon as we face a mathematical model — a relation obtained by physical laws or by experiments — in any field in science and technology, Gauss' formulae might be useful tools. Note that the distributions of the random quantities need not to be known, just the expectations and standard deviations. We give here an example from solid mechanics.

Example 8.8. Consider a beam of length $L = 3$ m. A random force P with expectation 25 000 N and standard deviation 5 000 N is applied at the midpoint of such a beam[2]. The modulus of elasticity E of a randomly chosen

---

[2]We neglect the fact that parameters at some stage have to be estimated.

beam has the expectation $2 \cdot 10^{11}$ Pa and the standard deviation $3 \cdot 10^{10}$ Pa. All beams share the same second moment of (cross-section) area $I = 1 \cdot 10^{-4}$ m^4. Then the vertical displacement of the midpoints is

$$U = \frac{PL^3}{48EI}.$$

Give approximately $\mathsf{E}[U]$ and $\mathsf{V}[U]$.

In the model, P and E are considered random variables; P being an external load and E describing material properties. Assume that P and E are uncorrelated. Introducing

$$h(P, E) = \frac{PL^3}{48EI}$$

we have

$$h_1(P, E) = \frac{\partial}{\partial P} h(P, E) = \frac{L^3}{48EI},$$

$$h_2(P, E) = \frac{\partial}{\partial E} h(P, E) = -\frac{PL^3}{48E^2I},$$

and Gauss' formulae yield

$$\mathsf{E}[U] = \frac{\mathsf{E}[P]L^3}{48\mathsf{E}[E]I} = \frac{25\,000 \cdot 3^3}{48 \cdot 2 \cdot 10^{11} \cdot 1 \cdot 10^{-4}} = 7.03 \cdot 10^{-4} \text{ m},$$

$$\mathsf{V}[U] = \mathsf{V}[P]\big[\,h_1(\mathsf{E}[P], \mathsf{E}[E])\,\big]^2 + \mathsf{V}[E]\big[\,h_2(\mathsf{E}[P], \mathsf{E}[E])\,\big]^2 = 1.11 \cdot 10^{-8} \text{ m}^2.$$

Hence $\mathsf{D}[U] = 1.06 \cdot 10^{-4}$ m and the coefficient of variation is 15%.

Suppose the vertical displacement must be smaller than 1.5 mm. Introducing

$$Z = 1.5 \cdot 10^{-3} - U,$$

we are able to use Eq. (8.11) to estimate the failure probability $\mathsf{P}(Z < 0)$. We have that Cornell's index $\beta_{\mathrm{C}} = (1.5 \cdot 10^{-3} - \mathsf{E}[U])/\mathsf{D}[U] = 7.52$ and hence an estimate is given as

$$\mathsf{P}(Z < 0) \leq \frac{1}{1 + \beta_{\mathrm{C}}^2} = 0.017.$$

$\square$

8.3.1 The delta method

Gauss' approximation gives, as we have seen, a way of estimating the variance for a non-linear function h of random variables. Here we use it to construct confidence intervals for quantities, which are functions of some parameters. In order to construct a confidence interval the distribution of the estimation error $\mathcal{E}$ needs to be found (see Section 4.5) for details. Here the error is of the form

$$\mathcal{E} = h(\theta) - h(\Theta^*).$$

If Θ^* are ML estimators then, by Theorem 4.3 and Example 5.5, the distribution of Θ^* is asymptotically normal. Next, by using Taylor expansion and estimating errors, it can be demonstrated that even the error $\mathcal{E} = h(\theta) - h(\Theta^*)$ is asymptotically normal with mean zero and variance $(\sigma_{\mathcal{E}}^2)^*$ computed using Gauss' formulae. Thus with approximately $1 - \alpha$ confidence, $h(\theta)$ is in

$$\left[h(\theta^*) - \lambda_{\alpha/2}\sigma_{\mathcal{E}}^*, \ h(\theta^*) + \lambda_{\alpha/2}\sigma_{\mathcal{E}}^* \right]. \tag{8.21}$$

An expression for the standard deviation $\sigma_{\mathcal{E}}^*$ is given in Eq. (8.23). This way of constructing approximative confidence intervals is called the *delta method*. (Actually, a special case of the method was given in Eq. (4.30) for the case of a one-dimensional parameter θ.)

Estimates of $\sigma_{\mathcal{E}}^2$ are computed by Gauss' approximation of the variance of $h(\Theta^*)$. Gauss' approximation formulae were presented earlier with explicit expressions in the two-dimensional case. We here state the general case of a d dimensional parameter $\theta = (\theta_1, \theta_2, \ldots, \theta_d)$. The ML estimator is a vector $\Theta^* = (\Theta_1^*, \Theta_2^*, \ldots, \Theta_d^*)$.

Let $h(\theta)$ be a scalar function and consider the vector of derivatives, called gradient and denoted by

$$\nabla h(\theta) = \left[\frac{\partial}{\partial \theta_1} h(\theta) \ldots \frac{\partial}{\partial \theta_d} h(\theta) \right]^{\mathsf{T}}.$$

Denote the covariance matrix of Θ^* with $\mathbf{\Sigma} = [\sigma_{ij}^2]$, where $\sigma_{ij}^2 = \mathsf{Cov}(\Theta_i^*, \Theta_j^*)$. Now if Θ^* is a vector of ML estimators then the covariance matrix is estimated by inverting the matrix with the second-order derivatives

$$\mathbf{\Sigma}^* = [(\sigma_{ij}^2)^*] = -[\ddot{l}(\theta^*)]^{-1}, \tag{8.22}$$

see Examples 4.11, 5.6 for explicit computation in the special case when $d = 2$. Gauss' formulae written using matrix notation give the estimate of the variance

$$(\sigma_{\mathcal{E}}^2)^* = \mathsf{V}[h(\Theta^*)] \approx \nabla h(\theta^*)^{\mathsf{T}} \mathbf{\Sigma}^* \nabla h(\theta^*)$$

$$= \sum_{i=1}^{d} \sum_{j=1}^{d} (\sigma_{ij}^2)^* \frac{\partial}{\partial \theta_i} h(\theta^*) \frac{\partial}{\partial \theta_j} h(\theta^*). \tag{8.23}$$

An illustration of a typical application of the delta method is given in Example 8.9.

Example 8.9 (Rating life of ball bearings). Recall Example 4.1 (page 70) where 22 lifetimes of ball bearings were presented, which we consider as independent observations of ball-bearing lifetime X. In this example we assume a parametric model for the distribution and study the uncertainty of the parameter estimates. In particular, we study the so-called *rating life*, L_{10}, a statistical measure of the life, which 90% of a large group of apparently identical ball bearings will achieve or exceed. In other words, L_{10} satisfies $P(X \leq L_{10}) = 1/10$.

ML estimates. Assume that a Weibull model is valid for the distribution of the lifetime:

$$F_X(x) = 1 - e^{-(x/a)^c}, \quad x \geq 0.$$

One can prove that the ML estimates of the parameters a and c are given by

$$a^* = \left(\frac{1}{n} \sum_{i=1}^{n} x_i^{c^*} \right)^{1/c^*},$$

$$\frac{1}{c^*} = \frac{\sum_{i=1}^{n} x_i^{c^*} \ln x_i}{\sum_{i=1}^{n} x_i} - \frac{1}{n} \sum_{i=1}^{n} \ln x_i.$$

From a computational point of view, c^* is first solved by iteration from the second equation; then a^* is calculated from the first equation. For our data set, one finds $a^* = 82.08$ and $c^* = 2.06$. (Thus, the distribution is close to a Rayleigh distribution ($c = 2$).)

The estimators are consistent and asymptotically two-dimensional normally distributed with variances and covariance

$$V[A^*] \approx 1.087 \frac{(a^*/c^*)^2}{n}, \quad V[C^*] \approx 0.608 \frac{(c^*)^2}{n}, \quad \text{Cov}[A^*, C^*] \approx 0.2545 \frac{a^*}{n},$$

and $E[A^*] \approx a^*$, $E[C^*] \approx c^*$. (Note that the correlation coefficient $\rho[A^*, C^*] \approx 0.313$.) The variances can be presented in the matrix form

$$\Sigma^* = \begin{bmatrix} 1.087 \dfrac{(a^*/c^*)^2}{n} & 0.2545 \dfrac{a^*}{n} \\[2ex] 0.2545 \dfrac{a^*}{n} & 0.608 \dfrac{(c^*)^2}{n} \end{bmatrix}$$

and are derived by inverting the matrix with second-order derivatives of the log-likelihood function evaluated at a^*, c^*, *i.e.* $-[\ddot{l}(a^*, c^*)]^{-1}$.

Studies of rating life. With our assumption of a Weibull distribution, an estimate of the rating life is given by the expression

$$L_{10}^* = a^* \cdot \left(-\ln(1 - \frac{1}{10}) \right)^{1/c^*}.$$

A point estimate for our data is thus $L_{10}^* = 27.53$ (10^6 revolutions).

As usual, we are interested in the uncertainty of this estimate. Gauss' approximation will be used to approximately compute $(\sigma_{\hat{\mathcal{E}}}^2)^* = V[L_{10}^*]$ by considering the random variables A^* and C^*. Introducing

$$h(a,c) = a \cdot \left(-\ln(1 - \frac{1}{10})\right)^{1/c}$$

we find components in the gradient vector $\nabla h(a,c)$

$$\frac{\partial}{\partial a}h(a,c) = \left(-\ln(1 - \frac{1}{10})\right)^{1/c},$$

$$\frac{\partial}{\partial c}h(a,c) = -\frac{a}{c^2} \cdot \left(\ln(-\ln(1 - \frac{1}{10}))\right) \cdot \left(-\ln(1 - \frac{1}{10})\right)^{1/c}.$$

Now we can compute the variance

$$(\sigma_{\hat{\mathcal{E}}}^2)^* = V[h(A^*, C^*)] \approx \nabla h(\theta^*)^T \mathbf{\Sigma}^* \nabla h(\theta^*)$$

$$= \left(\frac{\partial}{\partial a}h(a^*, c^*)\right)^2 V[A^*] + \left(\frac{\partial}{\partial c}h(a^*, c^*)\right)^2 V[C^*]$$

$$+ 2\frac{\partial}{\partial a}h(a^*, c^*)\frac{\partial}{\partial c}h(a^*, c^*)\, \mathrm{Cov}[A^*, C^*] = 43.1.$$

Since $\sigma_{\hat{\mathcal{E}}}^* = 6.57$, using Eq. (8.21) we conclude that with approximate confidence 95% the rating life L_{10} is in the interval

$$\left[L_{10}^* - \lambda_{\alpha/2}\sigma_{\hat{\mathcal{E}}}^*,\; L_{10}^* + \lambda_{\alpha/2}\sigma_{\hat{\mathcal{E}}}^*\right] = [\,27.53 - 1.96 \cdot 6.57,\; 27.53 + 1.96 \cdot 6.57\,]$$

$$= [\,14.66,\; 40.4\,],$$

millions of revolutions.

Problems

8.1. Let $X \in \mathrm{Po}(2)$ and $Y \in \mathrm{Po}(3)$ be two independent random variables. Define $Z = X + Y$ and give the distribution for Z.

8.2. Let $X \in \mathrm{N}(10, 3^2)$, $Y \in \mathrm{N}(6, 2^2)$ be independent random variables and define $Z = X - Y$.

(a) Give the distribution for Z.
(b) Calculate $P(Z > 5)$.

8.3. In a certain region, there are three powerplants, say A, B, and C. Let X_A, X_B, and X_C denote the number of (serious) interruptions in each individual powerplant during one year. Assume a Poisson distribution; from historical data, one then has $X_A \in \mathrm{Po}(0.05)$, $X_B \in \mathrm{Po}(0.42)$, $X_C \in \mathrm{Po}(0.37)$. Further, assume statistical independence.

Calculate the probability for at least one interrupt in the region during one year.

8.4. *An accelerated test.* Fifty ball bearings, subjected to a lifetime test, are divided into ten groups, in each of which there are five bearings. The lifetime of a single bearing is Weibull distributed with distribution function

$$F(t) = 1 - e^{-(t/a)^c}, \qquad t > 0,$$

where $a > 0$ and $c > 0$ are constants[3]. When the first bearing breaks down in each group, its lifetime is recorded. The group in question has therewith done its bit and is withdrawn from the test. Eventually, there will be 10 such observations, one from each group.

(a) Show that the distribution of the time until the first breakdown of a bearing in a certain group also is Weibull, and express its parameters — say a_1 and c_1 — in terms of a and c. Every bearing breaks down independent of the others.

(b) Now, estimates of a_1 and c_1 can be obtained from the data set of ten observations (for example by means of the maximum-likelihood method), and from those estimates we can, in turn, estimate a and c. Assume that we have used a routine from a software package[4] to get the estimates $a_1{}^* = 5.59 \cdot 10^6$ and $c_1{}^* = 1.56$ of a_1 and c_1, respectively. Estimate a and c.
At the cost of lost information, time has been saved, since we need not to wait for all bearings to break down.

In this exercise, the "lifetime" of a ball bearing is the number of revolutions covered before breakdown.

8.5. The water supply to a small town comes from two sources: from a reservoir and from pumping underground water. Suppose during a summer month, the amount of water available from each source is normally distributed $N(30, 9)$, $N(15, 16)$, million litres, respectively. Suppose the demand during the month is also variable and can be modelled as a normally distributed variable with mean 35 millions of litres and coefficient of variation 10%.

(a) Determine the probability P_f that there will be insufficient supply of water during the summer month. Assume that demand and supply vary independently.

(b) Determine the probability P_f that there will be insufficient supply of water during the summer months, under the assumption that demand and the total supply of water are negatively correlated with correlation coefficient $\rho = -0.8$.

8.6. A bus travels from the city A to a city C via the village B. The times are considered independent and exponentially distributed (quite unrealistic) with mean values (minutes) 40 $(A \rightarrow B)$; 40 $(B \rightarrow C)$. Calculate the probability that the route takes more than one and a half hour.

8.7. In an electric circuit of voltage U with three resistors R_1, R_2, R_3 in parallel, the current I is given by

$$I = U \left(\frac{1}{R_1} + \frac{1}{R_2} + \frac{1}{R_3} \right).$$

Consider U, R_1, R_2, and R_3 as independent random variables with expectations 120 V, 10 Ω, 15 Ω, and 20 Ω, respectively, and standard deviations 15 V, 1 Ω, 1 Ω, and 2 Ω, respectively. The four random variables are assumed to be independent. Give, approximately, $\mathsf{E}[I]$ and $\mathsf{D}[I]$.

---

[3]The location parameter b is $b = 0$.
[4]e.g. wweibfit.m from the Matlab toolbox WAFO.

8.8. Consider a situation with one strength variable R and one load variable S. The following expression for the failure probability is sometimes more convenient:

$$P_f = P\left(\frac{R}{S} - 1 < 0\right).$$

Let R and S be independent, log-normally distributed random variables, $\ln R \in N(m_R, \sigma_R^2)$, $\ln S \in N(m_S, \sigma_S^2)$. Derive an explicit expression for P_f in terms of m_R, σ_R, m_S, and σ_S.

8.9. Assume that in a design situation, one has log-normally distributed strength R and load S with $E[R] = 150$ MPa, $E[S] = 100$ MPa. The coefficient of variation of the load is known, $R[S] = 0.05$, *i.e.* $V[S] = 0.05^2 \cdot 100^2$ MPa^2. How large is the coefficient of variation of the strength allowed to be if the failure probability must be less than 0.001?

8.10. Small cracks of mode I (opening cracks, plane state) grow larger in metal when the specimen is subject to cyclic loads. The growth rate of a crack is given by

$$\frac{\Delta A}{\Delta N} = 2c_0 \left(\frac{\Delta\sigma\sqrt{\pi \cdot (A/2)}}{(\Delta K_I)_0}\right)^n,$$

where

A is the initial length of crack

$\Delta\sigma$ is the range of the varying stress applied

ΔN is a "small" number of load cycles

ΔA is the growth of the crack during the ΔN load cycles

c_0 is a constant, $c_0 = 1 \cdot 10^{-6}$ m

$(\Delta K_I)_0$ is a constant specific to the material

n is a constant specific to the material

For steel SIS 2309, we know that $(\Delta K_I)_0 = 61.6$ MN/m$^{3/2}$, and $n = 2.8$. If

$$E[A] = 2.5 \text{ mm}, \qquad D[A]/E[A] = 20 \text{ \%}$$
$$E[\Delta\sigma] = 250 \text{ MN/m}^2, \qquad D[\Delta\sigma]/E[\Delta\sigma] = 30 \text{ \%}$$

calculate the expectation and the coefficient of variation for the growth of the crack per applied cycle, i.e. for $\Delta A/\Delta N$. Assume that A and $\Delta\sigma$ are statistically independent.

8.11. The maximum electrical energy that can be delivered to a region during one certain day (the production capacity) is log normally distributed with a median of 6 GWh and coefficient of variation 0.1. The daily demand is also variable and depends on the economical activity, outdoor temperature, etc. The demand is also log normally distributed with coefficient of variation 0.2 and a median that is 60 % of the median of the production capacity.

(a) Compute the probability that the demand will not be satisfied on a certain day. Assume that demand and production capacity are independent. Calculate the related return period for the event "demand is not satisfied".

(b) Surveys indicate that the return period in (a) is incorrect. A deeper investigation showed that the logarithm of the demand and the logarithm of the production capacity are correlated with correlation coefficient -0.8. What is now the probability that the demand will not be satisfied on a certain day? And what about the return period?

8.12. In this exercise we prove the inequality in Eq. (8.11), that is

$$P(Z < 0) \leq \frac{1}{1 + \beta_{\mathrm{C}}^2},$$

where $Z = R - S$ and $\beta_{\mathrm{C}} = m_Z / \sigma_Z$.

(a) Let X be a random variable with $m = \mathsf{E}[X] > 0$ and $\sigma^2 = \mathsf{V}(X)$. Show that

$$P(X < 0) \leq \frac{\mathsf{E}[(X - a)^2]}{a^2} = \frac{\sigma^2 + (m - a)^2}{a^2}$$

for each $a > 0$.
(b) Use the inequality in (a) to obtain the bound in Eq. (8.11).

8.13. A beam is rigidly supported by a wall and simply supported at a distance ℓ from the wall according to Figure 8.3. The action P is assumed to be stochastic with $\mathsf{E}[P] = 4$ kN and $\mathsf{D}[P] = 1$ kN while the length ℓ is deterministic, $\ell = 5$ m. Further the beam is assumed to have a moment capacity M_F that is a random variable with $\mathsf{E}[M_F] = 20$ kNm, $\mathsf{D}[M_F] = 2$ kNm. Failure is given by $M > M_F$ where $M = P\ell/2$. The failure function becomes

$$h(M_F, \ell, P) = M_F - P\ell/2$$

and the probability of failure is given as $P_{\mathrm{f}} = \mathsf{P}(M_F - P\ell/2 < 0)$.

(a) We consider $R = M_F$ a strength variable and $S = P\ell/2$ a load variable. Calculate $\mathsf{E}[R]$, $\mathsf{D}[R]$, $\mathsf{E}[S]$, and $\mathsf{D}[S]$.
(b) Find an upper bound of the probability of failure (use Eq. (8.11)).
(c) Make an assumption of distribution; suppose R and S are normally distributed with parameters as found in (a). Compute the probability of failure. Compare with the result in (b).

8.14. In a mine, water is gathered at a rate S; $\mathsf{E}[S] = 0.05$, $\mathsf{V}[S] = 10^{-6}$. One plans to install n pumps with capacities R_i; $\mathsf{E}[R_i] = 0.0025$, $\mathsf{V}[R_i] = 10^{-7}$.

(a) Assume that the capacities of the pumps are independent. How many pumps should be installed in order for the safety index for the event that the water level is not increasing, is higher than 3.5.

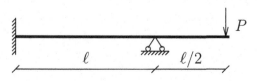

Fig. 8.3. Illustration for Problem 8.13.

(b) Assume now that there is a correlation between capacities of the pumps; the correlation coefficient is $\rho_{R_i, R_j} = 0.5$. Calculate n in this situation. (*Hint.* Use Eq. (5.12).)

8.15. In an industry, products are produced at a rate of on average 400 metric tons per day in the working week. The variation is quite large with a variance of 1000 ton^2; further, the amounts of goods for different days are strongly correlated with $\mathsf{Cov}[X_i, X_j] = \rho^{|j-i|}$ (where $\rho = 0.9$).

The work is scheduled on a weekly basis, in particular the number of lorries needed for the transports. Each lorry has a capacity of 10 ton per transport. From experience it is known that the number of transports per day for an individual lorry can be modelled as a Poisson distributed r.v. with intensity 1 hour^{-1}.

How many lorries are needed in order to have Cornell's index at least 3.5 for the whole production of one week being transported to customers?

Assume 7 hours of efficient work per day and 5 working days per week; further, assume that transports by different lorries can be modelled as independent r.v. (*Hint.* Use Eq. (5.12).)

Estimation of Quantiles

The notion of quantiles was introduced in Section 3.2: recall that a quantile x_α for an r.v. X is a constant such that

$$P(X \leq x_\alpha) = 1 - \alpha. \tag{9.1}$$

In this chapter we examine quantiles in somewhat more detail: we present methods for obtaining estimates of quantile values and also discuss techniques to assess the uncertainties of such estimates. In the previous chapter, Gauss approximation was used for that purpose. Here we present bootstrap methodology and Bayesian approaches through examples. Furthermore, we study particular applications of quantiles in reliability and engineering design, where analysis of the so-called characteristic strength is an important issue.

The chapter is organized as follows: first, the notion of characteristic strength is introduced and then examples are given where a parametric modelling is performed. In Section 9.2, the Peaks Over Threshold (POT) method is introduced. Finally, in Section 9.3, we present a type of problem where quality of components is concerned. Two methods are discussed, one based on asymptotic normality of estimation error, the other on Bayesian principles.

9.1 Analysis of Characteristic Strength

In Section 8.1, we discussed failure of a system in terms of variables of "strength" or "load" type. In designing systems in a wide sense (buildings, procedures), attention must be paid to gain statistical knowledge of the components of the system. We here analyse variables of "strength" type ("load" variables are modelled in the next chapter) and consider estimation of quantiles of the strength of a randomly chosen component.

Let the r.v. R model the strength of a randomly chosen component from a certain population. Using Eq. (9.1), the quantile r_α is the value of the load that will break $(1 - \alpha)100\%$ of the components in the population. In safety analysis, one is mostly interested in the fraction of weak components and hence α is chosen close to one (*e.g.* 0.95). That quantile r_α is called the *characteristic strength*. Thus, by definition, on average 5 out of 100 components will break when loaded by the load greater than the specified value of characteristic strength $r_{0.95}$. In practice the last statement cannot be true, since the quantile r_α is unknown and has to be estimated. The uncertainty in the estimate r_α^* of r_α therefore has to be analysed.

9.1.1 Parametric modelling

In many situations, strengths r_i, $i = 1, \ldots, n$, of tested components have been observed in experiments. To find a suitable model for the distribution function $F_R(r)$, experience and external information about the situation should be used to limit the class of possible distributions. For instance, we may assume that $F_R(r)$ belongs to some specific family of distributions, like Weibull, log-normal, etc. To indicate the parameter(s), we write $F_R(r; \theta)$. A restriction to a certain class of distributions can be assumed from previous knowledge in the actual field of application and earlier tests. Often observed strengths are plotted on different probability papers, and the family of distributions that gives the best fit to data is chosen. If the parameter θ were known, r_α can be computed by solving the equation

$$F_R(r_\alpha; \theta) = \mathsf{P}(R \leq r_\alpha) = 1 - \alpha.$$

Hence r_α is a function of θ and an estimate r_α^* is obtained by replacing θ by its estimate θ^*: $r_\alpha^* = r_\alpha(\theta^*)$. Uncertainties of the estimate r_α can be found by the delta method; another approach is bootstrap methodology (see Example 9.1 below).

In Example 4.23 we studied variability of a probability by bootstrap methodology. We now use that technique to investigate variability of a quantile. The parametric model is chosen to be Weibull.

Example 9.1 (Ball bearings: bootstrap study). Consider Example 4.1, where 22 lifetimes of ball bearings were presented. In this example we assume a parametric model for the distribution and study the uncertainty of an estimate $x_{0.9}^*$ of the quantile value $x_{0.9}$.

ML estimates. As in Example 8.9, page 210, we assume a Weibull model and thus the ML estimates are $a^* = 82.08$ and $c^* = 2.06$. The observations plotted in Weibull probability paper are shown in Figure 9.1, left panel.

Characteristic value. An estimate of the characteristic value $x_{0.9}$ based on the ML estimates is given by

$$x_{0.9}^* = a^*(-\ln 0.9)^{1/c^*} = 27.53.$$

Construction of a confidence interval for $x_{0.9}$ is not simple and we propose to use a bootstrap approach.

Bootstrap estimates. We resample from the original data set of 22 observations many times ($N_B = 5000$) and hence obtain the bootstrap estimates

$$(x_{0.9})_i^B = a_i^B(-\ln 0.9)^{1/c_i^B}, \quad i = 1, \ldots, N_B.$$

Note that in each of the simulations, an ML estimation is performed of a and c. The 5 000 pairs of bootstrapped parameter estimates are shown in Figure 9.1, right panel.

A histogram of bootstrapped quantile values $x_{0.9}^B$ is shown in Figure 9.2, left panel (the star on the abcissa indicates the ML estimate $x_{0.9}^*$). To construct a confidence interval, we need the bootstrap-estimation error $x_{0.9}^{\text{emp}} - (x_{0.9})_i^B$, where $x_{0.9}^{\text{emp}} = 33$ (obtained by considering the empirical distribution F_n and solving the equation $F_n(x_{0.9}) = 0.1$). The ring in Figure 9.2, left panel, indicates the estimate $x_{0.9}^{\text{emp}}$.

In Figure 9.2, right, the empirical distribution for the bootstrap-estimation error is given. We can see that the error distribution is skewed to the right

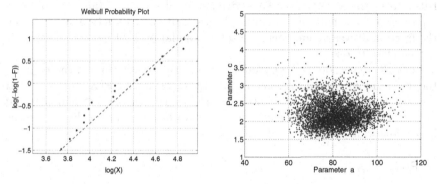

Fig. 9.1. *Left*: Ball-bearing data plotted in Weibull probability paper. *Right*: Simulated bootstrap sample, parameters a_i^B and c_i^B.

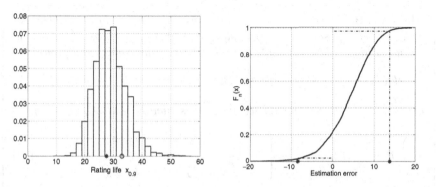

Fig. 9.2. *Left*: A histogram of $x_{0.9}$, based on $N_B = 5000$ resamplings on 22 original observations. Star: $x_{0.9}^*$; Ring: $x_{0.9}^{\text{emp}}$. *Right*: The empirical error distribution of $x_{0.9}^{\text{emp}} - (x_{0.9})_i^B$ with quantiles marked as stars.

and has larger positive errors than negative. Let us choose confidence level
0.95. We have marked with stars the quantiles $e^B_{1-\alpha/2}$, $e^B_{\alpha/2}$, left and right
star, respectively. Now Eq. (4.26) gives the following confidence interval for
$x_{0.9}$, [19.1, 41.4]. (Compare the interval with the one in Example 8.9.)   □

Our guess, based on some limited experience in fatigue analysis, is that
in engineering the parameters of type r_α are computed based on small numbers of observations. One of the reasons is that it can be very expensive to
test components or it may take long time to perform tests. However, since
tests have been performed on similar types of components one is quite sure
about the type of distribution of strength F_R and maybe even values of some
parameters. In such a situation a Bayesian approach is an option.

9.2 The Peaks Over Threshold (POT) Method

Hitherto in this chapter, we have given examples of uncertainty analysis of
a parametric estimate of a quantile. In applications to safety analysis, often
the fraction of weak components in some population is of interest. Then α is
chosen close to one.

The POT method, to be introduced next, can be used to find the α quan-
tile of a random variable X, *i.e.* a constant x_α such that $P(X \leq x_\alpha) = 1 - \alpha$
when α is close to zero (in its original formulation) or one (as demonstrated
in Remark 9.3). It is more convenient to write the definition of the α quantile
in the following alternative way

$$P(X > x_\alpha) = \alpha.$$

The method is based upon the following result (cf. [61]), which we summarize
as a theorem:

Theorem 9.1. *Under suitable conditions on the random variable X,
which are always satisfied in examples considered in this book, if the threshold u_0 is high (when u_0 tends to infinity), then the conditional probability*

$$P(X > u_0 + h \,|\, X > u_0) \approx 1 - F(h; a, c)$$

where $F(h; a, c)$ is a generalized Pareto distribution (GPD), given by

$$\text{GPD:} \qquad F(h; a, c) = \begin{cases} 1 - (1 - ch/a)^{1/c}, & \text{if } c \neq 0, \\ 1 - \exp(-h/a), & \text{if } c = 0, \end{cases} \qquad (9.2)$$

for $0 < h < \infty$ if $c \leq 0$ and for $0 < h < a/c$ if $c > 0$.

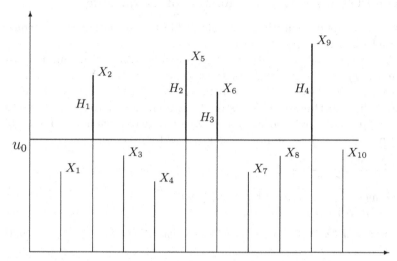

Fig. 9.3. Data $X_1, \ldots X_{10}$ and corresponding exceedances $H_1, \ldots, H_4$ over the threshold u_0.

Note that the GPD is used to model the *exceedances* over the threshold, hence the term Peaks Over Threshold. (For an illustration, see Figure 9.3.) In practice, the choice of a value for the threshold u_0 is not trivial. Several graphical methods have been developed; these are often combined with fitting a GPD to a range of thresholds, observing the stability of the corresponding parameter estimates.

There exists software to estimate the parameters in a GPD. The ML method can be applied, and also algorithms specifically derived for the purpose [38]. Moreover, also observe that when $c = 0$, $F(h; a, 0)$ is an exponential distribution with expected value a. For most distributions of X met in this book,

$$P(X > u_0 + h \,|\, X > u_0) \approx e^{-h/a}$$

with very good approximation when u_0 is sufficiently high, and consequently, one often assumes that $c = 0$. In case the exponential function does not model $P(X > u_0 + h \,|\, X > u_0)$ with sufficient accuracy, the more general Pareto distribution with $c \neq 0$ is used.

Remark 9.1. The standard Pareto distribution is defined by

$$F(x) = 1 - x^{-k}, \quad x \geq 1,$$

where $k > 0$ is a shape parameter. If X is Pareto distributed then, with $c = -1/k < 0$ and $a > 0$, $Y = -\frac{a}{c}(X - 1)$ is GPD with $F_Y(y) = F(y; a, c)$ as given in Eq. (9.2). □

9.2.1 The POT method and estimation of x_α quantiles

We turn now to the presentation of an algorithm for estimation of x_α, which can be used when α is close to zero.

Let X_i be independent identically distributed variables with common distribution $F_X(x)$. As before, let x_i be the observed values of X_i, $i = 1, \ldots, n$. Suppose for a fixed level u_0 the probability $p_0 = \mathsf{P}(X > u_0)$ can be estimated and the uncertainty of the estimate p_0^*, say, is not too large. Here a parametric approach could be taken involving choice of a particular family of distribution; alternatively, simply p_0^* is given as the fraction

$$p_0^* = \frac{\text{Number of } x_i > u_0}{n}.$$

The POT method can be used for $\alpha < p_0^*$, i.e. $x_\alpha > u_0$.

Now, if u_0 is high and $x > u_0$, then

$$\mathsf{P}(X > x) = \mathsf{P}(X > u_0) \cdot \mathsf{P}(X > u_0 + (x - u_0) \,|\, X > u_0) \qquad (9.3)$$

$$\approx \mathsf{P}(X > u_0)(1 - F(x - u_0; a, c)),$$

where F is a generalized Pareto distribution with a suitable scale parameter a and form parameter c (often taken to be zero) (see Theorem 9.1).

Let a^* and c^* be the estimates of a and c. Then x_α^* is the solution of the equation

$$p_0^*(1 - F(x_\alpha - u_0, a^*, c^*)) = \alpha$$

and the POT algorithm gives the following estimate:

$$x_\alpha^* = \begin{cases} u_0 + (a^*/c^*)\left[1 - (\alpha/p_0^*)^{c^*}\right], & \text{if } c \neq 0, \\ u_0 + a^* \ln{(p_0^*/\alpha)}, & \text{if } c = 0. \end{cases} \qquad (9.4)$$

Remark 9.2. An advantage with the POT method and the approximation in Eq. (9.3) is the capability to model the tails of the distribution: this in contrast to earlier chapters, where families of distributions were intended to model the central part of the unknown distribution.

The POT method consists of two steps. The first one is estimation of $p_0 = \mathsf{P}(X > u_0)$, basically by means of the method presented in the previous section; then the few extremely high values of x_i, are used to model the distribution at its tails. □

Remark 9.3. The POT method can also be used to find quantiles when α is close to one. Simply let $Y = -X$ and find the quantile $y_{1-\alpha}$. Using Eq. (9.4), since $1 - \alpha$ is close to zero, the x_α quantile is simply equal to $-y_{1-\alpha}$.

For example, let x_i be the observations of X and assume $\alpha = 0.999$ is of interest. Then define $y_i = -x_i$ and use the POT method to find the estimate $y_{0.001}^*$. Finally, let $x_{0.999}^* = -y_{0.001}^*$. □

The two following examples illustrate how the POT method can be used to find lower and higher quantiles, respectively.

9.2.2 Example: Strength of glass fibres

Consider experimental data of the breaking strength (GPa) of glass fibres of length 1.5 cm (see [72]). The empirical distribution, based on a sample of 63 observations, is shown in Figure 9.4. In this example, we want to estimate the lower 0.01 quantile. We use two methods: a parametric approach and the POT method. Note that we have less than 100 observations, hence finding the 0.01 quantile is a delicate issue. The lowest observed value (0.55) is probably close to the quantile.

Parametric model: Weibull. Strength of material is often modelled with a Weibull distribution,

$$F(x) = 1 - e^{-(x/a)^c}, \qquad x \geq 0$$

say. Statistical software return the ML estimates $a^* = 1.63$, $c^* = 5.78$. The quantile is then found as

$$x^*_{0.99} = a^*(-\ln(1 - 0.01))^{1/c^*} = 0.74.$$

By incident, the second smallest observed strength is 0.74; the third smallest 0.77. The Weibull is considered to fit the central parts of the distribution well but here our aim is to examine the tail, to find the quantile. The POT method is employed next.

POT method. To find the lower quantile, the data with opposite sign are investigated (cf. Remark 9.3). Investigating the stability of fitted shape parameter in a GPD for different choices of thresholds indicates a suitable threshold about -1.4. This threshold results in $p_0^* = 0.27$. The method of probability weighted moments (PWM) is used to estimate parameters and result in the estimates $a^* = 0.404$, $c^* = 0.248$, which gives the quantile of interest as $x^*_{0.99} = 0.49$.

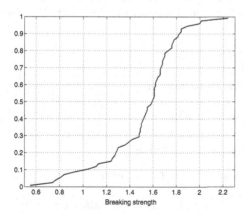

Fig. 9.4. Breaking strength of glass fibres. Empirical distribution.

Note that we presented only point estimates x_α^* in this example. Clearly, a full analysis should investigate the uncertainty of these as well. For instance, the standard error of the estimates when fitting a general GPD is larger compared to the case of exponential excursions (more parameters means higher uncertainty).

In the following example we perform a Bayesian analysis of the uncertainty of POT estimates.

9.2.3 Example: Accidents in mines

This is a continuation of Example 2.11, which finished with a question; how to compute the probability $P(K > 400)$, i.e. more than 400 perished people in a single mining accident. Here POT is used to estimate the probability.

Estimation using the POT method

Denote by k_i, $i = 1, 2, \ldots, 120$, the number of deaths in the ith accident. These form independent observations of K. Let us choose a threshold $u_0 = 75$, as we did in Example 2.11.

The first step of the POT method is to estimate the probability $p_0 = P(K > 75)$. Since there are 17 accidents with more than 75 deaths, we find the estimate

$$p_0^* = 17/120 = 0.142.$$

The 17 values are given next:

89 114 189 76 142 361 91 178 143 207 189 268

120 164 101 178 81

The second step in the POT method is to model the conditional distribution of excursions above $u_0 = 75$, $P(K - u_0 \leq h \mid K > u_0)$ by means of a GPD. Often one assumes that the shape parameter $c = 0$, i.e. that the distribution is exponential

$$P(K > u_0 + h \mid K > u_0) = e^{-h/a},$$

where a is the unknown parameter to be estimated. In Figure 9.5, left panel, we can see the exceedances plotted on exponential paper (see Example 4.3 for definition) follow a straight line and hence we have no reasons to reject the model. The ML estimate is $a^* = \sum k_i/n - u_0 = 83.3$. Consequently, the probability looked for is estimated as

$$P(K > 400) \approx p^* e^{-(400-u_0)/a^*} = 0.0029, \tag{9.5}$$

i.e. on average once in 333 mines accidents there will be more than 400 perished.

A quantile can also be estimated, by means of Eq. (9.4). For instance, for $\alpha = 0.001$

$$k_\alpha^* = u_0 + a^* \ln{(p^*/\alpha)} = 75 + 83.3 \ln{(0.142/0.001)} = 487.2.$$

Obviously both estimates are uncertain and in the following we use Bayesian modelling in order to investigate the size of the uncertainties. In the frequentistic approach, this could be done using confidence intervals and the delta method. However, as long as we do not have informative priors, both ways of analysing the uncertainty work equivalently well. Here we found it more illustrative to use the Bayesian approach.

Bayesian modelling of uncertainties

As usual, the unknown parameters $\theta_1 = p_0$ and $\theta_2 = 1/a$ are modelled by independent random variables Θ_1 and Θ_2. Note that it is more convenient to use the parameter $\theta_2 = 1/a$ than a since the family of gamma distributions forms conjugated prior for Θ_2, as demonstrated in the following remark.

Remark 9.4. Suppose the prior density for Θ_2 is Gamma(a,b). Let X be exponentially distributed with mean a, having the pdf

$$f_X(x) = \theta_2 \, e^{-\theta_2 x} \quad \text{with} \quad \theta_2 = \frac{1}{a},$$

and let x_i, $i = 1, \ldots, m$ be observations of independent experiments X. The likelihood function is given by

$$L(\theta_2) = \prod_{i=1}^{m} \theta_2 \, \exp(-\theta_2 x_i) = \theta_2^m \exp\left(-\theta_2 \sum_{i=1}^{m} x_i\right).$$

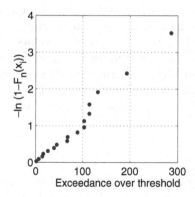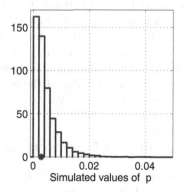

Fig. 9.5. *Left:* Exceedances over the threshold $u_0 = 75$, plotted on exponential paper. *Right:* Histogram; 10 000 simulated values of $p = \mathrm{P}(K > 400)$.

Following Eq. (6.9), *i.e.* $f^{\text{post}}(\theta_2) = c\,L(\theta_2)f^{\text{prior}}(\theta_2)$, we arrive at the posterior density

$$f^{\text{post}}(\theta_2) \in \text{Gamma}\left(a + m,\, b + \sum_{i=1}^{m} x_i\right). \tag{9.6}$$

$\square$

To model the unknown frequency Θ_1 we use beta priors. Suppose there is no prior information about the value p_0. Thus the prior density for θ_1 is Beta(1,1) (uniform prior) and by Eq. (6.22) with $n = 120$ and $k = 17$, the posterior density is Beta(18,104). We turn now to the choice of the prior density for Θ_2. Again suppose there is little experience of the size of θ_2 and hence the so-called "improper" prior $f^{\text{prior}}(\theta_2) = 1/\theta_2$ is proposed[1]. Now, with $x_i = k_i - u_0$, by Eq. (9.6), the posterior density for Θ_2 is

$$f^{\text{post}}(\theta_2) = \frac{c}{\theta_2} L(\theta_2) \in \text{Gamma}\left(m, \sum_{i=1}^{m}(k_i - u_0)\right) = \text{Gamma}\,(17, 1416).$$

(Note that θ_2, the inverse of the expected number of deaths in a mining accident, is an important parameter often estimated for other data sets; hence, there are reasons in using more informative priors. This could lower the uncertainty in the estimated value of the probability.)

The probability $P = \mathsf{P}(K > 400) = \Theta_1 \exp(-\Theta_2(400 - u_0))$ is a random variable. Since Θ_1, Θ_2 are independent the distribution of P could be computed using the version of the law of total probability in Eq. (5.22); this yields with $u_0 = 75$,

$$\mathsf{P}(P \le p) = \int_0^\infty \mathsf{P}\left(\Theta_1 \le p\,e^{\theta_2\,325}\right) f^{\text{post}}(\theta_2)\,d\theta_2.$$

This integral has to be computed numerically, giving the posterior distribution for P. Using this distribution, the predictive probability $\mathsf{E}[P] = 0.0044$ is found. A credibility interval is found as $[3.5 \cdot 10^{-4}, 0.016]$. As a complement to these findings, we use a Monte Carlo method and simulate a large number $N = 10\,000$, say, of independent values of θ_1, θ_2, compute N values of the probability p and present these in the form of a histogram (normalized to have integral one). The result is given in Figure 9.5 (right panel) and the star at the abscissa is the value of the estimate $p^* = 0.0029$, found in Eq. (9.5) by a frequentistic approach. We can see that the uncertainty is quite large.

9.3 Quality of Components

In this section we return to the study of a variable R describing strength of some kind of component. Suppose components with a prescribed quality

---

[1]This prior is not a pdf; see Section 6.5.1 for definition and interpretations.

$r_\alpha = \hat{r}$, say, are to be bought. The question is: Can one trust the value $\hat{r}$ assigned to the components — how can the accuracy of the characteristic strength be checked? Principally we wish to find the probability

$$p = \mathsf{P}(R \le \hat{r})$$

and compare this with the value $1 - \alpha$ claimed by a dealer.

9.3.1 Binomial distribution

Suppose we plan to test n components and check whether the strength exceeds a fixed threshold $\hat{r}$, say. A natural estimate of the probability $p = \mathsf{P}(R \le \hat{r})$ is the fraction of broken components divided by n. Using mathematical language, let K be the number of components that have strength below $\hat{r}$. Then, since the strengths of individual components are assumed to be independent, K is a binomially distributed variable:

$$K \in \mathrm{Bin}(n, p).$$

Suppose that k failures were observed in n tested components. From the table in Example 4.19, p. 90, the ML estimate of p is found as $p^* = k/n$ and if n is large $(np(1 - p) > 10)$ the error $\mathcal{E}$ is approximately normally distributed: $m_{\mathcal{E}} \approx 0$ and $(\sigma_{\mathcal{E}}^2)^* = p^*(1 - p^*)/n$. Hence with approximately $1 - \alpha$ confidence,

$$p \in \left[p^* - \lambda_{\alpha/2}\sigma_{\mathcal{E}}^*, \ p^* + \lambda_{\alpha/2}\sigma_{\mathcal{E}}^* \right]. \tag{9.7}$$

More often we are interested in the number of tests needed in order that the estimate p^* is sufficiently close to the unknown probability p. For example, we wish to find n such that, with confidence $1 - \alpha$, the relative error is less than 50%, *i.e.* n satisfying

$$\lambda_{\alpha/2} \frac{\sigma_{\mathcal{E}}^*}{p^*} \le 0.50, \quad \text{hence} \quad n \ge \frac{1 - p^*}{p^*} \left(\frac{\lambda_{\alpha/2}}{0.50} \right)^2.$$

An obvious generalization to the case of relative error less than $100q\%$ is

$$n \ge \frac{1 - p^*}{p^*} \left(\frac{\lambda_{\alpha/2}}{q} \right)^2. \tag{9.8}$$

If p^* is small we can simplify by replacing $1 - p^*$ in the numerator in the right-hand side of Eq. (9.8) by 1.

For example, let $\alpha = 0.10$, $q = 0.5$. In the case when $\hat{r} = r_\alpha$, then $p = 0.10$. (However, normally $\hat{r} \ne r_\alpha$.) We obtain that more than circa 108 components need to be tested in order to satisfy the accuracy requirements that the error of an estimate p^* is less than 50% with probability $1 - \alpha = 0.90$.

Since many experiments are needed, more elaborate methods have to be used. Measurements of the strengths will have to be made and possibly a parametric model for the strength employed.

Example 9.2 (Ball bearings). For simplicity only, assume that the components are the ball bearings from Example 4.1. The given observations were lifetimes. Suppose an expert claims that only 10% of such ball bearings have a lifetime less than 40 millions cycles. How reliable is this information?

In Example 4.1 we had $n = 22$ observations, which is a too small number to justify use of asymptotic normality of the error distribution if $p = 0.1$. The computed interval in Eq. (9.7) with confidence level 0.95 ($\alpha/2 = 0.025$) is not well motivated and in addition, we can see that it is quite wide ($p \in [0, 0.23]$). This is not surprising since more than 154 tested bearings are needed in order to get accuracy of the estimate of 50% with confidence 95% (see Eq. (9.8)).

□

9.3.2 Bayesian approach

As discussed in Chapter 6, conjugated beta priors are useful for estimation of a probability. We often assumed as prior distribution a Beta$(1,1)$, a uniform distribution corresponding to lack of knowledge. However, in the present application, the prior density must be determined. We illustrate with an example.

Example 9.3. Consider a production of components, coming in series all labelled with a quality expressed in characteristic strength $r_{0.9}$. Again, suppose we know that on average (taken over the series and based on many data) $p = 0.1$ but there is a variability between the series with coefficient of variation equal to 1. Using Eq. (6.11), we translate this information into the parameters a and b of the beta distribution. More precisely, we choose $a = 0.8$, $b = 6.2$.

Now, first assume that in the test of 25 components, 3 were weaker than the value $\hat{r}$. This information will change our beta priors and yield new parameters $\tilde{a} = 3.8$ and $\tilde{b} = 28.2$. Let $\theta = p$. The priors gives the probability $P(\Theta > 0.2) = 0.188$, which is quite high. The posterior distribution, on the other hand, renders $P(\Theta > 0.2) = 0.089$. If there were only two components that broke under the load $\hat{r}$, the posterior density has parameters $\tilde{a} = 2.8$ and $\tilde{b} = 29.2$, resulting in $P(\Theta > 0.2) = 0.029$.

In Figure 9.6, the prior density is shown as a solid curve. We further note that the posterior distribution resulting in the case of two broken components (dashed-dotted curve) has its mode[2] at a lower probability than the corresponding distribution for three broken components (dotted curve). (As in Example 9.1, the number of tested components (here: 25) is small.)

□

---

[2]If an r.v. X has a probability-density function $f(x)$, then $x = M$ is a *mode* of the distribution if $f(x)$ has a global maximum at M.

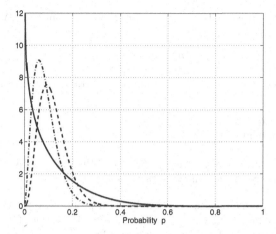

Fig. 9.6. Solid: Prior density. Dashed: Posterior density (three broken components). Dashed-dotted: Posterior density (two broken components).

Problems

9.1. A manufacturer of filters claims that only 4% of the products are of less quality than required in specifications. Let p be the probability that a filter is of poor quality. How many items need to be tested to get an estimate p^* of p having relative error less than 0.50 accuracy with confidence 95%? (*Hint.* See Eq. (9.8).)

9.2. Consider an a r.v. X. In the POT method, exceedances over a threshold u are analysed and the conditional probability

$$P(X > u + x \,|\, X > u)$$

is essential for the analysis.

(a) Show that

$$P(X > u + x \,|\, X > u) = \frac{1 - F(u + x)}{1 - F(u)}.$$

(b) Consider an exponential model, $F(x) = 1 - e^{-x}$, $x > 0$. Use the result in (a) to determine the distribution of the exceedances over a threshold u.

9.3. For a certain batch of ball bearings the expected value of the lifetime X (million revolutions) is $E[X] = 75$ and the coefficient of variation $R[X] = 0.40$. Make the assumption of a Weibull distribution and give an estimate of the rating life L_{10}, a quantile satisfying

$$P(X \le L_{10}) = 1/10.$$

Hint. Use Table 4 in appendix and the fact that for the scale parameter in a Weibull distribution,

$$a = \frac{E[X]}{\Gamma(1 + 1/c)}.$$

9.4. Consider strengths of fibres from Example 9.2.2. Assuming a Weibull distribution for the strength,

$$F(x) = 1 - e^{-(x/a)^c}, \qquad x \geq 0,$$

statistical software result in the ML estimates $a^* = 1.63$, $c^* = 5.78$ and the covariance matrix

$$\Sigma = \begin{pmatrix} 0.0014 & 0.0066 \\ 0.0066 & 0.3225 \end{pmatrix}$$

A point estimate of $x_{0.99}$ was found as 0.74. Use the delta method to construct a 95 %-approximate confidence interval for the quantile $x_{0.99}$.

9.5. Consider flipping a coin, giving "heads" with probability p. Assume that p is around 0.50. How many flips are needed to get a relative error of p^* less than 20% with high confidence? Perform the calculations for $\alpha = 0.05$ and $\alpha = 0.10$. (This problem was discussed in page 31.)

9.6. The strength of a storm at sea is measured by the peak value of significant wave height (H_s) that is observed during the storm. During 12 years of measurements at US NODC Buoy 46005, 576 storms have been recorded. Denote by X the peak value of H_s that is measured during a storm and assume that the values of H_s for different storm are iid r.v.

(a) Estimate the quantile for $\alpha = 0.001$, $P(X > x_\alpha) = 0.001$, if the following values of x for 40 serious storms are available (where the peak H_s exceeded 9 m):

$$
\begin{array}{cccccccccc}
9.6 & 9.6 & 9.5 & 11.0 & 11.9 & 9.2 & 9.6 & 9.1 & 9.9 & 9.3 \\
9.4 & 9.5 & 9.5 & 12.2 & 13.0 & 10.0 & 9.3 & 10.0 & 9.5 & 9.8 \\
9.2 & 10.8 & 9.9 & 9.8 & 10.1 & 12.3 & 10.8 & 9.4 & 9.1 & 11.1 \\
9.0 & 11.5 & 10.6 & 10.4 & 9.0 & 9.4 & 11.8 & 12.9 & 11.3 & 9.9
\end{array}
$$

Assume that the cdf of $H = X - 9$ is well approximated by an exponential distribution. (*Hint:* Use the POT method and $\sum_{i=1}^{40} h_i = 49.2$.)

(b) Use the delta method to derive approximative confidence intervals for $x_{0.001}$. (*Hint:* Assume that the estimators of p_0 and a are independent.)

(c) Estimate the intensity of storms and the expected number of storms during 100 years that are stronger than $x_{0.001}$.

Design Loads and Extreme Values

In the previous chapter, we presented methods to estimate quantiles of a random variable X. In the special case when X is an unknown strength of a component, the quantile is a measure, called *characteristic strength* (of material), or *production quality*. Often the 0.9 quantile is used; this has the interpretation that if a constant load is equal to the characteristic strength, then the likelihood of failure is 0.1.

The situation that a load is constant is often a crude approximation. If a load varies in time (or space) then the strength has to be chosen to accommodate for possible high values. A constant, used to describe the severity of a variable load, is the so-called *design load* s_T with *return period T*. Let for instance $T = 100$ years: a 100-year value s_{100} means that the load will exceed s_{100} on average once per 100 years if stationarity of the processes generating loads could be assumed. Usually variability of a load would change in such a long period and hence the above-mentioned interpretation should be rather seen as a more intuitive description of the severity of the load. Here we use another definition of design load s_T, namely that the probability that the load would exceed the design value s_T during a period $t = 1$ (in units of T) is $1/T$, *i.e.* $\mathsf{P}_t(A) = 1/T$ where $A =$"Load exceeds s_T". Since

$$\mathsf{P}_t(A) = \mathsf{P}(\text{The maximal load in the period } t > s_T)$$

this means that s_T is a $1/T$ quantile of a variable

$$M_t = \max_{s \in [0,t]} X(s),$$

where $X(s)$ is the value of the load at time instance s. Consequently in this chapter, we discuss estimation of quantiles for M_t, which is a maximal value of a sequence of random variables.

The parametric approach, presented in Section 9.1.1, will be used to estimate s_T. The generalized extreme-value distribution is introduced in Section 10.2 and analysed using extreme-value theory, this will be employed as a model for M_t. Estimation of design loads is very difficult and sometimes even questionable (we make statements about loads never observed) but such information is needed to construct safe structures. We also show by examples with data from real situations that the estimated values of the design load s_T are very uncertain.

10.1 Safety Factors, Design Loads, Characteristic Strength

Consider the simplest case with only one strength variable R and one load variable S, taking values r and s, respectively. A failure occurs when the load exceeds the strength, *i.e.* $s > r$. The true values of r and s are unknown. If the characteristic strength $r_{0.9}$ is available, we believe that

$$\mathsf{P}(R \leq r_{0.9}) = 0.1. \tag{10.1}$$

If we accept as high risk for failure as 0.1, we could allow to load the component (structure) with a load not higher than $s = r_{0.9}$.

Example 10.1. Consider again the wire in Example 8.7. The parameter $r_{0.9} = 100$ [kg] suggests is that the risk that the wire cannot support a load of 100 kg is 0.1. This can be interpreted as that 10% of the population of wires fail when loaded by the mass of 100 kg.   □

At the design stage one can choose the strength of a structure by specifying the value of the characteristic strength $r_{0.9}$. (Lower quantiles than $r_{0.9}$ could also be used to measure strength of materials, components or structures.) We show next how a simple analysis of the load can look like.

A deterministic fixed load level s, called the *design load*, which we require that the component (or structure) is able to carry without failing, is chosen. Depending on the intended safety level, a constant $c \geq 1$ (safety factor) is selected and it is required that the characteristic strength $r_{0.9} > cs$. How much $r_{0.9}$ should exceed s can be found in design norms and regulations. The exact value of c is decided by using a suitable safety analysis, usually employing safety indices.

Static and time variable loads

Often s is defined by means of a worst-case scenario. First we consider a *static* situation when the load S is more or less constant during the whole service time and deterioration of material strength can be neglected. For example, S can be the maximal pressure on a dam when it contains the maximum allowed amount of water, or it can be the weight that has to be supported by a beam. The static load S can be constant in time but still uncertain; for example, due to variability of geometry. In many practical design codes the static load is characterized by its expected value $m_S = \mathsf{E}[S]$. However, besides the static load the structure may experience a *time variable* (for example environmental) load often defined as the maximal load during a service time. The severity of the variable load is measured by the so-called return value; often a 50- or 100-year value s_{50} is used, however in some cases, *e.g.* dikes in the Netherlands or protection for high wave heights at offshore platforms, even 10 000-year values are employed. (Note that the service time of the structure does not need to be the same as the return period T.)

Remark 10.1. The notion of return period originated in the sciences of hydrology, when for instance analysis of severe floods was made, but the concept can be applied in other fields of science and technology as well. An early account of a statistical description of return periods for flood flows was made by Gumbel (one of the pioneer researchers in statistical extreme-value theory) in the early 1940s [31]. □

Design norm

Assume that we are interested in 50-year loads. The design norm gives two constants, c_1 and c_2, and indicates to us that we should have a characteristic strength $r_{0.9}$ that exceeds the design load $s = c_1 m_S + c_2 s_{50}$, *i.e.*

$$r_{0.9} > c_1 m_S + c_2 s_{50}, \tag{10.2}$$

in order to ensure sufficient safety during the service time.

Higher values of the constants c_1 and c_2 render a safer structure and also a more expensive one. The constants $c_1 \geq 1, c_2 \geq 0$, specified in design codes, define safety of a particular type of structure. In computations of c_1 and c_2 some typical distributions for strength R and load S are assumed. Hence, if the real strength, or load, has a distribution that differs from the one used in the derivation of c_1 and c_2, the true safety level can be lower than intended in the norm. Another reason for deviation of the true failure probability from the nominal value specified in the design norms is that the values of s_{50}, $r_{0.9}$, and m_S are estimated and hence uncertain.

10.2 Extreme Values

As mentioned in the introduction, finding 100-year loads is equivalent to finding the 0.01 quantile of the distribution of heights of yearly loads, *i.e.* the maximal load during one year. We return to this problem in Section 10.3, where we use extreme-value theory to give theoretical grounds for employing the generalized extreme-value distributions as models for the variability of yearly maxima.

However, the question of how high the maximum of some quantity X can be over a relative long period of time t has its own interest. For instance, the weather, rain, snow, wind, etc. change, and so do financial activities — stock prices, insurance claims, and many other quantities vary over time. Society is adapted to handle the *usual* variability of the environment. However, sometimes it rains or snows much more than usual, or it may be a very cold or warm period. Some accidents cause large losses that have been covered by insurance companies. Such extreme situations can lead to serious consequences, which one wishes to be prepared for and get an idea of the likelihood for them happening.

Examples of relevant questions are: What are the maximal values of claims related to a single storm that may happen during next 10 years, or what is the maximal daily rain observed at some meteorological measurement station during the next 20 years? Obviously, nobody knows the answers to these questions since these consider future values of variable quantities. Hence a more appropriate problem is to give a measure of *risk*, usually probability or safety indices, that maximal future losses (or amount of rain) S exceed the available resources R, *i.e.* $P(S > R)$. This section serves as an introduction to the problem of finding an appropriate class of distributions for the variable S, which represents "maximal demand". We hope that its reading will motivate the reader to deeper studies of extreme-value theory and its applications; for further reading, see for instance the seminal book by Gumbel [32], Leadbetter *et al.* [47], Coles [14], and the chapter by Smith [71].

10.2.1 Extreme-value distributions

Let $X_1, X_2, \ldots, X_n$ be iid random variables, each having distribution $F(x)$. Classical extreme-value theory deals principally with the distribution of the maximum

$$M_n(X) = \max(X_1, \ldots, X_n)$$

Similarly, if the quantity X is measured continuously during a period of length t, its maximal value will be denoted $M_t(X)$. Most often we shall use simpler notation and write M_n, M_t for $M_n(X), M_t(X)$, respectively.

Since X_i are independent the distribution function $F_{M_n}(x) = P(M_n \leq x)$ can be easily computed as follows

$$F_{M_n}(x) = P(X_1 \leq x, X_2 \leq x, \ldots, X_n \leq x) = P(X_1 \leq x)^n = F(x)^n. \quad (10.3)$$

In practice, a distribution $F(x)$ is assumed or assigned: from experience or based on analysis like presented in Chapter 4 (probability papers, etc.). The distributions are often fitted to the observations available and often describe the central parts well. However, the interest of extreme-value analysis is within the tails of the distribution. In Example 10.2, we see that the differences may be considerable.

Example 10.2. Suppose the maximal daily loads X_i are iid variables with distribution $F(x)$ and one is interested in the design load s_T, $T = 100$ years. The 100-year load is the 0.01 quantile of M_{365} with distribution given by Eq. (10.3). In other words, s_T is a solution of the equation $1 - F(x)^{365} = 0.01$, i.e. $F(x) = (1 - 0.01)^{1/365}$. Employing Taylor's formula, $(1 - x)^c \approx 1 - cx$, we find that $F(x) \approx 1 - 1/36500$, that is, s_T is close to the $1/36500$ quantile.

The $1/36500$ quantile is very sensitive to the exact shape of the distribution $F(x)$ for x where $F(x) \approx 1$ (in the tail). A model error, i.e. that $F(x)$ differs from the true distribution $P(X \leq x)$ in that region may result in big errors in the design load. We give now a numerical example:

Suppose a daily load X is log normally distributed with mean 1 and coefficient of variation 0.1, consequently

$$m = E[\ln(X)] = -0.005 \quad \text{and} \quad \sigma^2 = V[\ln(X)] = 0.01.$$

Hence the 100-year load is equal to 1.49, i.e. the mean plus five standard deviations. Now suppose we choose $F(x)$ to be the normal distribution $N(1, 0.01)$. Using $F(x)$, the 100-year load would be 1.40, i.e. mean plus four standard deviations.

As seen in Figure 10.1 (left panel), for mean one and coefficient of variation 0.1 the log-normal density can hardly be distinguished from the normal one. 100 simulated values from the log-normal distribution are shown in a normal probability paper in Figure 10.1 (right panel). Having observed 100 values x_i, one could erroneously assume that X is normally distributed.   □

In Example 10.2 a situation was presented when the use of Eq. (10.3) to find the low quantiles of M_n may lead to not-negligible errors due to uncertainty in the shapes of the tails for the cdf of X_i. Here the POT method, presented, in Chapter 9 could be used to solve the problem. However, there are also other practical problems and disadvantages to use (10.3) (or the

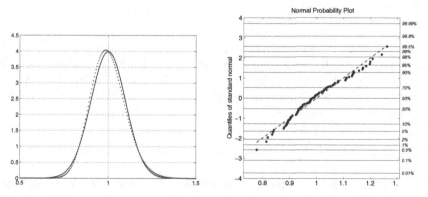

Fig. 10.1. *Left:* Solid line: normal pdf; Dotted line: log-normal pdf. *Right:* 100 simulated values from a log-normal distribution.

POT method) to find the design loads, namely, assumed independence of daily loads X_i may not be true; the distribution of X_i may vary (*e.g.* due to seasonal effects: loads in winter are different than in summer). There are specialized techniques to use POT in the situation of dependence, sometimes manifesting in clustering of large values, or seasonal dependence; however, these are outside the scope of this book and we refer the reader to specialized literature: Chapters 5-6 in [14] or for an application to ocean engineering, the report [2].

Here we present an alternative approach in which instead of estimating distribution X_i and using Eq. (10.3) (or the POT method) to find the design load, the distribution of yearly maxima is estimated directly if data over longer periods (several years) are available. The theoretical ground for this approach is an important result in extreme-value theory called the Extremal Types Theorem. This shows us that, under very general conditions, the distribution of M_n can be well approximated by the so-called Generalized Extreme-Value distribution (GEV). The accuracy of the approximation increases with n. Thus, this is a similar type of asymptotic result for maxima as the Central Limit Theorem was for sums or ML estimates.

Theorem 10.1. *If there are parameters $a_n > 0$, b_n and a non-degenerate probability distribution $G(x)$ such that*

$$\mathsf{P}\left(\frac{M_n - b_n}{a_n} \leq x\right) = \left[F(a_n x + b_n)\right]^n \to G(x) \tag{10.4}$$

then G is the Generalized Extreme Value distribution

$$GEV: \ G(x; a, b, c) = \begin{cases} \exp\left(-(1 - c(x - b)/a)_+^{1/c}\right), & if \ c \neq 0, \\ \exp\left(-\exp\{-(x - b)/a\}\right), & if \ c = 0, \end{cases} \tag{10.5}$$

where a is a scale parameter, b is a location parameter and c a shape parameter; $x_+ = \max(0, x)$.

The expression $(1 - c(x - b)/a)_+$ in Eq. (10.5) means that $1 - c(x - b)/a \geq 0$ and hence, if $c < 0$, the formula is valid for $x > b + (a/c)$ and if $c > 0$, it is valid for $x < b + (a/c)$. The case $c = 0$ is interpreted as the limit when $c \to 0$. Note that the Gumbel distribution is a GEV distribution with $c = 0$.

The consequence of Theorem 10.1 is that for large values of n

$$\mathsf{P}(M_n \leq x) \approx G\left(\frac{x - b_n}{a_n}\right), \tag{10.6}$$

which means that maximum of large number of iid variables X_i is well approximated by a distribution belonging to a class of generalized extreme-value distributions, see also Definition 3.3.

Remark 10.2. Note that there are situations when the maximum of iid variables is not asymptotically GEV, *i.e.* Eq. (10.4) does not hold for $G(x)$ defined in Eq. (10.5). Classical examples are when X_i are iid Poisson or geometrically distributed; see [47], p. 26.   □

Many real-world maximum loads belong to the GEV distributions with $c = 0$, *i.e.* the class of Gumbel distributions. For instance, if daily loads are normal, log-normal, exponential, Weibull (and some other distributions having so-called exponential tails) then the yearly (or monthly) maximum loads can be well modelled by a Gumbel distribution.

Let X_i be independent Gumbel distributed variables. Then an interesting result related to the distribution of the maximum can be derived; see the following example.

Example 10.3. Recall that a Gumbel distributed r.v. X with scale and location parameter a and b has the cdf

$$F(x) = \exp(-e^{-(x-b)/a}), \quad -\infty < x < \infty.$$

Now the maximum $M_n = \max_{1 \le i \le n} X_i$ has distribution

$$P(M_n \le x) = \left(\exp(-e^{-(x-b)/a})\right)^n = \exp(-ne^{-(x-b)/a})$$

$$= \exp(-e^{-(x-b)/a+\ln n}) = \exp(-e^{-(x-b-a\ln n)/a}). \quad (10.7)$$

Thus, the maximum of n independent Gumbel variables is also Gumbel with the same scale parameter and with location parameter changed from b to $b + a \ln n$. This property is sometimes referred to as the Gumbel distribution being max stable.   □

A numerical illustration of the derived result for Gumbel distributions is given in the following example. The application is related to design loads and operational time period.

Example 10.4. Assume that the maximum load on a construction during one year is given by a Gumbel distribution with expectation 1000 kg and standard deviation 200 kg. From the expressions for expectation and variance (given in appendix), is found $a = 156$, $b = 910$.

Suppose the construction will be used for 10 years. Then the maximum load over these 10 years is according to Eq. (10.7) given by a Gumbel distribution with mean $1000 + 156 \cdot \ln 10 = 1.4 \cdot 10^3$ kg and standard deviation 200 kg. The probability density functions for these two Gumbel distributions are shown in Figure 10.2. The solid line has mean $1.0 \cdot 10^3$ kg, the dashed-dotted $1.4 \cdot 10^3$ kg.   □

If $F(x)$ and $F(x)^n$ are of the same type, *i.e.* differ by values of location and scale parameters, then $F(x)$ is called *max stable*. As demonstrated in (10.7), the Gumbel distribution is max-stable. Actually the GEV distributions are the only max-stable distributions. We end this subsection with two remarks in which we give some properties of GEV distributions.

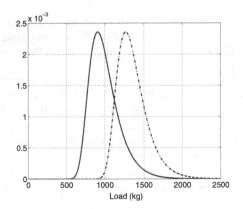

Fig. 10.2. Probability density function for Gumbel distributions with standard deviation 200 kg. Solid line: Gumbel distribution with mean $1.0 \cdot 10^3$ kg (maximum load, one year). Dashed-dotted line: Gumbel distribution with mean $1.4 \cdot 10^3$ kg (maximum load, ten years).

Remark 10.3 (Max stability). The GEV distribution is *max stable*, which means that maximum of n iid GEV distributed variables with parameters (a, b, c) is GEV distributed with the shape parameter c unchanged. While the scale parameter a is changed to a/n^c, the location parameter b is equal to

$$\begin{cases} b + \dfrac{a}{c}\left(1 - n^{-c}\right), & \text{if } c \neq 0, \\ b + a \ln n, & \text{if } c = 0. \end{cases}$$

□

Remark 10.4 (GEV – Random numbers). To simulate GEV-distributed Z with shape parameter $c \neq 0$, let U be a uniformly distributed random number, $U \in [0, 1]$. Then Z is the solution of the equation $U = F(Z)$, *i.e.*

$$Z = b + \frac{a}{c}\left(1 - (-\ln U)^c\right), \quad c \neq 0.$$

For $c = 0$ (Gumbel distribution), $Z = b - a \ln(-\ln U)$.   □

Choice between Gumbel and GEV

The Gumbel distribution $c = 0$ is often a natural model. This is because the distribution of maximum of independent (or even weakly dependent) normally, log-normally, gamma, Weibull loads is well approximated by a Gumbel distribution. Having one parameter fixed makes estimation simpler and at the same time the uncertainty of the estimated design load s_T smaller. However, before assuming that $c = 0$ it is recommendable to test whether data do not contradict the assumption.

The simplest test is to plot data on Gumbel probability paper and check if the observations lie reasonably close on a straight line. Next, a confidence interval for c can be computed (see Remark 10.5 for more details). If the confidence interval contains the value zero, then data do not contradict the assumption that yearly maxima are Gumbel distributed. The deviance can also be used to measure how much the GEV cdf better explains the variability of maxima compared to the Gumbel cdf. The deviance can be computed by means of

$$\text{DEV} = 2\big(l(a^*, b^*, c^*) - l(\tilde{a}^*, \tilde{b}^*)\big), \tag{10.8}$$

where $l(a^*, b^*, c^*)$ is the log-likelihood function and a^*, b^*, c^* are ML estimates of parameters in a GEV cdf, while $l(\tilde{a}^*, \tilde{b}^*)$ is the log-likelihood function and $\tilde{a}^*, \tilde{b}^*$ are ML estimates of parameters in a Gumbel cdf (see Remark 10.6 for computational details). If the deviance DEV is higher than $\chi_\alpha^2(1) = \lambda_{\alpha/2}^2$ then the Gumbel model should be rejected, $i.e.$ the GEV explains data significantly better.

Remark 10.5 (Confidence interval for shape parameter). The confidence interval for c can be derived using asymptotic normality of the ML estimators (see Theorem 4.3 and Section 8.3.1). If the number of observations is large (here fifty years or more), the asymptotic normality of estimators of the GEV cdf implies that with approximate confidence $1 - \alpha$

$$c \in \big[c^* - \lambda_{\alpha/2}\sigma_{\mathcal{E}}^*, \; c^* + \lambda_{\alpha/2}\sigma_{\mathcal{E}}^*\big]. \tag{10.9}$$

Here $\lambda_{\alpha/2}$ is the $\alpha/2$-quantile of an N(0,1) cdf, while $\sigma_{\mathcal{E}}^* \approx D[C^*]$. The standard deviation $\sigma_{\mathcal{E}}^*$ is one of the outputs of most programs used to estimate the parameters in a GEV cdf. It is computed by inverting the matrix $-[\ddot{l}(a^*, b^*, c^*)]$ (see Eq. (8.22) and Section 8.3.1 on the delta method). $\quad\square$

Remark 10.6 (Log likelihood for GEV and Gumbel distributions). Most of the programs used to estimate the parameters in a GEV (or Gumbel) cdf return the value of the log-likelihood as one of the outputs. If $z_1, \ldots, z_n$ are observed yearly maxima, then for a GEV pdf the log-likelihood

$$l(a, b, c) = \sum_{i=1}^{n} \ln\big(f(z_i; a, b, c)\big)$$

where

$$f(x; a, b, c) = \frac{1}{a}(1 - c(x - b)/a)_+^{1/c-1} \exp\left(-(1 - c(x - b)/a)_+^{1/c}\right),$$

when $c \neq 0$, and

$$f(x; a, b, c) = \frac{1}{a}\exp(-(x - b)/a)\exp\left(-\exp\{-(x - b)/a\}\right),$$

when $c = 0$. $\quad\square$

10.2.2 Fitting a model to data: An example

In this section we present a typical use of the GEV distribution. The parameters are estimated using statistical software. The data analysed represent 24 000 temperature readings, performed by the same person for 66 years (1919-1985). All observations were made at 8 a.m. outside Växjö in the Swedish province Småland. Suppose we are interested in the probability that the maximal temperature at 8 a.m. in the next 100 years exceeds $x = 27\,^\circ\text{C}$.

The observations are X_i, $i = 1, \ldots, 24\,000$. We face the problem that X_i are not equally distributed. Temperature in winter is obviously lower than in summer. This is a typical example of *seasonal variability* of the phenomenon. The solution is to combine the 24 000 observations into 66 yearly maxima Z_i, $i = 1, \ldots, 66$. It is then more reasonable to assume that Z_i have the same distribution and are independent (one should, however, check whether there are trends caused by climate change).

Distribution of yearly maximal temperature.

As mentioned before, the Gumbel distribution $c = 0$ is often a natural model. We first plot data on Gumbel paper (see Figure 10.3 (left)) and note that extreme temperature has shorter upper tail than the Gumbel model. We fit the GEV distribution and find the estimates $a^* = 1.67$, $b^* = 22.6$, $c^* = 0.323$. The estimated standard deviations are $\mathrm{D}[A^*] \approx 0.16$, $\mathrm{D}[B^*] \approx 0.20$, and $\mathrm{D}[C^*] \approx 0.04$. A 95 %-confidence interval for the shape parameter c is given by Eq. (10.9) as

$$[0.323 - 1.96 \cdot 0.04, \ 0.323 + 1.96 \cdot 0.04]$$

This does not contain the value $c = 0$, hence the Gumbel model should be rejected. In addition, the interval shows that the estimated c parameter is

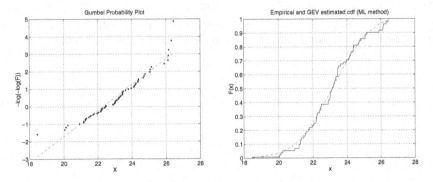

Fig. 10.3. Temperature measurements in Småland 1919-1985. *Left:* 66 yearly maxima plotted on Gumbel paper. *Right:* Comparison between the empirical distribution and the fitted GEV-model, dashed curve.

significantly positive. This is not surprising, since the positive c indicates that the maximum distribution has its upper bound estimated to be $b^* + (a^*/c^*) = 27.7°C$.

In Figure 10.3 (right), we see a comparison between the estimated GEV distribution and the empirical one. The agreement is good. Note that the estimated distribution describes variability of yearly maxima.

Distribution of M_{100}, maximal temperature during 100 years.

We are interested in the probability of the maximal temperature during 100 years. Obviously we do not have any observation of this random variable. The maximal temperature observed in 66 years was 26 °C. The calculation of the distribution of M_{100}, the maximum of 100 Z_i variables, is as follows:

$$P(M_{100} \leq x) = P(Z_i \leq x)^{100} = \exp\left\{-100(1 - c(x - b)/a)_+^{1/c}\right\}$$

$$\approx \exp\left\{-100(1 - c^*(x - b^*)/a^*)_+^{1/c^*}\right\}, \quad x \leq 27.7.$$

For $x = 27°C$, we have that during the following 100 years $P(M_{100} > 27) = 0.21$, while during the next year it is $P(M_1 > 27) = P(Z_1 > 27) = 0.002$.

Note that we could directly use the maximum stability of GEV to derive the distribution of M_{100}, which is GEV with the same parameters c, while a and b changed to a/n^c, $b + (a/c)(1 - n^{-c})$, respectively. Consequently, M_{100} is GEV with parameters a, b, c estimated to be 0.377, 26.6, 0.323, respectively, while, as found earlier, $M_1 = Z_i$ is GEV distributed with parameters a, b, c estimated to be 1.67, 22.6, 0.323, respectively.

Obviously the computed distribution for maximal temperature during 100 years assumes no changes in climate and independence of yearly maxima. Finally, note that M_{100} is a random variable and *not* a 100-year temperature s_{100} (the $1/100$ quantile of the M_1 distribution that is exceeded on average once in 100 years). In fact,

$$P(M_{100} \leq s_{100}) = \left(1 - \frac{1}{100}\right)^{100} \approx \frac{1}{e} = 0.37.$$

10.3 Finding the 100-year Load: Method of Yearly Maxima

In this section we employ the GEV distribution to estimate the T-year load. Suppose we have observed yearly loads M_t, $t = 1$ year, for a number of years. If the load varies relatively fast so that values of daily or weekly[1] loads can be considered independent then Theorem 10.1 tells us that the distribution of

---

[1]Here daily, weekly, monthly, or yearly loads mean the maximal value of the load during the specified period of time.

M_1 is well approximated by a cdf belonging to a GEV class. Since the design load s_T with return period T is equal to the level u solving the equation

$$\frac{1}{T} = P(M_1 > u),$$

and M_1 is modelled as GEV distribution then

$$s_T = b - a\ln(-\ln(1 - 1/T)), \quad \text{if } c = 0, \tag{10.10}$$

$$s_T = b + \frac{a}{c}\left(1 - (-\ln(1 - 1/T))^c\right), \quad \text{if } c \neq 0. \tag{10.11}$$

Next, using the observed yearly maxima a GEV cdf can be fitted to data, e.g. by means of the ML method (or other methods), and estimates $\theta^* = (a^*, b^*, c^*)$ found. An estimate of the design load s_T^* is then obtained by replacing a, b, c in Eqs. (10.10-10.11) by a^*, b^*, c^*.

Since in most cases long return periods are of interest, at least $T \geq 50$, and since we have that $-\ln(1 - 1/T) \approx 1/T$, we use the following, somewhat simpler, estimates of the design load:

$$s_T^* = b^* + a^*\ln T, \quad \text{if } c = 0, \tag{10.12}$$

$$s_T^* = b^* + \frac{a^*}{c^*}(1 - T^{-c^*}), \quad \text{if } c \neq 0. \tag{10.13}$$

Example 10.5. An estimate of the 100-year temperature in Växjö, is (by Eq. (10.13)) given by

$$s_{100}^* = 22.6 + \frac{1.67}{0.323}(1 - 100^{-0.323}) = 26.6 \ [°C].$$

$\square$

Remark 10.7. In some situations one may wish to use for example monthly maxima M_t, $t = 1/12$, to estimate s_T. If t is not equal to one year, by a moment of reflection is found that s_T is a t/T quantile of M_t. Hence, for $T \geq 50$, the estimate of the design load s_T is given by

$$s_T^* = b^* + a^*\ln(T/t), \quad \text{if } c = 0, \tag{10.14}$$

$$s_T^* = b^* + \frac{a^*}{c^*}\left(1 - (T/t)^{-c^*}\right), \quad \text{if } c \neq 0. \tag{10.15}$$

(Note that t and T must have the same units.) $\square$

10.3.1 Uncertainty analysis of s_T: Gumbel case

The Gumbel distribution, a special case of the GEV distribution with parameter $c = 0$, is often used to model M_1, yearly maxima. For $T \geq 50$, say, the estimate $s_T^* = b^* + a^*\ln T$, where a^*, b^* are ML estimates of the unknown parameters a, b. The ML estimators A^*, B^*, are asymptotically normally distributed (see Theorem 4.3) with variances

$$\mathsf{V}[A^*] \approx 0.61\frac{(a^*)^2}{n}, \quad \mathsf{V}[B^*] \approx 1.11\frac{(a^*)^2}{n}, \quad \mathsf{Cov}[A^*, B^*] \approx 0.26\frac{(a^*)^2}{n}.$$

(The variances and covariance are derived by inverting the matrix of second-order derivatives of the log-likelihood function.) Now using Eq. (5.11) we find

$$V[S_T^*] \approx 1.11 \frac{(a^*)^2}{n} + (\ln T)^2 \cdot 0.61 \frac{(a^*)^2}{n} + 2 \cdot 0.26 \cdot \ln T \frac{(a^*)^2}{n}$$

and hence with

$$\sigma_{\mathcal{E}}^* = a^* \sqrt{\frac{1.11 + 0.61(\ln T)^2 + 0.52 \ln T}{n}} \tag{10.16}$$

we have that with approximately $1 - \alpha$ confidence

$$s_T \in \left[s_T^* - \lambda_{\alpha/2}\sigma_{\mathcal{E}}^*, \ s_T^* + \lambda_{\alpha/2}\sigma_{\mathcal{E}}^* \right]. \tag{10.17}$$

Example 10.6 (Analysis of buoy data). We now study data from a buoy (US NODC Buoy 46005) situated in the NE Pacific (46.05 N, 131.02 W). The quantity called *significant wave height* (Hs) is important in ocean engineering and oceanography. This was calculated as the average of the highest one-third of all of the wave heights during the 20-minute sampling period. An alternative definition of Hs is as 4 times the standard deviation of the sea-surface level. Thus, in some sense, one can talk of Hs as representative of high values.

Obviously, there is a variability in Hs over the year; storms and hence high waves are *e.g.* less frequent in the summer. Again, we face the problem of *seasonality* as can be seen in Figure 10.4, left panel. If there are trends, for example due to global-warming effects (see [8], [9] for discussion of the wave climate in the North Atlantic) or periodic variability with longer periods than a few years, more advanced methods have to be used to study occurrences of high loads. These are not treated in this book and we refer, for examples in ocean engineering and oceanography, to the report by Anderson *et al.* [2].

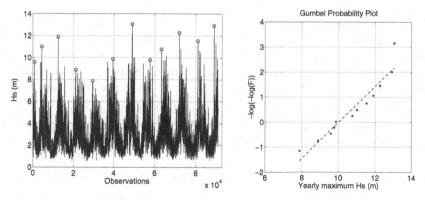

Fig. 10.4. *Left*: Time series of observations of Hs, 1st July 1993 – 1st July 2003. Yearly maxima indicated with rings. *Right*: Yearly maxima plotted in Gumbel probability paper.

Here we study data from 1993 to 2003 and assume that yearly maxima (where a year is defined as starting on July 1) are independent and identically distributed. These are marked as circles in the left plot. On the basis of extreme-value theory, we choose to model the yearly maximal Hs values using a Gumbel distribution. Only 12 yearly maxima of observations (unit: meters) are available:

$$9.6 \quad 11.0 \quad 11.9 \quad 8.9 \quad 7.9 \quad 9.9 \quad 13.0 \quad 9.8 \quad 10.8 \quad 12.3 \quad 11.5 \quad 12.9$$

Thus, it is hard to make a proper validation of the model and we only present the values on a Gumbel probability plot (Figure 10.4 (right panel)). The ML estimates of the parameters are $a^* = 1.5$ and $b^* = 10.0$, which gives the estimate of the 100-year significant wave height

$$s^*_{100} = b^* + a^* \ln(100) = 16.9 \ [\text{m}].$$

If we neglect the possibility of model error (namely that yearly maxima are not Gumbel distributed) and that the number of observations 12 is far too low to use asymptotic results like Theorem 4.3 (asymptotic normality of estimation errors) a confidence interval for s_{100} can be constructed. By means of (10.16)

$$\sigma^*_{\mathcal{E}} = 1.5 \sqrt{\frac{1.11 + 0.61(\ln(100))^2 + 0.52 \ln(100)}{12}} = 1.756$$

and hence, with approximately 95% confidence, s_{100} is bounded by $16.9 + 1.64 \cdot 1.756 = 19.8$ m. □

10.3.2 Uncertainty analysis of s_T: GEV case

In the case when data contradict the assumption that yearly maxima are Gumbel distributed, *e.g.* the confidence interval for c does not contain zero or the deviance DEV $> \chi^2_\alpha(1) = \lambda^2_{\alpha/2}$, then the GEV distribution is used to model the variability of yearly maxima and the design load s_T is estimated using Eq. (10.13).

If c is significantly negative then the predicted design load is usually very uncertain. One way of including the uncertainty into prediction of the design load is to estimate the confidence bound for s_T. The delta method (presented in Section 8.3.1) can be employed for this purpose. Some additional information needed for computations of the bounds is given in the following remark.

Remark 10.8. In order to use the delta method to evaluate the approximative confidence bound for s_T the gradient $\nabla s_T(a, b, c)$ first needs to be found, *i.e.* a vector containing the following partial derivatives:

$$\frac{\partial s_T}{\partial a} = \frac{1 - T^{-c}}{c}, \quad \frac{\partial s_T}{\partial b} = 1, \quad \frac{\partial s_T}{\partial c} = \frac{a}{c^2} \left(T^{-c} + cT^{-c} \ln(T) - 1 \right).$$

Now by Eq. (8.23) the approximate variance of the estimation error $\mathcal{E}$

$$(\sigma^2_{\mathcal{E}})^* = \nabla s_T(a^*, b^*, c^*)^{\mathsf{T}} \, \Sigma^* \, \nabla s_T(a^*, b^*, c^*)$$

where $\Sigma^* = \left[-\ddot{l}(a^*, b^*, c^*) \right]^{-1}$. Then with approximately $(1 - \alpha)$ confidence the design load is bounded by

$$s_T \leq s_T^* + \lambda_\alpha \sigma_{\mathcal{E}}^*. \tag{10.18}$$

□

10.3.3 Warning example of model error

This example has origin in an article by Coles and Pericchi [15], examining a series of daily rainfalls recorded at Maiquetia international airport, Venezuela. In Example 2.14, we already seen the data and concluded that daily rains exhibit seasonal variability. Here we wish to find the design value for the rainwater load and hence a natural analysis is first finding the distribution of yearly maxima M_t, t one year, and then finding the design value as described in Section 10.3. Let us first review the analysis presented in [15].

Model fit with Gumbel and GEV

First, the maximal daily rainfall observed during the years $1951, \ldots, 1998$ is computed. Thus we have 48 observations $z_1, \ldots, z_{48}$ of random variables $Z_1, \ldots, Z_{48}$ of the maximum amount of rain that falls during one day in each of the 48 years. Assume that Z_i are independent and choose the GEV class of distributions to model the data. ML estimates are found as $a^* = 19.9$, $b^* = 49.2$, and $c^* = -0.16$ and the standard deviation $D[C^*] \approx 0.14$. Hence by Eq. (10.9), with approximately 95% confidence, c lies in $[-0.16 - 1.96 \cdot 0.14, \ -0.16 + 1.96 \cdot 0.14]$. Since the interval contains $c = 0$ we conclude that the estimated parameter c^* does not significantly differ from zero.

In Figure 10.5, left, the observed maximal daily rainfall during the years $1951, \ldots, 1998$ are presented while in the right panel data are plotted on

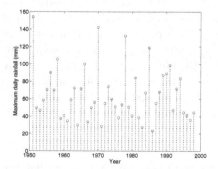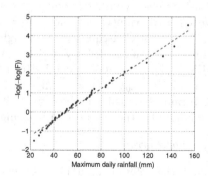

Fig. 10.5. *Left:* The observed yearly maximal rainfall in one day observed during the years 1951–1998 at Maiquetia international airport, Venezuela. *Right:* The data plotted on Gumbel paper.

Gumbel paper. We can see that the fit to the Gumbel distribution is acceptable. In engineering and statistical practice the Gumbel distribution would
be considered to be perfectly adequate to model the data. Consequently, the
estimation procedure is repeated under the assumption that Z_i are Gumbel
distributed, giving the ML estimates $\tilde{a}^* = 21.5$ and $\tilde{b}^* = 50.9$. Having the
estimates a^*, b^*, c^* and $\tilde{a}^*, \tilde{b}^*$ the deviance given in Eq. (10.8) can be applied. This measures how much better the GEV distribution explains the data
compared to the simpler Gumbel distribution. The obtained value DEV=1.67
should be compared with $\chi^2_{0.05}(1) = \lambda^2_{0.025} = 3.84$. Since DEV< 3.84, this
confirms our previous conclusions that the more complicated three-parameter
GEV distribution does not explain the variability of data significantly better
than the two-parameter Gumbel distribution does.

Estimation of design load

Suppose we wish to propose a design for a system that takes care of the large
amounts of rainwater in the tropical climate; thus a design of the rain fall
is needed. A quick glance at Table 8.1, page 205, indicates that we could
use the safety index $\beta_{\mathrm{HL}} = 3.7$, which corresponds to a risk[2] for failure
in one year to be 1 per 10 000. Using this piece of information, we look
for the quantile $z_{0.0001}$. For a Gumbel-distributed variable with parameters
$a^* = 21.5$ and $b^* = 50.9$ we get the design criterion that the system should
manage $s^*_{10000} = 249$ mm rain fall during one day.

We turn next to uncertainty analysis of the design load. Using Eq. (10.17)
with standard deviation computed using Eq. (10.16), we find that with approximately 95% confidence $s_{10000} \leq 249 + 1.64 \cdot 23.6 = 295$ mm. The confidence
level is achieved under the assumption that the Gumbel distribution is the
correct model for yearly maximal rain in one year. Now, in 1999 a catastrophe occurred with an accumulated rain in one day of 410 mm, causing around
50 000 deaths. The conclusion was that "the impossible had happened".

Let us also re-estimate the design load, including the observed 1999 year
disaster. The hypothesis that $c = 0$ has to be rejected. The parameters of
the GEV distribution are now $a^* = 20.8$, $b^* = 48.6$, and $c^* = -0.34$ and
$D[C^*] \approx 0.13$. Consequently, with high confidence we conclude that $c \neq 0$. In
addition, the deviance DEV$= 15.2 > \lambda^2_{\alpha/2} = 10.9$ for $\alpha = 0.001$, showing that
GEV explains the data much better than the Gumbel distribution does. The
design load s_{10000} is now estimated to be 1344 mm, far above the observed
1999 rain. The delta method gives that with approximately 95% confidence
the bound for the design load is as high as 3175 mm.

The model error

Before the 1999 maximum was observed, there were no indications that the
Gumbel model was not correct and a natural question is why not always

---

[2]When designing sea walls in the Netherlands, a return period of 10 000 years
was considered (Example 2.13).

use the GEV model to describe the variability of yearly maxima, instead of assuming that $c = 0$. Often in statistical practice, it is not recommended to use more complicated models than needed to describe data adequately. In the case studied here, including one more parameter c to the model would not explain better the variability of data but made the design value more uncertain causing additional costs to meet the required safety level. On the other hand, using the GEV distribution as a model for the yearly maxima instead of assuming that $c = 0$ is a way of including model uncertainty in the analysis (see the following subsection for some more detailed discussions on uncertainties in the design-load estimation).

Let us now compute s_T^* using the GEV model estimated for the data from the years 1951-1998, $i.e.$ $a^* = 19.9$, $b^* = 49.2$, and $c^* = -0.16$. The design load $s_{10000}^* = 468$ mm and, with approximately 95% confidence, it is smaller than 1030 mm. Clearly, using the design load 468 mm, one could be better prepared for the catastrophe that occurred 1999.

Remark 10.9. Coles and Pericchi proposed to use a Bayesian approach to predict future rainfall. The model for Z conditionally that the parameters a, b, and c are known was a GEV distribution. Then they used suitable priors for the parameters. (Seasonality was also included in their model.) The prior density was then updated using the available observations. Since there are some technical problems in finding the normalization constant in the updating procedure, the so-called MCMC (Markov Chain Monte Carlo) procedure was employed to get the posterior distribution for the parameters. The theory behind the MCMC algorithm is beyond the scope of this book. The algorithm is very useful in the Bayesian updating scheme when many parameters are uncertain.   □

10.3.4 Discussion on uncertainty in design-load estimates

As seen in Remark 10.8, the design load s_T is a strictly decreasing function of c, having large negative derivative for $c < 0$. Consequently the uncertainty in the value of parameter c will heavily influence the uncertainty of s_T (except the case when c is significantly positive). The main reason for the uncertainty of c is the notorious lack of data. In practice, cases can be found where 100-year design loads are estimated using less than 20 years of measurements. Sometimes there are reliable observations for a period of about 50 years and seldom more than 100 years of reliable data are available. Note that even if data for longer periods were available, new problems could appear, namely, changes in environment that would require more parameters to model and hence the uncertainty may not be smaller. Our conclusion is that the uncertainty of s_T^*, for $T > 100$ years, is hard to avoid and should not be neglected.

Since the confidence bounds for s_T are extremely large, these are seldom used as design values. Instead longer return periods T are used for definition of design loads. The dikes in the Netherlands should withstand a 10 000-year sea-level and offshore platforms have to be designed so that a 10 000-year wave is not hitting their decks. In the following example we demonstrate that the design load with a return period of 10 000 years can with non-negligible probability have a return period of only 100 years.

Example 10.7. The difficulty in estimating the design load is illustrated by the following experiment. Suppose the yearly maxima Z_i are GEV distributed with parameters $(a, b, c) = (20, 50, -0.2)$ (the parameters are chosen to be close to the ones valid for the Venezuela rain data). The 100-year and 1000-year loads are found by Eq. (10.13) as $s_{100} = 201$, $s_{1000} = 348$ (suitable units).

Now suppose 50 yearly maxima have been observed from the distribution. This is achieved by simulating 50 random numbers from GEV distribution with the parameters. Next, one checks whether the Gumbel distribution fits well the data and computes the *estimate s_T^**, $T = 10\,000$ years.

The numerical experiment was repeated 1000 times in order to get an idea of the uncertainty. The following result was found:

$s_{10\,000}^*$ was lower than s_{100} in about 5% of the cases

$s_{10\,000}^*$ was lower than s_{1000} in about 25% of the cases

Hence, the probability that "the impossible would happen" is non-negligible, due to the huge uncertainty of the parameter c and the limited time of observation (50 years).

Finally, it is worth noting that in about 65% of the cases, the Gumbel distribution fitted well the data (could not be rejected at 95%-confidence level). ☐

The topic discussed in the last example has importance when evaluating safety of a structure during its service time T_s, say. Assume that T_s is much shorter than the return period T (typically $T_s = 50$ while $T = 10\,000$ years); then the probability that the load exceeds the design loading the service period is close to T_s/T.

For example, if the service time $T_s = 50$ years while $T = 10\,000$ years then the chances of observing the 10 000-year load in 50 years is 1:200, *i.e.* negligible, while if $T = 100$ years, the probability is about $1/2$ (more precisely $1 - \exp(-0.5) = 0.39$). Thus one should not be surprised in observing the 100 year-load in a 50-year period. The last example shows that even the design of 10 000-year load may be observed during such a period.

Problems

10.1. Consider the random variables X_i, $i = 1, \dots, 5$, each of which is uniformly distributed on $(-1, 1)$. Find an expression for the distribution of the random variable $Y = \max(X_1, \dots, X_5)$.

10.2. Consider wind speeds of storms. Assume that the speed of the wind is varying according to some unknown distribution. The maximum wind speed U during a time interval, 30 minutes say, is often well modelled by a Gumbel distribution, that is

$$F_U(u) = \exp\left(-e^{-(u-b)/a}\right),$$

where $a > 0$ and b are constants. Since wind speeds are positive, b will be so high that the probability of obtaining negative values is negligible.

Assume that a storm lasts for 3 hours and that we measure the maximum wind speed during these six 30-minute periods, resulting in the random variables $U_1, U_2, \ldots, U_6$. You may assume that $U_1, U_2, \ldots, U_6$ are independent.

(a) Give the distribution function for $U_{\max} = \max(U_1, \ldots, U_6)$, expressed in terms of $F_U(u)$.
(b) Assume that $a = 4$ m/s. During a hurricane, the maximum wind speed was at some places higher than 40 m/s. Give the value of b corresponding to a probability of 50 % that the maximum wind speed will exceed 40 m/s during 3 hours. In other words, what is the value b such that the median value of $U_{\max}$ is 40 m/s?

10.3. Consider the exponential distribution,

$$F(x) = 1 - e^{-x}, \quad x \geq 0.$$

Use Eq. (10.4) with $a_n = 1$ and $b_n = \ln n$ to prove that the limiting distribution of the sample extremes is the Gumbel distribution.

10.4. In the following data set are found 19 observations of X, yearly maximum of one-hour averages of concentration of sulphur dioxide SO_2 (pphm), Long Beach, California ([66]). The observations were recorded in 1956-1974.

$$47 \quad 41 \quad 68 \quad 32 \quad 27 \quad 43 \quad 20 \quad 27 \quad 25 \quad 18 \quad 33 \quad 40 \quad 51 \quad 55 \quad 40$$

$$55 \quad 37 \quad 28 \quad 34$$

A statistician decides after plotting in probability paper that a Gumbel distribution might fit the annual maxima:

$$F_X(x) = \exp(-e^{-(x-b)/a}), \quad -\infty < x < \infty.$$

The parameters a and b are estimated by the ML method and estimates are returned by statistical software as $a^* = 10.6$, $b^* = 31.9$.

(a) Estimate the 100-year one-hour average, x_{100}, i.e. the 0.01 quantile of X.
(b) For a Gumbel distributed r.v., one can show that the estimators A^* and B^* are asymptotically normally distributed. The covariance matrix is given by

$$\begin{pmatrix} V(A^*) & \mathrm{Cov}(A^*, B^*) \\ \mathrm{Cov}(A^*, B^*) & V(B^*) \end{pmatrix} \approx \frac{(a^*)^2}{n} \begin{pmatrix} 0.61 & 0.26 \\ 0.26 & 1.11 \end{pmatrix}.$$

Note that the estimates are correlated! Calculate the coefficient of correlation.

(c) Use the covariance matrix to find an approximative distribution of the estimation error $\mathcal{E} = x_{100} - (B^* + A^* \ln(100))$.

(d) Calculate an approximative confidence interval for x_{100}.

10.5. Starting from Eq. (10.13), derive the expression for the gradient $\nabla s_T(a, b, c)$ found in Remark 10.8.

10.6. Recall the example from Section 10.3.3, the situation with data from years 1951-1998. A software package returns the following GEV estimates and covariance matrix:

$$a^* = 19.8931, \qquad b^* = 49.1592, \qquad c^* = -0.1648,$$

$$\mathbf{\Sigma}^* = \begin{pmatrix} 7.0099 & 5.0433 & 0.0848 \\ 5.0433 & 11.2277 & 0.1791 \\ 0.0848 & 0.1791 & 0.0191 \end{pmatrix}.$$

Use the delta method to compute a 95% upper bound for the 10 000-year design load.

10.7. Consider a Weibull distributed r.v. X:

$$F_X(x) = 1 - e^{-(x/a)^c}, \qquad x > 0.$$

Show that $Y = \ln X$ is Gumbel distributed and find its scale parameter. (This fact can be used when constructing test statistics for the Weibull distribution, see [24].)

A

Some Useful Tables

In the following pages, we first present a list of some common distributions discussed in this book, including expressions for expectations and variances. Thereafter follow:

Table 1. A table of the standard normal distribution, $N(0,1)$.
Table 2. A table with quantiles for Student's t distribution.
Table 3. A table with quantiles for the χ^2 distribution.
Table 4. Coefficient of variation for a Weibull distributed random variable.

Distribution		Expectation	Variance
Beta distribution, Beta(a,b)	$f(x) = \frac{\Gamma(a+b)}{\Gamma(a)\Gamma(b)}x^{a-1}(1-x)^{b-1},\ 0<x<1$	$\frac{a}{a+b}$	$\frac{ab}{(a+b)^2(a+b+1)}$
Binomial distribution, Bin(n,p)	$p_k = \binom{n}{k}p^k(1-p)^{n-k},\ k=0,1,2,\ldots,n$	np	$np(1-p)$
First success distribution	$p_k = p(1-p)^{k-1},\ k=1,2,3,\ldots$	$\frac{1}{p}$	$\frac{1-p}{p^2}$
Geometric distribution	$p_k = p(1-p)^k,\ k=0,1,2,\ldots$	$\frac{1-p}{p}$	$\frac{1-p}{p^2}$
Poisson distribution, Po(m)	$p_k = e^{-m}\frac{m^k}{k!},\ k=0,1,2,\ldots$	m	m
Exponential distribution, Exp(a)	$F(x) = 1-e^{-x/a},\ x\geq 0$	a	a^2
Gamma distribution, Gamma(a,b)	$f(x) = \frac{b^a}{\Gamma(a)}x^{a-1}e^{-bx},\ x\geq 0$	a/b	a/b^2
Gumbel distribution	$F(x) = e^{-e^{-(x-b)/a}},\ x\in\mathbb{R}$	$b + a\cdot 0.5772\ldots$	$a^2\pi^2/6$
Normal distribution, N(m,σ^2)	$f(x) = \frac{1}{\sigma\sqrt{2\pi}}e^{-(x-m)^2/2\sigma^2},\quad x\in\mathbb{R}$ $F(x) = \Phi((x-m)/\sigma),\quad x\in\mathbb{R}$	m	σ^2
Log-normal distribution, $\ln X \in$ N(m,σ^2)	$F(x) = \Phi\left(\frac{\ln x-m}{\sigma}\right),\quad x>0$	$e^{m+\sigma^2/2}$	$e^{2m+2\sigma^2} - e^{2m+\sigma^2}$
Uniform distribution, U(a,b)	$f(x) = 1/(b-a),\ a\leq x\leq b$	$\frac{a+b}{2}$	$\frac{(a-b)^2}{12}$
Weibull distribution	$F(x) = 1-e^{-\left(\frac{x-b}{a}\right)^c},\quad x\geq b$	$b + a\Gamma(1+1/c)$	$a^2\left[\Gamma(1+\tfrac{2}{c}) - \Gamma^2(1+\tfrac{1}{c})\right]$

Table 1. Standard-normal distribution function

If $X \in N(0,1)$, then $P(X \leq x) = \Phi(x)$, where $\Phi(\cdot)$ is a non-elementary function given by

$$\Phi(x) = \int_{-\infty}^{x} \frac{1}{\sqrt{2\pi}}\, e^{-\xi^2/2} \, d\xi.$$

This table gives function values of $\Phi(x)$. For negative values of x, use that $\Phi(-x) = 1 - \Phi(x)$.

x	0.00	0.01	0.02	0.03	0.04	0.05	0.06	0.07	0.08	0.09
0.0	0.5000	0.5040	0.5080	0.5120	0.5160	0.5199	0.5239	0.5279	0.5319	0.5359
0.1	0.5398	0.5438	0.5478	0.5517	0.5557	0.5596	0.5636	0.5675	0.5714	0.5753
0.2	0.5793	0.5832	0.5871	0.5910	0.5948	0.5987	0.6026	0.6064	0.6103	0.6141
0.3	0.6179	0.6217	0.6255	0.6293	0.6331	0.6368	0.6406	0.6443	0.6480	0.6517
0.4	0.6554	0.6591	0.6628	0.6664	0.67600	0.6736	0.6772	0.6808	0.6844	0.6879
0.5	0.6915	0.6950	0.6985	0.7019	0.7054	0.7088	0.7123	0.7157	0.7190	0.7224
0.6	0.7257	0.7291	0.7324	0.7357	0.7389	0.7422	0.7454	0.7486	0.7517	0.7549
0.7	0.7580	0.7611	0.7642	0.7673	0.7704	0.7734	0.7764	0.7794	0.7823	0.7852
0.8	0.7881	0.7910	0.7939	0.7967	0.7995	0.8023	0.8051	0.8078	0.8106	0.8133
0.9	0.8159	0.8186	0.8212	0.8238	0.8264	0.8289	0.8315	0.8340	0.8365	0.8389
1.0	0.8413	0.8438	0.8461	0.8485	0.8508	0.8531	0.8554	0.8577	0.8599	0.8621
1.1	0.8643	0.8665	0.8686	0.8708	0.8729	0.8749	0.8770	0.8790	0.8810	0.8830
1.2	0.8849	0.8869	0.8888	0.8907	0.8925	0.8944	0.8962	0.8980	0.8997	0.9015
1.3	0.9032	0.9049	0.9066	0.9082	0.9099	0.9115	0.9131	0.9147	0.9162	0.9177
1.4	0.9192	0.9207	0.9222	0.9236	0.9251	0.9265	0.9279	0.9292	0.9306	0.9319
1.5	0.9332	0.9345	0.9357	0.9370	0.9382	0.9394	0.9406	0.9418	0.9429	0.9441
1.6	0.9452	0.9463	0.9474	0.9484	0.9495	0.9505	0.9515	0.9525	0.9535	0.9545
1.7	0.9554	0.9564	0.9573	0.9582	0.9591	0.9599	0.9608	0.9616	0.9625	0.9633
1.8	0.9641	0.9649	0.9656	0.9664	0.9671	0.9678	0.9686	0.9693	0.9699	0.9706
1.9	0.9713	0.9719	0.9726	0.9732	0.9738	0.9744	0.9750	0.9756	0.9761	0.9767
2.0	0.9772	0.9778	0.9783	0.9788	0.9793	0.9798	0.9803	0.9808	0.9812	0.9817
2.1	0.9821	0.9826	0.9830	0.9834	0.9838	0.9842	0.9846	0.9850	0.9854	0.9857
2.2	0.9861	0.9864	0.9868	0.9871	0.9875	0.9878	0.9881	0.9884	0.9887	0.9890
2.3	0.9893	0.9896	0.9898	0.9901	0.9904	0.9906	0.9909	0.9911	0.9913	0.9916
2.4	0.9918	0.9920	0.9922	0.9925	0.9927	0.9929	0.9931	0.9932	0.9934	0.9936
2.5	0.9938	0.9940	0.9941	0.9943	0.9945	0.9946	0.9948	0.9949	0.9951	0.9952
2.6	0.9953	0.9955	0.9956	0.9957	0.9959	0.9960	0.9961	0.9962	0.9963	0.9964
2.7	0.9965	0.9966	0.9967	0.9968	0.9969	0.9970	0.9971	0.9972	0.9973	0.9974
2.8	0.9974	0.9975	0.9976	0.9977	0.9977	0.9978	0.9979	0.9979	0.9980	0.9981
2.9	0.9981	0.9982	0.9982	0.9983	0.9984	0.9984	0.9985	0.9985	0.9986	0.9986
3.0	0.9987	0.9987	0.9987	0.9988	0.9988	0.9989	0.9989	0.9989	0.9990	0.9990
3.1	0.9990	0.9991	0.9991	0.9991	0.9992	0.9992	0.9992	0.9992	0.9993	0.9993
3.2	0.9993	0.9993	0.9994	0.9994	0.9994	0.9994	0.9994	0.9995	0.9995	0.9995
3.3	0.9995	0.9995	0.9996	0.9996	0.9996	0.9996	0.9996	0.9996	0.9996	0.9997
3.4	0.9997	0.9997	0.9997	0.9997	0.9997	0.9997	0.9997	0.9997	0.9997	0.9998
3.5	0.9998	0.9998	0.9998	0.9998	0.9998	0.9998	0.9998	0.9998	0.9998	0.9998
3.6	0.9998	0.9998	0.9999	0.9999	0.9999	0.9999	0.9999	0.9999	0.9999	0.9999

Table 2. Quantiles of Student's t-distribution

If $X \in t(n)$, then the α quantile $t_\alpha(n)$ is defined by

$$P\big(X > t_\alpha(n)\big) = \alpha, \quad 0 < \alpha < 1.$$

This table gives the α quantile $t_\alpha(n)$. For values of $\alpha \geq 0.9$, use that

$$t_{1-\alpha}(n) = -t_\alpha(n), \quad 0 < \alpha < 1.$$

n	α						
	0.1	0.05	0.025	0.01	0.005	0.001	0.0005
1	3.078	6.314	12.706	31.821	63.657	318.309	636.619
2	1.886	2.920	4.303	6.965	9.925	22.327	31.599
3	1.638	2.353	3.182	4.541	5.841	10.215	12.924
4	1.533	2.132	2.776	3.747	4.604	7.173	8.610
5	1.476	2.015	2.571	3.365	4.032	5.893	6.869
6	1.440	1.943	2.447	3.143	3.707	5.208	5.959
7	1.415	1.895	2.365	2.998	3.499	4.785	5.408
8	1.397	1.860	2.306	2.896	3.355	4.501	5.041
9	1.383	1.833	2.262	2.821	3.250	4.297	4.781
10	1.372	1.812	2.228	2.764	3.169	4.144	4.587
11	1.363	1.796	2.201	2.718	3.106	4.025	4.437
12	1.356	1.782	2.179	2.681	3.055	3.930	4.318
13	1.350	1.771	2.160	2.650	3.012	3.852	4.221
14	1.345	1.761	2.145	2.624	2.977	3.787	4.140
15	1.341	1.753	2.131	2.602	2.947	3.733	4.073
16	1.337	1.746	2.120	2.583	2.921	3.686	4.015
17	1.333	1.740	2.110	2.567	2.898	3.646	3.965
18	1.330	1.734	2.101	2.552	2.878	3.610	3.922
19	1.328	1.729	2.093	2.539	2.861	3.579	3.883
20	1.325	1.725	2.086	2.528	2.845	3.552	3.850
21	1.323	1.721	2.080	2.518	2.831	3.527	3.819
22	1.321	1.717	2.074	2.508	2.819	3.505	3.792
23	1.319	1.714	2.069	2.500	2.807	3.485	3.768
24	1.318	1.711	2.064	2.492	2.797	3.467	3.745
25	1.316	1.708	2.060	2.485	2.787	3.450	3.725
26	1.315	1.706	2.056	2.479	2.779	3.435	3.707
27	1.314	1.703	2.052	2.473	2.771	3.421	3.690
28	1.313	1.701	2.048	2.467	2.763	3.408	3.674
29	1.311	1.699	2.045	2.462	2.756	3.396	3.659
30	1.310	1.697	2.042	2.457	2.750	3.385	3.646
40	1.303	1.684	2.021	2.423	2.704	3.307	3.551
60	1.296	1.671	2.000	2.390	2.660	3.232	3.460
120	1.289	1.658	1.980	2.358	2.617	3.160	3.373
∞	1.282	1.645	1.960	2.326	2.576	3.090	3.291

Table 3. Quantiles of the χ^2 distribution

If $X \in \chi^2(n)$, then the α quantile $\chi_\alpha^2(n)$ is defined by

$$P\big(X > \chi_\alpha^2(n)\big) = \alpha, \quad 0 < \alpha < 1$$

This table gives the α quantile $\chi_\alpha^2(n)$.

n	0.9995	0.999	0.995	0.99	0.975	0.95	0.05	0.025	0.01	0.005	0.001	0.0005
1	—	—	$<10^{-2}$	$<10^{-2}$	$<10^{-2}$	$<10^{-2}$	3.841	5.024	6.635	7.879	10.83	12.12
2	$<10^{-2}$	$<10^{-2}$	0.0100	0.0201	0.0506	0.1026	5.991	7.378	9.210	10.60	13.82	15.20
3	0.0153	0.0240	0.0717	0.1148	0.2158	0.3518	7.815	9.348	11.34	12.84	16.27	17.73
4	0.0639	0.0908	0.2070	0.2971	0.4844	0.7107	9.488	11.14	13.28	14.86	18.47	20.00
5	0.1581	0.2102	0.4117	0.5543	0.8312	1.145	11.07	12.83	15.09	16.75	20.52	22.11
6	0.2994	0.3811	0.6757	0.8721	1.237	1.635	12.59	14.45	16.81	18.55	22.46	24.10
7	0.4849	0.5985	0.9893	1.239	1.690	2.167	14.07	16.01	18.48	20.28	24.32	26.02
8	0.7104	0.8571	1.344	1.646	2.180	2.733	15.51	17.53	20.09	21.95	26.12	27.87
9	0.9717	1.152	1.735	2.088	2.700	3.325	16.92	19.02	21.67	23.59	27.88	29.67
10	1.265	1.479	2.156	2.558	3.247	3.940	18.31	20.48	23.21	25.19	29.59	31.42
11	1.587	1.834	2.603	3.053	3.816	4.575	19.68	21.92	24.72	26.76	31.26	33.14
12	1.934	2.214	3.074	3.571	4.404	5.226	21.03	23.34	26.22	28.30	32.91	34.82
13	2.305	2.617	3.565	4.107	5.009	5.892	22.36	24.74	27.69	29.82	34.53	36.48
14	2.697	3.041	4.075	4.660	5.629	6.571	23.68	26.12	29.14	31.32	36.12	38.11
15	3.108	3.483	4.601	5.229	6.262	7.261	25.00	27.49	30.58	32.80	37.70	39.72
16	3.536	3.942	5.142	5.812	6.908	7.962	26.30	28.85	32.00	34.27	39.25	41.31
17	3.980	4.416	5.697	6.408	7.564	8.672	27.59	30.19	33.41	35.72	40.79	42.88
18	4.439	4.905	6.265	7.015	8.231	9.390	28.87	31.53	34.81	37.16	42.31	44.43
19	4.912	5.407	6.844	7.633	8.907	10.12	30.14	32.85	36.19	38.58	43.82	45.97
20	5.398	5.921	7.434	8.260	9.591	10.85	31.41	34.17	37.57	40.00	45.31	47.50
21	5.896	6.447	8.034	8.897	10.28	11.59	32.67	35.48	38.93	41.40	46.80	49.01
22	6.404	6.983	8.643	9.542	10.98	12.34	33.92	36.78	40.29	42.80	48.27	50.51
23	6.924	7.529	9.260	10.20	11.69	13.09	35.17	38.08	41.64	44.18	49.73	52.00
24	7.453	8.085	9.886	10.86	12.40	13.85	36.42	39.36	42.98	45.56	51.18	53.48
25	7.991	8.649	10.52	11.52	13.12	14.61	37.65	40.65	44.31	46.93	52.62	54.95
26	8.538	9.222	11.16	12.20	13.84	15.38	38.89	41.92	45.64	48.29	54.05	56.41
27	9.093	9.803	11.81	12.88	14.57	16.15	40.11	43.19	46.96	49.64	55.48	57.86
28	9.656	10.39	12.46	13.56	15.31	16.93	41.34	44.46	48.28	50.99	56.89	59.30
29	10.23	10.99	13.12	14.26	16.05	17.71	42.56	45.72	49.59	52.34	58.30	60.73
30	10.80	11.59	13.79	14.95	16.79	18.49	43.77	46.98	50.89	53.67	59.70	62.16
40	16.91	17.92	20.71	22.16	24.43	26.51	55.76	59.34	63.69	66.77	73.40	76.09
50	23.46	24.67	27.99	29.71	32.36	34.76	67.50	71.42	76.15	79.49	86.66	89.56
60	30.34	31.74	35.53	37.48	40.48	43.19	79.08	83.30	88.38	91.95	99.61	102.7
70	37.47	39.04	43.28	45.44	48.76	51.74	90.53	95.02	100.4	104.2	112.3	115.6
80	44.79	46.52	51.17	53.54	57.15	60.39	101.9	106.6	112.3	116.3	124.8	128.3
90	52.28	54.16	59.20	61.75	65.65	69.13	113.1	118.1	124.1	128.3	137.2	140.8
100	59.90	61.92	67.33	70.06	74.22	77.93	124.3	129.6	135.8	140.2	149.4	153.2

Table 4. Coefficient of variation of a Weibull distribution

The distribution function is given by

$$F_X(x) = 1 - e^{-(x/a)^c}, \quad x > 0,$$

and then the coefficient of variation is

$$R(X) = \frac{\sqrt{\Gamma(1 + 2/c) - \Gamma^2(1 + 1/c)}}{\Gamma(1 + 1/c)}.$$

c	$\Gamma(1 + 1/c)$	$R(X)$
1.00	1.0000	1.0000
2.00	0.8862	0.5227
2.10	0.8857	0.5003
2.70	0.8893	0.3994
3.00	0.8930	0.3634
3.68	0.9023	0.3025
4.00	0.9064	0.2805
5.00	0.9182	0.2291
5.79	0.9259	0.2002
8.00	0.9417	0.1484
10.00	0.9514	0.1203
12.10	0.9586	0.1004
20.00	0.9735	0.0620
21.80	0.9758	0.0570
50.00	0.9888	0.0253
128.00	0.9956	0.0100

Short Solutions to Problems

Problems of Chapter 1

1.1

(a) Possible values: $X = 0, 1, 2, 3$.

(b) $P(X = 0) = (1 - 0.5) \cdot (1 - 0.8) \cdot (1 - 0.2) = 0.08$.

$P(X = 1) = 0.5 \cdot (1-0.8) \cdot (1-0.2) + (1-0.5) \cdot 0.8 \cdot (1-0.2) + (1-0.5) \cdot (1-0.8) \cdot 0.2 = 0.42$.

(c) $P(X < 2) = P(X = 0) + P(X = 1) = 0.08 + 0.42 = 0.50$.

1.2 $A \cup B = A \cup (A^c \cap B)$, $\quad B = (A \cap B) \cup (A^c \cap B)$. The events A and $A^c \cap B$ are excluding, and so are $A \cap B$ and $A^c \cap B$. Hence $P(A \cup B) = P(A) + P(A^c \cap B)$, $\quad P(B) = P(A \cap B) + P(A^c \cap B)$. Subtraction gives the result. Alternatively: Deduce from the so-called Venn diagrams.

1.3 $P(A \cap B) = [\text{independence}] = P(A) P(A) > 0$, hence $P(A \cap B) \neq 0$ and the events are not excluding.

1.4 $P(A) = p$, $P(A^c) = 1 - p$. Since $A \cap A^c = \emptyset$, $P(A \cap A^c) = 0$. But $P(A)P(A^c) = p(1 - p) > 0$ if $p > 0$. Hence the events are not independent. If $p = 0$ then the events are independent.

1.5

(a) $\binom{12}{3} 0.05^3 0.95^9 = 0.017$.

(b) $0.95^{12} = 0.54$.

1.6

(a) $57/(57 + 53) = 57/110$.

(b) $32/50$.

(c) $P(\text{"Vegetarian"}) = \frac{57}{110}$, $P(\text{"Woman"}) = \frac{50}{110}$,
$P(\text{"Vegetarian"} \cap \text{"Woman"}) = \frac{32}{110}$. But $\frac{57}{110} \cdot \frac{50}{110} \neq \frac{32}{110}$, hence the events are dependent.

1.7

$$p = P(\text{"At least one light functions after 1000 hours"})$$
$$= 1 - P(\text{"No light functions after 1000 hours"}) = 1 - (1 - 0.55)^4 = 0.96.$$

1.8 P("Circuit functions") $= 0.8 \cdot 0.8 + 0.8 \cdot 0.2 + 0.2 \cdot 0.8 = 0.96$. Alternatively, reasoning with complementary event: $1 - 0.2 \cdot 0.2 = 0.96$.

1.9 $A =$"Lifetime longer than one year", $B =$"Lifetime longer than five years". $P(B|A) = P(A \cap B)/P(A) = P(B)/P(A) = 1/9$.

1.10 Law of total probability: $0.6 \cdot 0.04 + 0.9 \cdot 0.01 + 0.01 \cdot 0.95 = 0.024 + 0.009 + 0.0095 = 0.0425$.

1.11 Let $N =$"Number of people with colour blindness". Then $N \in \text{Bin}(n,p)$, $P(N > 0) = 1 - P(N = 0) = 1 - (1-p)^n$. Since for p close to zero, $1 - p \approx \exp(-p)$, we have $P(N > 0) \approx 1 - \exp(-np)$; hence $n \geq 75$. (Alternatively, p is close to zero, hence $N \in \text{Po}(np)$, etc.)

1.12 $N =$ "Number of erroneous filters out of n". Model: $N \in \text{Bin}(n,p)$, where $n = 200$, $p = 0.01$. As $n > 10$, $p < 0.1$, Poisson approximation is used: $N \in \text{Po}(np)$, i.e. $N \in \text{Po}(0.2)$. $P(N > 2) = 1 - P(N \leq 2) \approx 1 - (e^{-0.2} + 0.2 \cdot e^{-0.2} + \frac{0.2^2}{2} e^{-0.2}) = 0.0011$.

Problems of Chapter 2

2.1

(a) $P(X \leq 2) = P(X = 0) + P(X = 1) + P(X = 2) = e^{-3}\frac{3^0}{0!} + e^{-3}\frac{3^1}{1!} + e^{-3}\frac{3^2}{2!} = \frac{17}{2}e^{-3} = 0.423$

(b) $P(0 \leq X \leq 1) = P(X = 0) + P(X = 1) = e^{-3}\frac{3^0}{0!} + e^{-3}\frac{3^1}{1!} = 4e^{-3} = 0.199$

(c) $P(X > 0) = 1 - P(X \leq 0) = 1 - P(X = 0) = 1 - e^{-3}\frac{3^0}{0!} = 0.950$

(d)

$$P(5 \leq X \leq 7 | X \geq 3) = \frac{P(5 \leq X \leq 7 \cap X \geq 3)}{P(X \geq 3)} = \frac{P(5 \leq X \leq 7)}{P(X \geq 3)} =$$

$$= \frac{P(5 \leq X \leq 7)}{1 - P(X \leq 2)} = \frac{e^{-3} \cdot \left(\frac{3^5}{5!} + \frac{3^6}{6!} + \frac{3^7}{7!}\right)}{1 - \frac{17}{2}e^{-3}} = 0.300.$$

2.2

(a) By independence, $p = 0.92^6 \cdot 0.08 = 0.049$ (see also geometric distribution).

(b) $1/0.08 = 12.5$ months.

2.3 Bayes' formula gives $P(A|B) = 0.33$.

2.4 Introduce the events $A_1 =$ "Fire-emergency call from industrial zone", $A_2 =$"Fire-emergency call from housing area", $F =$"Fire at arrival". Further, $P(A_1) = 0.55$, $P(A_2) = 0.45$, $P(F|A_1) = 0.05$, $P(F|A_2) = 0.90$. Thus

$$P(A_1 | F) = \frac{P(F|A_1)P(A_1)}{P(F)} = \frac{P(F|A_1)P(A_1)}{P(F|A_1)P(A_1) + P(F|A_2)P(A_2)}$$

$$= \frac{0.05 \cdot 0.55}{0.05 \cdot 0.55 + 0.90 \cdot 0.45} = 0.064$$

2.5 Introduce $A_1 =$ "Dot sent", $A_2 =$ "Dash sent", $B =$ "Dot received". From the text, $P(B|A_2) = 1/10$, $P(B^c|A_1) = 1/10$. Asked for: $P(A_1|B)$.

Odds for A_1, A_2: $q_1^{\text{prior}} = 3$, $q_2^{\text{prior}} = 4$. Posterior odds, given B is true: $q_1^{\text{post}} = (1-1/10)\cdot 3$, $q_2^{\text{post}} = (1/10)\cdot 4$. Hence $P(A_1|B) = q_1^{\text{post}}/(q_1^{\text{post}}+q_2^{\text{post}}) = 0.87$.

2.6

(a) Solution 1: There are four possible gender sequences: BB, BG, GB, and GG. All sequences are equally likely. We know that there is at least one girl, hence the sequence BB is eliminated and three cases remain. The probability that the other child is also a girl is hence $1/3$.

Solution 2: The odds for the four gender combinations are equal: $q_i^{\text{prior}} = 1$. $A = $"The Smiths tell you that they have 2 children and at least one is a girl". We wish to find $P(GG\,|\,A)$. Since $P(A|BB) = 0$, $P(A|BG) = P(A|GB) = P(A|GG) = 1$, the posterior odds given A is true are $0 : 1 : 1 : 1$. Hence $P(GG|A) = 1/3$.

(b) $A = $"You see the Smiths have a girl". $P(A|BB) = 0$, $P(A|BG) = P(A|GB) = 1/2$, $P(A|GG) = 1$. Thus the posterior odds are $0 : 1/2 : 1/2 : 1$ and hence

$$P(GG|A) = \frac{q_4^{\text{post}}}{q_1^{\text{post}} + \cdots + q_4^{\text{post}}} = \frac{1}{2}.$$

2.7 $A = $ "A person is infected", $B = $ "Test indicates person infected". Bayes' formula: $P(A|B) = \frac{0.99\cdot 0.0001}{0.99\cdot 0.0001+0.001\cdot(1-0.0001)} \approx 0.09$.

2.8 We have

$$P(3 \text{ leakages} \,|\, \text{Corr}) = \left((\lambda_{\text{Corr}}5)^3/3!\right)\exp(-\lambda_{\text{Corr}}5) = 0.05$$

and similarly

$$P(3 \text{ leakages} \,|\, \text{Thermal}) = 0.20, \quad P(3 \text{ leakages} \,|\, \text{Other}) = 1.7\cdot 10^{-7}.$$

Hence the posterior odds are $q_{\text{Corr}}^{\text{post}} = 4\cdot 0.05 = 0.2$, $q_{\text{Therm}}^{\text{post}} = 1\cdot 0.20 = 0.2$, $q_{\text{Other}}^{\text{post}} = 95\cdot 1.7\cdot 10^{-7} = 2\cdot 10^{-5}$. In other words, the odds are roughly 1:1:0. The two reasons for leakage are now equally likely.

2.9 $p = $ P("A certain crack is detected"), $(p = 0.8)$; $N = $ the number of cracks along the distance inspected; $K = $ the number of detected cracks along the distance inspected.

(a) $P(K = 0 \,|\, N = 2) = (1 - p)(1 - p) = 0.04$.

(b) Since $P(N = 0) + P(N = 1) + P(N = 2) = 1$, there are never more than two cracks. Law of total probability: $P(K = 0) = P(K = 0|N = 0)P(N = 0)+P(K = 0|N = 1)P(N = 1) + P(K = 0|N = 2)P(N = 2) = P(N = 0) + (1 - p)P(N = 1) + (1 - p)^2 P(N = 2) = 0.42$.

(c) Bayes' formula: $P(N = 0|K = 0) = P(K = 0|N = 0)P(N = 0)/P(K = 0) = 1\cdot 0.3/0.424 = 0.71$.

2.10

(a) $1 - (1 - p)^{24\,000} \approx 1 - (1 - 24\,000\cdot p) = 24\,000 p = 1.2\cdot 10^{-3}$, where $p = 5\cdot 10^{-8}$.

(b) On average, $n = 1/p$ street crossings to the first accident. One year has $6\cdot 200$ street crossings, giving a return period of $1.7\cdot 10^4$ years.

2.11

(a) $\lambda \approx 5/10 = 1/2$ [year^{-1}]

(b) $T \approx 2$ [years]
(c) $P_t(A) \approx \frac{1}{2} \cdot \frac{1}{12} = \frac{1}{24}$ and hence $p = 1 - P_t(A) \approx 23/24$.

2.12 Introduce A_1, A_2: fire ignition in hospital No. 1 and No. 2, respectively. Asked for:

$$p = P(N_{A_1}(t) > 0 \cap N_{A_2}(t) > 0) = P_t(A_1) \cdot P_t(A_2),$$

$t = 1/12$ year. By Eq. (2.11),

$$p = \left[1 - \exp\left(\frac{1}{12}\exp(-7.1 + 0.75 \cdot \ln(6000))\right)\right]$$

$$\cdot \left[1 - \exp\left(\frac{1}{12}\exp(-7.1 + 0.75 \cdot \ln(7500))\right)\right] = 0.0025.$$

2.13

(a) $\lambda_A \approx (48 + 26 + 44)/3 = 39.3$ year^{-1}.
(b) $N \in \text{Po}(m)$, $m = \lambda_A \cdot P(B) \cdot 1/12$. Since $P(B) \approx (37 + 41 + 49)/(1108 + 1089 + 1192) = 0.0345$, we find $P_t(A \cap B) = 1 - \exp(-m) \approx 0.11$.

2.14

(a) The factors given lead to the following intensities of fires in the town: $\lambda_1 = 2.5$, $\lambda_2 = 5$, $\lambda_3 = 7.5$, $\lambda_4 = 10$ (year^{-1}). Choose a uniform prior odds: $q_i^0 = 1$, $i = 1, \dots, 4$.
(b) $C =$ "No fire start during two months". Poisson assumption: $P(C|\Lambda = \lambda_i) = e^{-\lambda_i/6}$ and hence $P(C|\Lambda = \lambda_1) = 0.66$, $P(C|\Lambda = \lambda_2) = 0.43$, $P(C|\Lambda = \lambda_3) = 0.27$, $P(C|\Lambda = \lambda_4) = 0.19$. The posterior odds are given as $q_i^{\text{post}} = P(C|\Lambda = \lambda_i)q_i^0$ and thus $q_1^{\text{post}} = 0.66$, $q_2^{\text{post}} = 0.43$, $q_3^{\text{post}} = 0.27$, $q_4^{\text{post}} = 0.19$.
(c) Theorem 2.2 yields $P^{\text{post}}(\Lambda = \lambda_i) = q_i^{\text{post}} / \sum_j q_j^{\text{post}}$, giving 0.43, 0.28, 0.17, 0.12. $B =$ "No fire starts next month". With $P(B|\Lambda = \lambda_i) = \exp(-\lambda_i t)$, $t = 1/12$, the law of total probability gives:

$$P^{\text{post}}(B) = \sum P(B|\Lambda = \lambda_i)P^{\text{post}}(\Lambda = \lambda_i) = 0.68.$$

Problems of Chapter 3

3.1

(a) $e^{-0.2 \cdot 3} = 0.549$.
(b) $E[T] = 1/0.2 = 5$ (hours).

3.2 Alternatives (i) and (iii). The function in (ii) does not integrate to one, the function in (iv) takes negative values.

3.3 $x_{0.95} = 10(-\ln(0.95))^{1/5} = 5.52$.

3.4 $F_X(x) = \exp(-e^{-(x-b)/a}) \Rightarrow F_Y(y) = P(e^X \le y) = P(X \le \ln y) = F_X(\ln y) = \exp(-y^{-1/a}e^{b/a})$, $y > 0$.

3.5

(a) $F_Y(y) = \begin{cases} 1 - e^{-y^2/a^2} & y > 0 \\ 0, & y \le 0 \end{cases}$.

(b) $f_Y(y) = \dfrac{d}{dy} F_Y(y) = \begin{cases} \frac{2}{a^2} \cdot y e^{-y^2/a^2} & y > 0 \\ 0, & y \le 0 \end{cases}$.

3.6 $E[T] = \int_0^\infty u\, f_T(u)\, du = [-u(1 - F_T(u))]_0^\infty + \int_0^\infty (1 - F_T(u))\, du$. We show that the first term is equal to zero. Consider $t(1 - F_T(t)) = t \int_t^\infty f_T(u)du < \int_t^\infty u f_T(u)\, du$. Since $E[T]$ exists, $\int_t^\infty u f_T(u)\, du \to 0$ as $t \to 0$, thus $t(1 - F_T(t)) \to 0$.

3.7 $E[Y] = \int_0^\infty e^{-y^2/a^2}\, dy = \frac{a}{2} \int_{-\infty}^\infty e^{-u^2}\, du = \frac{a}{2}\sqrt{\pi}$.

3.8

(a) $x_{0.50} = 0$ by symmetry of the pdf around zero.
(b) $\int_{-\infty}^\infty \frac{|x|}{\pi(1+x^2)}\, dx = \infty$.

3.9 $x_{0.01} = b - a\ln(-\ln(1 - 0.01)) = 67 \text{ m}^3/\text{s}$

3.10 Table: $x_{0.01} = \lambda_{0.01} = 2.33$; $x_{0.025} = \lambda_{0.025} = 1.96$, and $x_{0.95} = -x_{0.05} = -\lambda_{0.05} = -1.64$.

3.11 Table: $\chi^2_{0.001}(5) = 20.52$; $\chi^2_{0.01}(5) = 15.09$; $\chi^2_{0.95}(5) = 1.145$.

3.12

(a) $P(X > 200) = 1 - \Phi(\frac{200-180}{7.5}) = 1 - \Phi(2.67) = 0.0038$.
(b) Use Eq. (3.11): $x_{0.01} = 180 + 7.5\lambda_{0.01} = 197.5$. Thus 1% of the population of men is longer than 197.5 cm.

3.13 Table in appendix gives for the gamma distribution $E[X] = 10/2 = 5$, $V[X] = 10/2^2 = 2.5$. $E[Y] = 3E[X] - 5 = 10$, $V[Y] = 3^2V[X] = 22.5$.

3.14 $E[X] = m$, $D[X] = m$; hence $R[X] = 1$.

Problems of Chapter 4

4.1

(a) $E[M_1^*] = m$, $E[M_2^*] = 3m/2$, $E[M_3^*] = m$. Thus M_1^* and M_3^* are unbiased.
(b) $V[M_1^*] = \sigma^2/2$, $V[M_2^*] = 5\sigma^2/4$, $V[M_3^*] = \sigma^2/4$. Thus M_3^* has the smallest variance (and is unbiased).

4.2

(a) $m^* = \frac{1}{n}\sum_{i=1}^{70} \ln x_i = 0.99$, $(\sigma^2)^* = \frac{1}{n-1}\sum_{i=1}^{70} (\ln x_i - m^*)^2 = 0.0898$, $\sigma^* = 0.3$.
(b) We have $1/1000 = P(X > h_{1000}) = 1 - \Phi((\ln h_{1000} - m)/\sigma)$, thus $\lambda_{0.001} = (\ln h_{1000} - m)/\sigma \iff h_{1000} = \exp(m + \sigma\lambda_{0.001}) \Rightarrow h_{1000}^* = \exp(m^* + \sigma^*\lambda_{0.001}) = 6.8$ m.

4.3

(a) Log-likelihood function and its derivative:

$$l(p) = k \ln p + (n - k) \ln(1 - p) + \ln \binom{n}{k}$$

$$\dot{l}(p) = \frac{k}{p} - \frac{n - k}{1 - p}$$

Solving $\dot{l}(p) = 0$ yields the ML estimate $p^* = k/n$, which can be shown to maximize the function.

(b)

$$\ddot{l}(p) = -\frac{k}{p^2} - \frac{n - k}{(1 - p)^2} = -\frac{k(1 - p)^2 + (n - k)p^2}{p^2(1 - p)^2}$$

$$= -\frac{k - 2kp + np^2}{p^2(1 - p)^2}$$

Now, with $p^* = k/n$, we find $\ddot{l}(p^*) = -n/(p^*(1 - p^*))$ and hence $(\sigma_{\hat{\varepsilon}}^2)^* = p^*(1 - p^*)/n$.

4.4

(a) $L(a) = \prod_{i=1}^{n} f(x_i; a) = \prod_{i=1}^{n} \frac{2x_i}{a^2} e^{-\frac{x_i^2}{a^2}}$. Log-likelihood function:

$$l(a) = \ln L(a) = \sum_{i=1}^{n} \ln\left(\frac{2x_i}{a^2} e^{-\frac{x_i^2}{a^2}}\right) = \sum_{i=1}^{n} \left(\ln 2x_i - 2 \ln a - \frac{x_i^2}{a^2}\right)$$

with derivative

$$\dot{l}(a) = -\frac{2n}{a} + 2\sum_{i=1}^{n} \frac{x_i^2}{a^3}.$$

Hence $a^* = \sqrt{\sum_{i=1}^{n} x_i^2/n} = 2.2$.

(b) Since

$$\ddot{l}(a) = \frac{2n}{a^2} - \frac{6}{a^4}\sum x_i^2$$

we find $\ddot{l}(a^*) = -4n/(a^*)^2$ and hence $(\sigma_{\hat{\varepsilon}}^2)^* = (a^*)^2/4n = 0.15$. An asymptotic 0.9 interval is then

$$[2.2 - 1.64 \cdot \sqrt{0.15}, \ 2.2 + 1.64 \cdot \sqrt{0.15}] = [1.56, 2.84]$$

(c) $[1.72, 3.28]$.

4.5 Tensile strength $X \in N(m, 9)$.

(a) $m^* = 20$, $n = 9$, $(\sigma_{\hat{\varepsilon}}^2)^* = \sigma^2/n = 1$; thus with 95 % confidence $m \in \left[m^* \pm \lambda_{0.05}\sigma_{\hat{\varepsilon}}^*\right] = [18.4, 21.6]$.

(b) $2 \cdot \lambda_{0.05} \cdot \sigma/\sqrt{9} = 2 \cdot \lambda_{0.025} \cdot \sigma/\sqrt{n} \Rightarrow n = 9(\lambda_{0.025}/\lambda_{0.05})^2 = 12.8$. Thus, the number must be $n = 13$ and one needs $13 - 9 = 4$ observations more.

4.6 $Q = 0.024$, $\chi^2_{0.05}(1) = 3.84$. Do not reject the hypothesis about a fair coin.

4.7

(a) $X \in \text{Bin}(3, 1/4)$.

(b) Since $X \in \text{Bin}(3, 1/4)$, $P(X = 0) = (3/4)^3$, $P(X = 1) = 3 \cdot (1/4) \cdot (3/4)^2 = 27/64$, $P(X = 2) = 9/64$, $P(X = 3) = 1/64$. It follows that $Q = 11.5$ and since $Q > \chi^2_{0.01}(4 - 1) = 11.3$ we reject the hypothesis. (It seems that the frequency of getting 3 spades is too high.)

4.8 Minimize $g(a) = \text{V}[\Theta_3^*] = a^2 \sigma_1^2 + (1 - a)^2 \sigma_2^2$; $g'(a) = 2a\sigma_1^2 - 2(1 - a)\sigma_2^2 = 0 \Leftrightarrow a = \sigma_2^2/(\sigma_1^2 + \sigma_2^2)$ (local minimum since $g''(a) = 2\sigma_1^2 + 2\sigma_2^2$).

4.9

(a) $m^* = \bar{x} = 33.1$.

(b) $\underset{\sim}{\mathcal{E}} \in \text{N}(0, (\sigma_{\mathcal{E}}^2)^*)$, where $(\sigma_{\mathcal{E}}^2)^* = \bar{x}/n$. Hence $[29.5, 36.7]$.

(c) Eq. (4.28) gives

$$\chi^2_{0.975}(2 \cdot 331) = 662 \left(\sqrt{\frac{2}{9 \cdot 662}}(-1.96) + 1 - \frac{2}{9 \cdot 662} \right)^3 = 592.6.$$

In a similar manner follows $\chi^2_{0.025}(2 \cdot 331 + 2) = 737.3$. Now Eq. (4.29) gives the interval $[\chi^2_{0.975}(662)/20, \chi^2_{0.025}(664)/20] = [29.6, 36.9]$.

4.10

(a) Since high concentrations are dangerous, is to find a *lower* bound of interest.

(b) $m^* = \bar{x} = 9.0$; $n = 12$; $\sigma_{\mathcal{E}}^* = \sqrt{s_n^2/n} = 6.15/\sqrt{12}$; $\alpha = 0.05$.
Since with approximate confidence $1 - \alpha$, $m \geq \bar{x} - \lambda_\alpha \sigma_{\mathcal{E}}^*$, we find $m \geq 6.0$.

4.11 The interval presented by B is wider; hence, B used a higher confidence level $(1 - \alpha = 0.95)$ as opposed to A $(1 - \alpha = 0.90)$.

4.12

(a) There are $r = 9$ classes in which the $n = 55$ observations are distributed as 1, 7, 10, 6, 8, 8, 6, 5, 4; $m^* = 334/55 = 6.1$. Further, $p_1^* = \exp(-m^*)(1 + m^* + (m^*)^2/2) = 0.0577$, $p_2^* = \exp(-m^*)(m^*)^3/3! = 0.0848$, $p_3^* = 0.1294$, $p_4^* = 0.1579$, $p_5^* = 0.1605$, $p_6^* = 0.1399$, $p_7^* = 0.1066$, $p_8^* = 0.0723$, $p_9^* = 1 - \sum_{i=1}^{8} p_i^* = 0.0909$. One finds $Q = 5.21$ which is smaller than $\chi^2_{0.05}(r-1-1) = 14.07$. Hence do not reject hypothesis about Poisson distribution.

(b) With $\sigma_{\mathcal{E}}^* = \sqrt{m^*/55} = 0.33$ and $\lambda_{0.05} = 1.64$ it follows that with approximate confidence 0.95, $m \leq m^* + \lambda_{0.05}\sigma_{\mathcal{E}}^* = 6.64$.

Problems of Chapter 5

5.1

(a) $P(X = 2, Y = 3) = [\text{independence}] = P(X = 2)P(Y = 3) = 0.60 \cdot 0.25 = 0.15$.

(b) $P(X \leq 2, Y \leq 3) = [\text{independence}] = P(X \leq 2)P(Y \leq 3) = 0.80 \cdot 0.75 = 0.60$.

5.2 Multinomial probability: $\frac{5!}{3!1!1!}0.73^3 \cdot 0.20 \cdot 0.07 = 0.11$.

5.3

(a) Using multinomial probabilities (or independence) $p_{X_A,X_B}(0,0) = (1 - p_A - p_B)^2 = 0.16$, $p_{X_A,X_B}(0,1) = 2p_B(1 - p_A - p_B)^2 = 0.20$. $p_{X_A,X_B}(1,0) = 0.28$, $p_{X_A,X_B}(1,1) = 0.175$, $p_{X_A,X_B}(0,2) = 0.0625$, $p_{X_A,X_B}(2,0) = 0.1225$.

(b) $X_A \in \text{Bin}(2, p_A)$, $X_B \in \text{Bin}(2, p_B)$. Use of formulae for mean and variance for binomial variables gives the results:

$$E[X_A] = 0.70, \quad E[X_A^2] = 2p_A^2 + 2p_A = 0.945,$$

$$E[X_B] = 2p_B = 0.50, \quad E[X_B^2] = 2p_B^2 + 2p_B = 0.625.$$

$E[X_A X_B] = \sum x_A x_B p_{x_A x_B}(x_A x_B) = 1 \cdot 1 \cdot p(1,1) = 0.175$.
$V[X_A] = E[X_A^2] - (E[X_A])^2 = 0.455$. $V[X_B] = E[X_B^2] - (E[X_B])^2 = 0.375$.
$\text{Cov}[X_A, X_B] = E[X_A X_B] - E[X_A]E[X_B] = -0.175$. $\rho(X_A, X_B) = \frac{\text{Cov}[X_A X_B]}{\sqrt{V[X_A]V[X_B]}} = -0.42$.

5.4

(a) Marginal distributions by Eq. (5.2):

j	1	2	3		k	1	2	3
p_j	0.10	0.35	0.55		p_k	0.20	0.50	0.30

(b) $P(Y = 3|X = 2) = P(X = 2, Y = 3)/P(X = 2) = 0.2/0.35 = 0.57$.

(c) The probability that give two interruption, the expert is called three times.

5.5 $\int_{x=0}^{0.3} \int_{y=0}^{0.4} f_{X,Y}(x,y)\, dx dy = 0.12$.

5.6 $E[2X + 3Y] = 2E[X] + 3E[Y] = 2 \cdot \frac{7}{2} + 3 \cdot \frac{6}{4} = 11.5$.

5.7 $V[N_1] = 4.2$, $V[N_2] = 2.5$, $\text{Cov}[N_1, N_2] = 0.85 \Rightarrow$
$V[N_1 - N_2] = V[N_1] + V[N_2] - 2\text{Cov}[N_1, N_2] = 5$.

5.8

(a) $E[Y_1] = E[Y_2] = 0$, $V[Y_1] = 1$, $V[Y_2] = \varrho^2 V[X_1] + (1 - \varrho^2)V[X_2] = 1$, $\text{Cov}[Y_1, Y_2] = E[Y_1 Y_2] - E[Y_1]E[Y_2]$, where $E[Y_1 Y_2] = \varrho E[X_1^2] + \sqrt{1 - \varrho^2}E[X_1 X_2] = \varrho$; since here $E[X_1 \cdot X_2] = E[X_1]E[X_2] = 0$. Hence $\text{Cov}[Y_1, Y_2] = \varrho$ and $\rho_{Y_1 Y_2} = \varrho$.

(b) $(Y_1, Y_2) \in N(0, 0, 1, 1, \varrho)$ and hence

$$f_{Y_1,Y_2}(y_1, y_2) = \frac{1}{2\pi\sqrt{1 - \varrho^2}} e^{-\frac{1}{2(1-\varrho^2)}(y_1^2 + y_2^2 - 2\varrho y_1 y_2)}.$$

5.9

(a) $F_{X|X>0}(t) = \frac{P(X \le t \cap X > 0)}{P(X > 0)} = \frac{F_X(t) - F_X(0)}{1 - F_X(0)}$, $t > 0$.

(b) $F_X(x) = \Phi(\frac{x-m}{\sigma})$. From (a) it follows, using $1 - \Phi(-m/\sigma) = \Phi(m/\sigma)$, that
$$F_T(t) = P(X \le t \mid X > 0) = \frac{\Phi(\frac{t-m}{\sigma}) + \Phi(\frac{m}{\sigma}) - 1}{\Phi(\frac{m}{\sigma})}, \quad t > 0.$$

(c) Differentiating the distribution function in (b) yields
$$f_T(t) = \frac{\frac{1}{\sigma}\Phi'(\frac{t-m}{\sigma})}{\Phi(\frac{m}{\sigma})} = \frac{1}{\Phi(m/\sigma)} \cdot \frac{1}{\sigma\sqrt{2\pi}} e^{-(t-m)^2/2\sigma^2}, \quad t > 0.$$

5.10

$$P(X = k \mid X + Y = n) = \frac{P(X = k, X + Y = n)}{P(X + Y = n)} = \frac{P(X = k, Y = n - k)}{P(X + Y = n)}$$

$$= \frac{P(X = k)P(Y = n - k)}{P(X + Y = n)} = \frac{\frac{e^{-m_1} m_1^k}{k!} \cdot \frac{e^{-m_2} m_2^{n-k}}{(n-k)!}}{\frac{e^{-(m_1+m_2)}(m_1+m_2)^n}{n!}}$$

$$= \binom{n}{k} \left(\frac{m_1}{m_1 + m_2}\right)^k \left(1 - \frac{m_1}{m_1 + m_2}\right)^{n-k},$$

i.e. the probability-mass function for $\mathrm{Bin}(n, \frac{m_1}{m_1+m_2})$.

5.11

$$P(X = x) = \sum_{y=0}^{\infty} P(X = x | Y = y) P(Y = y) = \sum_{y=x}^{\infty} \left[\binom{y}{x} p^x (1-p)^{y-x}\right]\left[\frac{e^{-m} m^y}{y!}\right]$$

$$= \frac{(mp)^x e^{-m}}{x!} \sum_{y=x}^{\infty} \frac{((1-p)m)^{y-x}}{(y-x)!} = \frac{(mp)^x e^{-m}}{x!} \sum_{k=0}^{\infty} \frac{((1-p)m)^k}{k!}$$

$$= \frac{(mp)^x e^{-m}}{x!} e^{(1-p)m} = \frac{(mp)^x}{x!} e^{-mp}.$$

Hence, $X \in \mathrm{Po}(mp)$ and $E[X] = mp$.

Problems of Chapter 6

6.1 $a = b = 1 \Rightarrow f(\theta) = c\theta^{1-1}(1-\theta)^{1-1} = c, \quad 0 < \theta < 1$, and hence with $c = 1$, $\Theta \in \mathrm{U}(0,1)$.

6.2 $a = 1 \Rightarrow f(\theta) = c\theta^{1-1}e^{-b\theta} = ce^{-b\theta}, \quad \theta \geq 0$, and hence, for $c = b$, Θ is an exponentially distributed r.v. with expectation $1/b$.

6.3 Let the intensity of imperfections be described by the r.v. Λ.

(a) $E[\Lambda] = 1/100 \ \mathrm{km}^{-1}$.
(b) $\mathrm{Gamma}(13, 600)$.
(c) $E[\Lambda] = 13/600 = 0.022 \ [\mathrm{km}^{-1}]$.

6.4

(a) $\underset{\sim}{\Lambda} \in \mathrm{N}(\lambda^*, (\sigma_{\mathcal{E}}^2)^*)$, where $\lambda^* = n/\sum t_i$, $(\sigma_{\mathcal{E}}^2)^* = (\lambda^*)^2/n = n/(\sum t_i)^2$. For the data, $\lambda^* = 0.0156$, $\sigma_{\mathcal{E}}^* = 0.0032$, hence $\underset{\sim}{\Lambda} \in \mathrm{N}(0.0156, 0.0032^2)$.

(b) Let $t = 24$. Since $P = \exp(-\Lambda t)$ and $-\Lambda t \in \mathrm{N}(-0.0156 \cdot t, (0.0032 \cdot t)^2)$, P is lognormally distributed and

$$E[P] = \exp(-24 \cdot 0.0156 + (24 \cdot 0.0032)^2/2) \approx \exp(-24 \cdot 0.0156) = 0.69,$$

i.e. the same as in the frequentistic approach, $P = \exp(-\lambda^* t)$.

6.5

(a) For example, one has called once and waited for 15 min, got no answer, and then rang off immediately.
(b) Gamma $(4, 32)$.
(c) $4/32 = 1/8 = 0.125$ min^{-1}.
(d)

$$\mathsf{P}^{\mathrm{pred}}(T > t) = \mathsf{E}[e^{-\Lambda t}] = \int_0^\infty e^{-\lambda t} f^{\mathrm{post}}(\lambda)\, d\lambda = \frac{32^4}{\Gamma(4)} \int_0^\infty e^{-\lambda t} \lambda^3 e^{-32\lambda}\, d\lambda$$

$$= \frac{32^4}{\Gamma(4)} \int_0^\infty \lambda^3 e^{-\lambda(32+t)}\, d\lambda = \left(\frac{32}{32+t}\right)^4$$

Thus $\mathsf{P}(T > 1) = 0.88$, $\mathsf{P}(T > 5) = 0.56$, $\mathsf{P}(T > 10) = 0.34$.

6.6

(a) $p^* = 5/5 = 1$
(b) Posterior distribution: Beta$(6, 1)$.
(c) A = "The man will win in a new game". Since $\mathsf{P}(A|P = p) = p$, $\mathsf{P}^{\mathrm{pred}}(A) = \mathsf{E}[P] = 6/7$.

6.7

(a) Dirichlet(1,1,1)
(b) Dirichlet(79,72,2)
(c) $72/153 = 0.47$.

6.8

(a) $\Lambda \in$ Gamma(a, b); $\mathsf{R}[\Lambda] = 2$ yields $a = 1/4$ and since $a/b = 1/4$, we find $\Lambda \in$ Gamma$(1/4, 1)$. Predictive probability: $\mathsf{E}[\Lambda]t = \frac{1}{4} \cdot \frac{1}{2} = 1/8 = 0.125$.
(b) Updating the distribution in (a) yields $\Lambda^{\mathrm{post}} \in$ Gamma$(5/4, 3)$. Predictive probability:

$$\mathsf{E}[\Lambda]t = \frac{5}{4} \cdot \frac{1}{3} \cdot \frac{1}{2} = \frac{5}{24} = 0.21$$

(about twice as high as in (a)).

6.9 With $t = 1/52$, $p = (10.25/(10.25 + 1/52))^{244.25} = 0.63$. The approximate predictive probability is $1 - (244.25/10.25)/52 = 0.54$.

6.10 Since $f_T(t) = \lambda \exp(-\lambda t)$, the likelihood function is $L(\lambda) = \lambda^n \exp(-\lambda \sum_{i=1}^n t_i)$. If $f^{\mathrm{prior}}(\lambda) \in$ Gamma(a, b), i.e. $f^{\mathrm{prior}}(\lambda) = c \cdot \lambda^{a-1} \exp(-b\lambda)$, then

$$f^{\mathrm{post}}(\lambda) = c \cdot \lambda^{a+n-1} e^{-(b+\sum_{i=1}^n t_i)\lambda},$$

i.e. a Gamma$(a + n, b + \sum_{i=1}^n t_i)$.

6.11

(a) With $\Theta = m$, we have $\Theta \in$ N$(m^*, m^*/n)$. Hence with $m^* = 33.1$, $n = 10$, $\Theta \in$ N$(33.1, 3.3)$.
(b) $[m^* - 1.96\sqrt{m^*/n},\ m^* + 1.96\sqrt{m^*/n}]$, i.e. [29.5, 36.7] (the same answer as in Problem 4.9 (b)).

6.12

(a) $\Lambda \in \mathrm{Gamma}(1, 1/12)$, hence $\mathsf{P}(C) \approx \Lambda t$ and $\mathsf{P}^{\mathrm{pred}}(C) = \mathsf{E}[P] = 12/365$. Further, $\mathsf{R}[P] = 1$.

(b) $\Lambda \in \mathrm{Gamma}(5, 3+1/12)$; $\mathsf{P}^{\mathrm{pred}}(C) \approx (5/37)(12/365) = 0.0044$. $\mathsf{R}[P] = 1/\sqrt{5} = 0.45$.

(c) $\Theta_1 =$ Intensity of accidents involving trucks in Dalecarlia;
$\Theta_2 = \mathsf{P}(B) = $ A truck is a tank truck.
Data and use of improper priors yields $\Theta_1 \in \mathrm{Gamma}(118, 3)$. With a uniform prior for Θ_2 is obtained $\Theta_2 \in \mathrm{Beta}(37 + 41 + 39 + 1, 1108 + 1089 + 1192 - 37 - 41 - 39 + 1)$, $i.e.$ $\mathrm{Beta}(118, 3273)$. Hence

$$\mathsf{P}^{\mathrm{pred}}(C) \approx \mathsf{E}[\Theta_1 \Theta_2 \, t] = \frac{118}{3} \frac{118}{118 + 3273} \frac{1}{365} = 0.0037,$$

a similar answer as in (b). Uncertainty: For the posterior densities $\mathsf{R}[\Theta_1] = 1/\sqrt{118}$, $\mathsf{R}[\Theta_2] = 1/\sqrt{3392}\sqrt{(1-p)/p} = 1/\sqrt{3392} \cdot \sqrt{27.73}$ (with $p = 0.0348$) and hence with Eq. (6.42), $\mathsf{R}[P] = \sqrt{(1 + 1/118)(1 + 27.73/3392) - 1} = 0.13$. (Compare with the result in (b)).

Problems of Chapter 7

7.1

(a) $\mathsf{P}(T > 50) = \exp(-\int_0^{50} \lambda(s)\,ds) = 0.79$.

(b) $\mathsf{P}(T > 50 \,|\, T > 30) = \exp(-\int_{30}^{50} \lambda(s)\,ds) = 0.87$.

7.2 Application of the Nelson–Aalen estimator results in

t_i	276	411	500	520	571	672	734	773	792
$\Lambda^*(t_i)$	0.1111	0.2361	0.3790	0.5456	0.7456	0.9956	1.3290	1.8290	2.8290

7.3 Constant failure rate means exponential distribution for life time. $F_{T_1}(t) = F_{T_2}(t) = 1 - \exp(-\lambda t)$, $t \geq 0$. The life time T of the whole system is given by $T = \max(T_1, T_2)$:

$$F_T(t) = \mathsf{P}(T \leq t) = \mathsf{P}(T_1 \leq t, T_2 \leq t) = F_{T_1}(t)\,F_{T_2}(t).$$

It follows that $\lambda_T(t) = f_T(t)/(1 - F_T(t)) = 2\lambda(1 - \exp(-\lambda t))/(2 - \exp(-\lambda t))$.

7.4 Let $Z \in \mathrm{Po}(m)$. $\mathsf{R}[Z] = \mathsf{D}[Z]/\mathsf{E}[Z] = 1/\sqrt{m}$. Thus $0.50 = 1/\sqrt{m} \Rightarrow m = 4$; $\mathsf{P}(Z = 0) = \exp(-4) = 0.018$.

7.5 $\mathsf{P}(N(2) > 50) \approx 1 - \Phi\big((50.5 - 20 \cdot 2)/\sqrt{20 \cdot 2}\big) = 1 - \Phi(1.66) = 0.05$.

7.6

(a) $N(1) \in \mathrm{Po}(\lambda \cdot 1) = \mathrm{Po}(1.7)$; $\mathsf{P}(N(1) > 2) = 1 - \mathsf{P}(N(1) \leq 2) = 1 - \exp(-1.7)(1 + 1.7 + (1.7)^2/2) = 0.24$.

(b) X (distance between imperfections) is exponentially distributed with mean $1/\lambda$; hence $\mathsf{P}(X > 1.2) = \exp(-1.2\lambda) = 0.13$.

7.7

(a) Barlow–Proschan's test; Eq.(7.19), $(n = 24)$ gives $z = 11.86$ and with $\alpha = 0.05$ results in the interval $[8.8, 14.2]$; hence, no rejection of the hypothesis of a PPP.

(b) T_i = Distance between failures, $T_i \in \exp(\theta)$; $\theta^* = \bar{t} = 64.13$. Since $\lambda^* = 1/\theta^*$, $\lambda^* = 0.016$ [hour^{-1}].

7.8 $m_1^* = 21/30$; $(\sigma_{\mathcal{E}_1}^2)^* = m_1^*/30$; $m_2^* = 16/45$; $(\sigma_{\mathcal{E}_2}^2)^* = m_2^*/45$. With $m^* = m_1^* - m_2^*$ we have $\sigma_{\mathcal{E}}^2 = V[M^*]$ and an estimate is found as $(\sigma_{\mathcal{E}}^2)^* = (\sigma_{\mathcal{E}_1}^2)^* + (\sigma_{\mathcal{E}_2}^2)^*$. Numerical values: $m^* = 0.34$, $\sigma_{\mathcal{E}}^* = 0.177$ which gives the confidence interval $[0.34 - 1.96 \cdot 0.177, 0.34 + 1.96 \cdot 0.177]$, *i.e.* $[-0.007, 0.69]$. The hypothesis that $m_1 = m_2$ cannot be rejected but we suspect that $m_1 > m_2$.

7.9

(a) Let $N(A) \in \mathrm{Po}(\lambda A)$. Let A be a disc with radius r. Then $P(R > r) = P(N(A) = 0) = e^{-\lambda \pi r^2}$, that is, a Rayleigh distribution with $a = 1/\sqrt{\lambda \pi}$.
(b) $E[R] = 1/2\sqrt{\lambda}$ (cf. Problem 3.7).
(c) $E[R] = 1/2\sqrt{2 \cdot 10^{-5}} = 112$ m.

7.10 Let N be the number of hits in the region: $N \in \mathrm{Po}(m)$. We find $m^* = 537/576 = 0.9323$, ($n = 576$). With $p_k^* = P(N = k) = \exp(-m^*)(m^*)^k/k!$, the following table results

k	0	1	2	3	4	> 5
n_k	229	211	93	35	7	1
$n \cdot p_k^*$	226.74	211.39	98.54	30.62	7.14	1.57

We find $Q = 1.17$. Since $\chi_{0.05}^2(6 - 1 - 1) = 9.49$, we do not reject the hypothesis about Poisson distribution.

The two last groups should be combined. Then $Q = 1.018$ found, which should be compared to $\chi_{0.05}^2(5 - 1 - 1) = 7.81$. Hence, even here, one should not reject the hypothesis about Poisson distribution.

7.11

(a) The intensity: $334/55 = 6.1$. $p = 1 - \Phi((10.5 - 6.1)/\sqrt{6.1}) = 0.038$. Expected number of years: $p \cdot t = 0.038 \cdot 55 = 2.1$ (the observed data had 3 such years).
(b) DEV$= 2(-123.8366 - (-123.8374)) = 0.0017$. Since $\chi_{0.01}^2(1) = 6.63$, we do not reject the hypothesis $\beta_1 = 0$. There is no sufficient statistical evidence that the number of hurricanes is increasing over the years.

7.12 We have 25 observations ($n = 25$) from $\mathrm{Po}(m)$, where $m^* = 71/25 = 2.84$. The statistics of the number of pines in a square is as follows:

0	1	2	3	4	5	6
1	4	5	8	4	1	2

We combine groups in order to apply a χ^2 test and with $p_k^* = \exp(-m^*)(m^*)^k/k!$, the following table results:

k	< 2	2	3	4	> 4
n_k	5	5	8	4	3
$n \cdot p_k^*$	5.6	5.9	5.6	4.0	4.0

We find $Q = 1.48$; since $\chi_{0.05}^2(5 - 1 - 1) = 7.8$, the hypothesis about a Poisson process is not rejected.

7.13

(a) $N^{\text{Tot}}(t) =$ The total number of transports; $N^{\text{Tot}}(t) \in \text{Po}(\lambda t)$, where $\lambda = 2000$ day^{-1}. It follows that

$$P(N^{\text{Tot}}(5) > 10300) = 1 - P(N^{\text{Tot}}(5) \leq 10300) \approx 1 - \Phi\left(\frac{10300 - 2000 \cdot 5}{\sqrt{2000 \cdot 5}}\right)$$

$$= 1 - \Phi(3.0) = 0.0013,$$

where we used normal approximation.

(b) $N^{\text{Haz}}(t) =$ The number of transports of hazardous material during period t. $N^{\text{Haz}}(t) \in \text{Po}(\mu)$ with $\mu = p\lambda t = 160t$. For a period of $t = 5$ days, $\mu = 800$. Normal approximation yields

$$P(N^{\text{Haz}}(5) > 820) = 1 - \Phi\left(\frac{820 - 800}{\sqrt{800}}\right) = 0.24.$$

Problems of Chapter 8

8.1 $X + Y \in \text{Po}(2 + 3) = \text{Po}(5)$.

8.2

(a) $Z \in \text{N}(10 - 6, 3^2 + 2^2)$, i.e. $Z \in \text{N}(4, 13)$.
(b) $P(Z > 5) = 1 - P(Z \leq 5) = 1 - \Phi\left(\frac{5-4}{\sqrt{13}}\right) = 0.39$.

8.3 Let $X = X_A + X_B + X_C$. Then $X \in \text{Po}(0.84)$ and $P(X \geq 1) = 1 - P(X = 0) = 1 - \exp(-0.84) = 0.57$.

8.4 Let $T = \min(T_1, \ldots, T_n)$, where T_i are independent Weibull distributed variables. Then

(a)

$$F_T(t) = 1 - (1 - F(t))^n = 1 - (1 - 1 + e^{-(t/a)^c})^n = 1 - e^{-n(t/a)^c}$$

$$= 1 - e^{-\left(t/(an^{-1/c})\right)^c}$$

This is a Weibull distribution with scale parameter $a_1 = a \cdot n^{-1/c}$, location parameter $b_1 = 0$, and shape parameter $c_1 = c$.

(b) $c^* = c_1^* = 1.56$; $a^* = a_1^* \cdot n^{1/c_1^*} = 1.6 \cdot 10^7$ $(n = 5)$.

8.5

(a) Let $S_r \in \text{N}(30, 9)$, $S_p \in \text{N}(15, 16)$. Water supply: $S = S_r + S_p \in \text{N}(45, 25)$. Demand: $D \in \text{N}(35, (35 \cdot 0.10)^2)$. Hence $S - D \in \text{N}(10, 25 + 3.5^2)$. $P_f = P(S - D \leq 0) = 1 - \Phi(10/\sqrt{25 + 3.5^2}) = 0.051$.

(b) $V[S - D] = 25 + 3.5^2 + 2 \cdot (-1) \cdot (-0.8) \cdot 5 \cdot 3.5 = 65.25$ and $P_f = 0.11$. The risk of insufficient supply of water has doubled!

8.6 $T = T_1 + T_2$; $T_1 \in \text{Gamma}(1, 1/40)$, $T_2 \in \text{Gamma}(1, 1/40)$, $T \in \text{Gamma}(2, 1/40)$; $P(T > 90) = 1 - P(T \leq 90) = \exp(-90/40)(1 + 90/40) = 0.34$ using Eq. (8.6).

8.7 Gauss formulae give $E[I] \approx 26$ A, $D[I] \approx 3.6$ A.

8.8 $P_f = P(R/S < 1) = P(\ln R - \ln S < 0) = \Phi\left(\frac{m_S - m_R}{\sqrt{\sigma_R^2 + \sigma_S^2}}\right)$

8.9 $\sigma_S^2 = \ln(1 + 0.05^2) \approx 0.0025$, $m_S = \ln 100 - \sigma_S^2/2 \approx 4.604$, $m_R = \ln 150 - \sigma_R^2/2 \approx 5.01 - \sigma_R^2/2$. Since $0.001 \geq P(\text{"Failure"}) = \Phi\left(\frac{m_S - m_R}{\sqrt{\sigma_R^2 + \sigma_S^2}}\right)$ (cf. Problem 8.8), we get the condition $\frac{m_S - m_R}{\sqrt{\sigma_R^2 + \sigma_S^2}} \geq \lambda_{0.999} = -3.09$ and hence $\sigma_R^2 \leq 0.014$, i.e. $R(R) = \sqrt{\exp(\sigma_R^2) - 1} \leq 0.12$. The coefficient of variation must be less than 0.12.

8.10 Gauss' formulae give $E[\frac{\Delta A}{\Delta N}] \approx 43.3$ nm, $V[\frac{\Delta A}{\Delta N}] = 1.321 \cdot 10^{-15} + 1.5 \cdot 10^{-17} = 1.34 \cdot 10^{-15}$ and hence $R[\frac{\Delta A}{\Delta N}] \approx 0.85$.

8.11

(a) R : Production capacity, S: maximum demand during the day. Wanted: $P_f = P(R < S) = P(\ln R - \ln S < 0)$. Independence $\Rightarrow Z = \ln R - \ln S \in N(m, \sigma^2)$, where $m = m_R - m_S = \ln 6 - \ln 3.6 = 0.5108$, $\sigma^2 = \sigma_R^2 + \sigma_S^2 = \ln(1 + R(R)^2) + \ln(1 + R(S)^2)$. It follows that $P_f = P(Z < 0) = 0.0107$, hence return period $1/P_f = 93.5$ days.

(b) Correlation $\Rightarrow \sigma^2 = \sigma_R^2 + \sigma_S^2 + 2 \cdot 1 \cdot (-1)\rho\sigma_R\sigma_S = 0.0809$. It follows that $P_f = 0.0363$ and return period $1/P_f = 27.6$ days.

8.12

(a) $P(X < 0) = \int_{x=-\infty}^0 f_X(x)\,dx \leq \int_{x=-\infty}^0 \frac{(x-a)^2}{a^2} f_X(x)\,dx \leq \int_{-\infty}^\infty \frac{(x-a)^2}{a^2} f_X(x)\,dx = \frac{E[(X-a)^2]}{a^2} = \frac{\sigma^2 + (m-a)^2}{a^2}$.

(b) Let $X = R - S$. Then $P(R < S) \leq \frac{1}{a^2}(\sigma_R^2 + \sigma_S^2 + (m_R - m_S - a)^2)$ for all $a > 0$. The right-hand side has minimum for $a = \frac{\sigma_R^2 + \sigma_S^2 + (m_R - m_S)^2}{m_R - m_S} > 0$ and the minimum value is $\frac{\sigma_R^2 + \sigma_S^2}{\sigma_R^2 + \sigma_S^2 + (m_R - m_S)^2} = \frac{1}{1 + \beta_C^2}$. The inequality is shown.

8.13

(a) $m_R = E[M_F] = 20$ kNm, $\sigma_R^2 = 2^2$ (kNm)^2, $m_S = \frac{\ell}{2}E[P] = 10$ kNm, $\sigma_S^2 = (\ell/2)^2 V[P] = 2.5^2$ (kNm)^2.

(b) $P_f \leq \frac{1}{1 + \beta_C^2} = \frac{2^2 + 2.5^2}{2^2 + 2.5^2 + (20 - 10)^2} = 0.093$. ($\beta_C = 3.12$).

(c) $1 - \Phi(3.12) = 0.001$.

8.14

(a) Failure probability: $P(Z < 0)$, where $Z = h(R_1, \ldots, R_n, S) = \sum_{i=1}^n R_i - S$. Safety index: $\frac{E[Z]}{\sqrt{V[Z]}} = \frac{nE[R_i] - E[S]}{\sqrt{nV[R_i] + V[S]}}$, from which it is found $n = 23$.

(b) Introduce $R = R_1 + \cdots + R_n$. Then

$$V[R] = \sum_{i=1}^n V[R_i] + 2\sum_{i<j} \text{Cov}[R_i, R_j] = nV[R_i] + 2\sum_{i<j} \rho V[R_i]$$

$$= V[R_i]\left[n + 2\frac{n(n-1)}{2}\rho\right] = nV[R_i]\left[1 + \rho(n-1)\right]$$

and hence the safety index $\frac{nE[R_i] - E[S]}{\sqrt{nV[R_i](1 + \rho(n-1)) + V[S]}}$ from which it is found $n = 30$. Higher correlation required more pumps.

8.15 Production. $X =$"Total production during a working week (tons)". Then

$$E[X] = 5 \cdot 400 = 2000,$$

$$V[X] = V[X_1 + \cdots + X_5] = V[X_1] \sum_{i,j}^{5} \rho^{|i-j|} = 1000(5 + 8\rho + 6\rho^2 + 4\rho^3 + 2\rho^4)$$

$$= 21\,300.$$

Transportation. Let N_i be the number of transportations of one lorry in a week; $N_i \in Po(m)$ where $m = \lambda t = 1 \cdot 7 \cdot 5 = 35$. Let $Y_i =$"Capacity (tons) of one lorry during a week", $Y =$ "Total capacity during a week (ton) using n lorries". We have that $Y_i = 10\,N_i$ and $Y = \sum_{i=1}^{n} Y_i = 10 \sum_{i=1}^{n} N_i$. Now $\sum N_i \in Po(35m)$ and hence

$$E[Y] = 350n, \qquad V[Y] = 3500n.$$

Solving for n in

$$\frac{350n - 2000}{\sqrt{3500n + 21300}} > 3.5$$

yields $n = 8$ lorries are needed.

Problems of Chapter 9

9.1 $1/0.04(1.96/0.5)^2 = 384.16$, hence 384 items need to be tested.

9.2

(a) Use the definition of conditional probability.
(b)

$$\frac{1 - F(u + x)}{1 - F(u)} = \frac{e^{-(u+x)}}{e^{-u}} = e^{-x}.$$

for $x > 0$. Hence, exceedances are again exponentially distributed.

9.3 Table 4 in appendix gives $c = 2.70$; hence $a = 84.3$ and $L_{10}^* = 36.6$.

9.4 Introduce

$$h(a, c) = a \cdot \left(-\ln(1 - \frac{1}{100})\right)^{1/c} = a \cdot \left(-\ln(0.99)\right)^{1/c}.$$

The quantities

$$\frac{\partial}{\partial a} h(a, c) = \left(-\ln(0.99)\right)^{1/c},$$

$$\frac{\partial}{\partial c} h(a, c) = -\frac{a}{c^2} \cdot \left(\ln(-\ln(0.99))\right) \cdot \left(-\ln(0.99)\right)^{1/c}.$$

evaluated at the ML estimates are 0.451 and 0.101, respectively. The delta method results in the approximate variance 0.0042 and since $x_{0.99}^* = 0.74$, with approximate 0.95 confidence

$$x_{0.99} \in \left[0.74 - 1.96 \cdot \sqrt{0.0042},\ 0.74 + 1.96 \cdot \sqrt{0.0042}\right], \quad \text{i.e.} \quad x_{0.99} \in [0.61, 0.87].$$

9.5 With $p = 0.5$, Eq. (9.8) gives $n \geq (1 - p)/p(\lambda_{\alpha/2}/q)^2$, where $q = 0.2$. $\alpha = 0.05:\ n \geq 96.0$; $\alpha = 0.10:\ n \geq 67.2$. Cf. the discussion at page 31.

9.6

(a) $p_0^* = 40/576 = 0.069$ and $a^* = 49.2/40 = 1.23$, hence by Eq. (9.4) $x_{0.001}^* = 9 + 1.23\ln(0.069/0.001) = 14.2$ m.

(b) Let $\theta_1 = p_0$ and $\theta_2 = a$. From the table in Example 4.19 the estimates of variances are found: $(\sigma_{\mathcal{E}_1}^2)^* = p_0^*(1 - p_0^*)/n = 0.0001$ $(n = 576)$, $(\sigma_{\mathcal{E}_2}^2)^* = (a^*)^2/n = 0.0378$ $(n = 40)$. The gradient vector is equal to $[a^*/p_0^*\ \ \ln(p_0^*/\alpha)] = [17.83\ 4.23]$, hence $(\sigma_{\mathcal{E}}^2)^* = 17.83^2 \cdot 0.0001 + 4.23^2 \cdot 0.0378 = 0.708$ giving an approximate 0.95 confidence interval for $x_{0.001}$; $[14.2 - 1.96\sqrt{0.708}, 14.2 + 1.96\sqrt{0.708}] = [12.6, 15.8]$.

(c) With $\lambda^* = 576/12$ [year]$^{-1}$, we find $\mathsf{E}[N] = \lambda \cdot \mathsf{P}(B) \cdot t \approx \lambda^* \cdot 0.001 \cdot 100 = 4.8$. (Thus, the value $x_{0.001}$ is approximately the 20-year storm.)

Problems of Chapter 10

10.1 $F_Y(y) = (F_X(y))^5$, where $X \in U(-1, 1)$ and thus $F_X(x) = \int_{-1}^{x} \frac{1}{2}\,d\xi = \frac{1}{2}(x + 1)$, $-1 < x < 1$. Hence $F_Y(y) = \frac{1}{2^5}(y + 1)^5$, $-1 < y < 1$.

10.2

(a) Let $n = 6$ be the number of observations. Due to independence, we have

$$F_{U_{\max}}(u) = \left(F_U(u)\right)^n = \left(\exp(-e^{-(u-b)/a})\right)^n = \exp(-n \cdot e^{-(u-b)/a})$$
$$= \exp(-e^{\ln n - (u-b)/a}) = \exp\left(-e^{-(u-(b+a\ln n))/a}\right).$$

Thus, $U_{\max}$ is also Gumbel distributed with scale parameter a and location parameter $b + a\ln n$.

(b) Let $a = 4$ m/s, $u_0 = 40$ m/s, $p = 0.50$. Find b such that $\mathsf{P}(U_{\max} > u_0) = p$: $b = u_0 + a\ln(-\ln(1 - p)) - a\ln n = 31.4$ m/s.

10.3

$$F^n(a_n x + b_n) = (1 - e^{-x - \ln n})^n = \left(1 - \frac{e^{-x}}{n}\right)^n \to \exp(-e^{-x}) \quad \text{as} \quad n \to \infty.$$

10.4

(a) $x_{100}^* = 31.9 - 10.6 \cdot \ln(-\ln(0.99)) = 80.7$ pphm.

(b) $0.26/\sqrt{0.62 \cdot 1.11} = 0.32$.

(c) $(\sigma_{\mathcal{E}}^2)^* = \mathsf{V}[B^* + \ln(100)A^*] = 9.9^2$, hence approx. $\mathcal{E} \in N(0, 9.9^2)$.

(d) $[80.66 - 1.96 \cdot 9.9, 80.66 + 1.96 \cdot 9.9] = [61.3, 100.1]$.

10.5 Use common rules for differentiation, for instance $\frac{d}{dx}(a^x) = a^x\,\ln a$.

10.6 We find $\nabla s_T(a^*, b^*, c^*) = [21.6231\ 1\ {-2.46 \cdot 10^3}]^\mathsf{T}$ and hence by Remark 10.8 $\sigma_{\mathcal{E}}^* = 330.8$. With $s_{10000}^* = 479.3$ follows the upper bound: $479.3 + 1.64 \cdot 330.8 = 1022$.

10.7

$$P(Y \leq y) = P(\ln X \leq y) = P(X \leq e^y) = F_X(e^y)$$
$$= 1 - \exp(-(e^y/a)^c) = 1 - \exp(-e^{-cy/a^c}).$$

The scale parameter is a^c/c.

References

1. O. O. Aalen. Nonparametric inference for a family of counting processes. *The Annals of Statistics*, 6:701–726, 1972.
2. C. W. Anderson, D. J. T. Carter, and D. Cotton. Wave climate variability and impact on offshore design extremes. Report for Shell International and the Organization of Oil & Gas Producers, 2001.
3. A. H-S. Ang and W. H. Tang. *Probability Concepts in Engineering Planning and Design*. J. Wiley & Sons, New York, 1984.
4. F. J. Anscombe and R. J. Aumann. A definition of subjective probability. *The Annals of Mathematical Statistics*, 34:199–205, 1964.
5. L Bortkiewicz von. *Das Gesetz der Kleinen Zahlen*. Teubner, Leipzig, 1898.
6. L. D. Brown and L. H. Zhao. A test of the Poisson distribution. *Sankhya*, 64:611–625, 2002.
7. U. Brüde. Basstatisik över olyckor och trafik samt andra bakgrundsvariabler. Technical Report VTI notat 27-2005, VTI, 2005.
8. D. J. T. Carter. Variability and trends in the wave climate of the North Atlantic: A review. In *Proceedings of the 9th ISOPE Conference*, volume III, pages 12–18, 1999.
9. D. J. T. Carter and L. Draper. Has the north-east Atlantic become rougher? *Nature*, 332:494, 1988.
10. G. Casella and R. L. Berger. *Statistical Inference*. Duxbury Press, second edition, 2002.
11. E. Çinlar. *Introduction to Probability Theory and its Applications*. Prentice Hall, Eaglewood Cliffs, New Jersey, 1975.
12. R. D. Clarke. An application of the Poisson distribution. *Journal of the Institute of Actuaries*, 72:48, 1946.
13. W. G. Cochran. Some methods for strengthening the common χ^2 tests. *Biometrics*, 10:417–451, 1954.
14. S. Coles. *An Introduction to Statistical Modelling of Extreme Values*. Springer-Verlag, New York, 2001.
15. S. Coles and L. Pericchi. Anticipating catastrophes through extreme value modelling. *Appl. Statist.*, 52:405–416, 2003.
16. H. Cramér. Richard von Mises' work in probability and statistics. *The Annals of Mathematical Statistics*, 24:657–662, 1953.

17. H. Cramér and M. R. Leadbetter. *Stationary and Related Stochastic Processes.* Wiley (republication by Dover 2004), New York, 1967.

18. C. Dean and J. F. Lawless. Tests for detecting overdispersion in Poisson regression models. *Journal of the American Statistical Association*, 84:467–471, 1989.

19. P. Diaconis and B. Efron. Computer-intensive methods in statistics. *Scientific American*, 248:96–109, 1983.

20. P. J. Diggle. *Statistical Analysis of Spatial Point Patterns.* Arnold Publishers, 2003.

21. O. Ditlevsen and H. O. Madsen. Structural reliability methods. Internet edition 2.2.5, Department of Mechanical Engineering, DTU, Lyngby, 2005.

22. N. R. Draper and H. Smith. *Applied Regression Analysis.* Wiley, New York, third edition, 1998.

23. B. Efron and R. Tibshirani. *An Introduction to the Bootstrap.* Chapman and Hall, New York, 1993.

24. J. W. Evans, R. A. Johnson, and D. W. Green. Two- and three parameter Weibull goodness-of-fit tests. Technical Report FPL-RP-493, United States Department of Agriculture, Forest Service, Forest Products Laboratory, 1989.

25. W. Feller. *Introduction to Probability Theory and its Applications*, volume I. Wiley, New York, third edition, 1968.

26. F. Garwood. Fiducial limits for the Poisson distribution. *Biometrika*, 28: 437–442, 1936.

27. H. G. Gauch Jr. *Scientific Method in Practice.* Cambridge University Press, Cambridge, 2003.

28. A. Gelman, J. B. Carlin, H. S. Stern, and D. B. Rubin. *Bayesian Data Analysis.* Chapman & Hall, 1995.

29. I. J. Good. Some statistical applications of Poisson's work. *Statistical Science*, 1:157–180, 1986.

30. M. Greenwood and G. U. Yule. An inquiry into the nature of frequency distributions representative of multiple happenings with particular reference to the occurrence of multiple attacks or disease of repeated accidents. *Journal of the Royal Statistical Society*, 83:255–279, 1920.

31. E. J. Gumbel. The return period of flood flows. *The Annals of Mathematical Statistics*, 12:163–190, 1941.

32. E. J. Gumbel. *Statistics of Extremes.* Columbia University Press (republication by Dover 2004), New York, 1958.

33. A. Gut. *An Intermediate Course in Probability.* Springer-Verlag, New York, 1995.

34. D. J. Hand, F. Daly, A. D. Lunn, K. J. McConway, and E. Ostrowski. *A Handbook of Small Data Sets.* Chapman & Hall, 1994.

35. A. M. Hasofer and N. C. Lind. Exact and invariant second-moment code format. *Journal of the Engineering Mechanics Division, ASCE*, 100:111–121, 1974.

36. U. Hjorth. *Computer Intensive Statistical Methods. Validation, Model Selection and Bootstrap.* Chapman & Hall, London, 1994.

37. T. Hodgkiess. *Materials Performance*, pages 27–29, 1984.

38. J. R. M. Hosking and J. R. Wallis. Parameter and quantile estimation for the generalised Pareto distribution. *Technometrics*, 29:339–349, 1987.

39. C. Howson. *Hume's Problem: Induction and the Justification of Belief.* Oxford University Press, Oxford, 2000.

40. R. G. Jarrett. A note on the intervals between coal mining accidents. *Biometrika*, 66:191–193, 1979.

41. N. L. Johnson, S. Kotz, and N. Balakrishnan. *Continuous Univariate Distributions, Volume 1. Second Edition.* John Wiley & Sons, 1994.

42. S. Kaplan and B.J. Garrick. On the use of Bayesian reasoning in safety and reliability decisions – three examples. *Nuclear Technology*, 44:231–245, 1979.

43. J. P. Klein and M. L. Moeschberger. *Survival Analysis: Techniques for Censored and Truncated Data.* Springer-Verlag, 1997.

44. N. Kolmogorov. *Grundbegriffe der Warscheinlichkeitsrechnung.* Springer-Verlag, 1933.

45. P. S. Laplace. Mémoire sur les approximations des formules qui sont fonctions de très grands nombres, et sur leurs applications aux probabilités (originally published in 1810). In *Oeuvres complètes de Laplace*, volume XII, pages 301–348. Gauthier-Villars, Paris, 1898.

46. L. Le Cam. On some asymptotic properties of maximum likelihood estimates and related bayes estimates. *University in California Publications in Statistics*, 1:277–330, 1953.

47. M. R. Leadbetter, G. Lindgren, and H. Rootzén. *Extremes and Related Properties of Random Sequences and Processes.* Springer-Verlag, 1983.

48. L. M. Leemis. Relationships among common univariate distributions. *The American Statistician*, 40:143–146, 1986.

49. E. L. Lehmann and G. Casella. *Theory of Point Estimation.* Springer-Verlag, New York, 1998.

50. I. Lerche and E. K. Paleologos. *Environmental Risk Analysis.* McGraw-Hill, 2001.

51. G. Lindgren. Hundraårsvågen – något nytt? In G. Grimvall and O. Lindgren, editors, *Risker och riskbedömningar*, pages 53–72. Studentlitteratur, Lund, 1995.

52. D. V. Lindley. Theory and practice of Bayesian statistics. *The Statistician*, 32:1–11, 1983.

53. W. M. Makeham. On the law of mortality. *J. Inst. Actuar*, 13:325–367, 1867.

54. R. Mises von. On the correct use of Bayes' formula. *The Annals of Mathematical Statistics*, 13:156–165, 1942.

55. J. P. Morgan, N. R. Chaganty, R. C. Dahiya, and M. J. Doviak. Let's make a deal: The player's dilemma (with discussion). *The American Statistician*, 45:284–289, 1991.

56. J Nelder and R. W. M. Wedderburn. Generalized linear models. *Journal of the Royal Statistical Society A*, 135:370–384, 1972.

57. W. Nelson. Theory and applications of hazard plotting for censored failure data. *Technometrics*, 14:945–965, 1972.

58. M. Numata. Forest vegetation in the vicinity of Choshi. Coastal flora and vegetation at Choshi, Chiba prefecture IV. *Bulletin of Choshi Marine Laboratory, Chiba University*, 3:28–48, 1961.

59. H. J. Otway, M. E. Battat, R. K. Lohrding, R. D. Turner, and R. L. Cubitt. A risk analysis of the Omega West reactor. Technical Report LA 4449, Los Alamos Scientific Laboratory, Univ. of California, 1970.

60. Y. Pawitan. *In All Likelihood: Statistical Modelling and Inference Using Likelihood.* Oxford University Press, Oxford, 2001.

61. J Pickands. Statistical inference using extreme order statistics. *Annals of Statistics*, 3:119–131, 1975.

62. D. A. Preece, G. J. S. Ross, and S. P. J. Kirkby. Bortkewitsch's horse-kicks and the generalised linear model. *The Statistician*, 37:313–318, 1988.
63. F. Proschan. Theoretical explanation of observed decreasing failure rate. *Technometrics*, 5:373–383, 1963.
64. T. Pynchon. *Gravity's Rainbow*. Jonathan Cape, London, 1973.
65. G. Ramachandran. Statistical methods in risk evaluation. *Fire Safety Journal*, 2:125–145, 1979-80.
66. E. M. Roberts. Review of statistics of extreme values with applications to air quality data. Part ii: Applications. *Journal of Air Pollution Control Association*, 29:733–740, 1979.
67. S. M. Ross. *Introduction to Probability Models*. Academic Press, seventh edition, 2000.
68. J. Rydén and I. Rychlik. A note on estimation of intensities of fire ignitions with incomplete data. *Fire Safety Journal*, Accepted for publication, 2006.
69. M. Sandberg. Statistical determination of ignition frequency. Master's thesis, Mathematical statistics, Lund University, Lund, 2004.
70. R. L. Scheaffer and J. T. McClave. *Probability and Statistics for Engineers*. Duxbury Press, fourth edition, 1995.
71. R. L. Smith. Statistics of extremes, with applications in environment, insurance and finance. In B. Finkenstadt and H. Rootzén, editors, *Extreme Values in Finance, Telecommunications and the Environment*, pages 1–78. Chapman and Hall/CRC Press, 2003.
72. R. L. Smith and J. C. Naylor. A comparison of maximum likelihood and Bayesian estimators for the three-parameter Weibull distribution. *Applied Statistics*, 36:358–369, 1987.
73. E. Sparre. Urspårningar, kollisioner och bränder på svenska järnvägar mellan åren 1985 och 1995. Master's thesis, Mathematical statistics, Lund University, Lund, 1995.
74. Statistics Sweden, Örebro. *Energy statistics for non-residential premises in 2002*, 2003.
75. I. Stewart. The interrogator's fallacy. *Scientific American*, 275:172–175, 1996.
76. Swedish Rescue Services Agency, Karlstad. *Räddningstjänst i siffror*, 2003.
77. P. Thoft-Christensen and M. J. Baker. *Structural Reliability Theory and its Applications*. Springer-Verlag, 1982.
78. British Standards Institute. *Application of Fire Safety Engineering Principles to the Design of Buildings*, 2003.
79. JCSS Committee, www.jcss.ethz.ch. *Probabilistic Model Code*, 2001.
80. W. Weibull. A statistical theory of the strength of materials. Ingenjörsvetenskapsakademiens handlingar No. 151, Royal Swedish Institute for Engineering Research. Stockholm, Sweden, 1939.
81. W. Weibull. A statistical distribution function of wide applicability. *Journal of Applied Mechanics*, 18:293–297, 1951.
82. D. Williams. *Weighing the Odds. A Course in Probability and Statistics*. Cambridge University Press, 2001.
83. E. B. Wilson and M. M. Hilferty. The distribution of chi-square. *Proceedings of the National Academy of Sciences*, 17:684–688, 1931.
84. R. Wolf. *Vierteljahresschrift Naturforsch. Ges. Zürich*, 207:242, 1882.
85. J. K. Yarnold. The minimum expectation in X^2 goodness of fit tests and the accuracy of approximations for the null distribution. *Journal of the American Statistical Association*, 65:864–886, 1970.

Index